BL^{S}S

The Berkeley
Interactive
Statistical System

D. Mark Abrahams
& Fran Rizzardi

DEPARTMENT OF STATISTICS

UNIVERSITY OF CALIFORNIA · BERKELEY

BLSS

The Berkeley
Interactive
Statistical System

W·W·NORTON & COMPANY

NEW YORK · LONDON

To our families

QA
276
.4
A271
1988
c.3
MATH

Library of Congress Cataloging-in-Publication Data

Abrahams, D. Mark.
   BLSS: the Berkeley interactive statistical system.

   Bibliography: p.
   Includes index.
   1. BLSS (Computer program)   2. Statistics — Data
processing.   I. Rizzardi, Fran.   II. Title.
QA276.4.A27 1988   519.5′028′55369   88-15110

ISBN 0-393-95586-9

W. W. Norton & Company, Inc., 500 Fifth Avenue, New York, N.Y. 10110
W. W. Norton & Company, Ltd., 37 Great Russell Street, London WC1B 3NU

Published simultaneously in Canada by Penguin Books Canada Ltd., 2801 John Street,
Markham, Ontario L3R 1B4.

Printed in the United States of America

1 2 3 4 5 6 7 8 9 0

CONTENTS

FOREWORD

The BLSS Project's existence is due to the willingness of the University of California, Berkeley—and in particular, Leonard Kuhi, Provost of the College of Letters and Science—to agree with our conviction that the mission of the University includes supporting the development of good instructional software.

But the program was well on its way before the official University funding started. It grew through the efforts of a sequence of graduate students and a young faculty member, all working without any official sponsorship or reward. The most dedicated of this sequence of talented and energetic young statisticians are the authors of this book.

BLSS is aimed at undergraduate and early graduate instructional use. It was designed for simplicity of use and playfulness, so that users can easily experiment and get quick responses showing the effects of their experimentation. The capability of working interactively with data can bring a concreteness and a sense of reality to the subject matter that is difficult to get by paper-and-pencil textbook problems alone.

At Berkeley BLSS is currently used by well over one thousand students a year—in the natural sciences, in the social sciences, and in statistics. Its usage is rapidly increasing. In particular, our department is using BLSS in all of our large lower division service courses.

One of the main purposes of the BLSS Project is to produce a statistical software system satisfying a fairly universal range of instructional needs. To do this, we carried out a widespread survey of universities to find what statistical capabilities would be important to add. As a result, we have a planned growth in mind for BLSS to make it still more capable and cosmopolitan.

One final small sermon: BLSS should not make any user's job easier—it should make it harder. At best, the computer is a tool to relieve humans from large amounts of tedious drudge work and let them focus on what humans are best at—hard and critical thinking about the underlying issues. The intelligence required cannot be artificial.

LEO BREIMAN
Director, Statistical Computing Facility
Department of Statistics
March 1988   University of California, Berkeley

PREFACE

BLSS (pronounced *bliss*) is a highly interactive statistical software system which runs on UNIX-based computers and workstations. It provides a wide range of capabilities to a wide range of people: students and computer novices find it easy to learn, so that they can start doing statistics quickly; experienced computer users find it a flexible interface with reliable tools for getting work done efficiently.

BLSS has grown out of the instructional needs of the Department of Statistics at the University of California, Berkeley. It is used at all levels within the Department, from large introductory undergraduate classes for nonmajors to specialized graduate classes; it is used in classes throughout the campus in fields such as anthropology, chemistry, electrical engineering, and forestry; and it is available at an increasing number of other institutions, both academic and commercial.

BLSS provides:

- A *simple, command-oriented* user interface which also recognizes *assignments* and algebraic *expressions*. For example, the command:

    . regress x y

  forms the regression of *y* on *x*; the command:

    . a = sqrt(b^2 + c^2)

  assigns to *a* the value of the expression $\sqrt{b^2+c^2}$. (The '.' in these examples is the prompt character which BLSS types when it is ready to accept commands.)

- An *intuitive way of storing data:* datasets can be scalars, row vectors, column vectors, matrices, or three-way arrays (if you need them). The work area can contain as many datasets as necessary, each with its own name, shape, and dimensions.

- *Standard statistical routines* such as descriptive statistics, plots, multiple regression, frequency distributions, cross-tabulation, confidence intervals, hypothesis tests (t-tests, z-tests, χ^2-tests), probability functions, random number generation, EDA routines, etc.

- Selected *advanced statistical routines,* including univariate and multivariate spectral analysis and ACE methods (estimation of optimal nonlinear transforms for linear models).

- *Matrix manipulation* including standard matrix arithmetic and decompositions and a variety of nonstandard—but natural—matrix operators and commands which allow datasets to be operated on as a whole instead of one element at a time.

- *Educational features* including instructional commands (such as coin tossing, drawing tickets from a box, random sampling, an elementary demonstration of the central limit theorem, and an illustration of confidence intervals via simulation), the ability to customize the environment on a per-class basis, a number of interactive demonstration modules ('movies') which can be easily modified or extended, and a collection of datasets for use in examples.

- *Convenience features* such as the ability to abbreviate commands (using *aliases* and *strings*), change their default behavior, customize output, repeat commands, and review and edit a history of commands run in the current session.

- *Extensibility features* such as the ability to write macros (command scripts), add commands which call favorite local subroutines, and interface to UNIX at both the command and programming language levels.

- *On-line documentation* for all commands and library datasets.

About this book. This book is a comprehensive introduction, user's guide, and reference manual for BLSS.

Part One—the *BLSS User's Guide* or *BUG*—teaches BLSS itself using common statistical techniques as examples. It begins at the novice level and progresses to a reasonably advanced level. There are many possible paths through Part One; Section 1.1 discusses them further.

However, BLSS provides too many commands, with too many options, to cover them all in the expository fashion of Part One. Part Two of the book—the *BLSS Reference Manual* or *BRM*—contains detailed documentation on all commands and features of BLSS, including many statistical techniques not mentioned in Part One. Part Three—the *BLSS Data Library* or *BDL*—describes the datasets which are included with BLSS for use in examples. The manual entries in Parts Two and Three are all available on-line within BLSS.

What this book does not cover. Unlike some books on statistical systems, this one does not attempt to explain, except when necessary for describing the software, any statistical concepts or methodology. There are two reasons. First, there are already many excellent statistics texts at all levels and we see no need to duplicate these efforts. (See the Bibliography at the end of the book for a very short list.) Second, statistical techniques cannot properly be applied cookbook style—adequate understanding cannot be obtained from a cursory overview. (Nonetheless, we do include many examples of statistical procedures as part of our explanation of BLSS itself.)

On this subject, we want to issue a warning. No statistical software is magic; all statistical techniques can be misused or misinterpreted. We hope that users will take the initiative to understand the assumptions behind the statistical procedures being used and to check that they apply to the problem under consideration—and to realize the consequences of not doing so.

We further limited the size of this book by omitting much technical information: We give only a basic discussion of macros and no discussion of installation, administration, or adding new BLSS commands (compiled programs). Such information is covered in separate documents which are available from the BLSS Project (see the address at the end of this preface).

To the student. Learning to use statistics is like learning any other practical art or science. It cannot be learned out of a textbook alone, with a bit of theory and a few classroom examples. Practical experience on real datasets is necessary; BLSS provides the means for this.

One of our primary goals in developing BLSS has been to make the mechanics of computing easy, so that you can concentrate your thoughts on statistics instead. To this end, many parts of this book, and quite a few commands in BLSS, were written with the student in mind. Nonetheless, the computer cannot 'do statistics' for you automatically. *You* must provide the guiding intelligence behind your statistical analyses. Before you use BLSS with an assignment, you should read the relevant material in your textbook. When you use BLSS, you will want to have both your textbook and this book at hand.

To the instructor. This book is written at the level of an undergraduate text. Because courses which are more advanced make more use of BLSS, we assume increasing sophistication as the chapters progress. Chapters 1, 2, and 3 are pitched at students in lower-level courses with no previous computer experience and are very easy. Chapters 4, 5, and 6 can be read by most lower-level students (although some may need to pause for thought at places) and all intermediate students. Chapter 7 is aimed at a variety of audiences: Section 7.1 is very easy but successive sections assume increasing knowledge of undergraduate probability theory. Chapters 8, 9, 10, and 11 require a certain ability to abstract and are pitched at advanced undergraduates. The chapters in Part One are reasonably independent of each other (although everybody should read Chapter 1) so you can select material from various chapters according to your class's level and needs.

For elementary classes which start with statistics, the usual way to introduce the software is through the material in Chapters 2 and 3, which cover arithmetic on individual numbers and lists of numbers and simple data analysis on univariate datasets.

For elementary classes which start with probability, the coin-tossing simulation, **coin** (in Section 7.1), is a good introduction. The commands **box**, **demoiv**, **sample** (all in Chapter 7), and **confid** may be used to illustrate concepts from probability and theoretical statistics at an elementary level, especially when used in conjunction with a text such as *Statistics* by Freedman, Pisani, and Purves. Some of these commands use the term *box model* from that text. This term is nothing more than a simple, concrete description of a *finite discrete random variable;* we explain it in Section 7.1.2.

Chapters 5 through 8 cover the commands and techniques most often used by elementary and intermediate classes. Consult Part Two of this book for additional commands and techniques not covered in Part One—the manual entry *topics(BRM)* contains a list of manual entries organized by topic and is a good place to start.

However, the real power of BLSS for instruction at intermediate and advanced levels comes from being able to assemble individual commands into larger sequences. Students learn to build up complex statistical procedures out of their elementary parts. They can examine the intermediate results, experiment with their own variations, and thus build up their intuitions. Chapter 9 provides examples of such extended command sequences.

Why UNIX? Why have we chosen UNIX as the operating system on which to run BLSS? By limiting ourselves to a single operating system, we can concentrate on developing better software rather than on satisfying the idiosyncrasies of many different operating systems; and from our standpoint, UNIX is the most important operating system for the foreseeable future. It is the only vendor-independent operating system which runs on a wide variety of hardware (from desktop machines to mainframes and even supercomputers), and its networking and resource-sharing capabilities far exceed those of any other we know of. Our department has found that UNIX provides an excellent environment for research (because of the powerful CPU's on which it runs and the rich variety of available software), for communicating with colleagues worldwide, and for teaching. For example, administering classes is much easier on UNIX than on PC's, because it eliminates the need for multiple copies of software, datasets, help files, and other resources, and because it provides easy communication between instructor, teaching assistants, and students.

Knowledge of UNIX is not necessary to use BLSS. However, for people who will use UNIX tools in conjunction with BLSS, we have provided two appendices—one on UNIX itself and one on **vi**, its most popular editor. These serve as introductions for the curious and as quick references for people already learning them.

Software versions. BLSS software is under continuing development. Capabilities are added and changes are made with each release. This book documents BLSS Release 4.0 of May 1988. Most of it also applies to earlier releases, but to get the exact results described in this book, you must use Release 4.0. The release number is displayed each time you start up BLSS, so you always know what version you are using. A list of software changes between different releases is available by typing 'news' when within BLSS; the on-line documentation provided with BLSS is the final authority on your release.

Obtaining BLSS. BLSS is distributed by the Department of Statistics at the University of California, Berkeley. For information on obtaining and running it, or current information on its capabilities, contact:

> BLSS Project
> Department of Statistics
> University of California
> Berkeley, CA   94720
>
> 415-642-5258

or send electronic mail to:

> blss@bach.berkeley.edu   (INTERNET)
> blss@ucbbbach.bitnet   (BITNET)
> ucbvax!bach!blss   (UUCP)

Comments on BLSS documentation and software are welcome and can be sent to us at the addresses above. Many improvements in both software and documentation have originated as comments from users; we hope that this pattern will continue.

HISTORY AND
ACKNOWLEDGMENTS

Many people and several institutions have contributed to what is now BLSS; we gratefully acknowledge them here.

BLSS evolved from the original version of ISP written at the Princeton University Department of Statistics by Dave Donoho and Ken Birman in 1974-78 under the direction of Peter Bloomfield. Donoho wrote the original version of the ISP shell and the following commands which are still in use: **biweight**, **medmad**, **medpolish**, **oneway**, **robust**, **smooth**, **twoway**, and—in conjunction with Donald McNeil—**boxplot** and **compare**. Birman wrote the original version of the **let** command; McNeil assembled the original data library.

James Reeds oversaw the installation of ISP at the University of California, Berkeley, in 1979; he, Steve Hubert, Ross Ihaka, and Brian Yandell maintained and enhanced ISP at the Berkeley Department of Statistics from 1979 through 1982. We mention here the major enhancements which are still in use. Reeds greatly expanded the capabilities of the ISP shell, adding many of the convenience features discussed in Chapter 10. The string expansion mechanisms were adapted, with permission, from those written by William Joy for the UNIX C shell. Hubert wrote the **code** and **xvalid** commands and portions of the **redo** mechanism. Ray Faith (then of Stanford University) wrote the original version of the **scatter** command; Reeds adapted it to use the terminal-independent *libcurses.a* library; Yandell made various improvements to it. Stanley Freisen wrote the **cluster** command. In 1982, this system was renamed *Berkeley ISP* in order to distinguish it from versions elsewhere which were evolving independently.

The present authors assumed primary responsibility for Berkeley ISP in fall 1982. Since then, Keith Haycock wrote the time series commands **autcor**, **autcov**, **autreg**, **demod**, **dft**, **filter**, **imprsk**, **pgrml**, **pgrmk**, **spec1**, **speck**, **trnfrk**, and **tsop**; he also wrote the current version of the time series subroutine library which these use, based on the original version written by David Brillinger. William Krebs wrote the **box**, **confid**, and **demoiv** commands; the latter was adapted from a long line of Berkeley demoiv commands, the first written by David Freedman. The **ace** and **mdrace** commands use subroutines written by Jerome Friedman; the **mdrace** command also uses the mdrace subroutines written by Robert Koyak. The **freq** and **xtab** commands are based on earlier versions written by Jon Dart at the University of California, San Diego. A number of commands draw upon the high quality public domain libraries

EISPACK (for eigenvalue-eigenvector problems) and LINPACK (for linear algebra, matrix decompositions, etc.).

In 1987, Berkeley ISP was renamed *BLSS* (an acronym for *B*erke*L*ey Interactive *S*tatistical *S*ystem), in order to avoid confusion with at least two other statistical software systems being marketed under the name ISP (some derived from the original Princeton ISP and some not).

Partial support for BLSS software development has been provided by the University of California, Berkeley, College of Letters and Science since fall 1985. Leo Breiman is—in more ways than one—the godfather of this project; without him it would not exist in its present form. Ani Adhikari and Charles Stone have provided extensive and invaluable comments on both the book and the software. Dan Coster, Jon Dart, Kjell Doksum, Roger Purves, John Rasmussen, Aram Thomasian, and many other users and instructors have been generous with their comments and suggestions. James Blakly has handled the administrative aspects of the BLSS project with grace and aplomb; Leslie Leonard provided additional administrative guidance; Claire LeDonne of the Berkeley Campus Software Office has been a wonderful help in dealing with the many faces of software distribution. Ed Moy made the monospace font available on the typesetter; Zhishun Alex Liu supplied occasional but crucial UNIX wizardry. John Hawkins, our editor at W. W. Norton, made many useful suggestions and helped develop the overall shape of this book.

Finally, we thank the Berkeley Department of Statistics as a whole for providing the ideal environment for this project.

PART ONE

BLISS User's Guide

CHAPTER 1

Introduction

Part One of this book is an introduction to BLSS, the Berkeley Interactive Statistical System.[1] We do not explain any statistical concepts (such as *histogram, standard deviation* or *regression*), but assume that you have learned or are learning these elsewhere. If you are using BLSS in conjunction with a statistics class, you should study the relevant material in the textbook *before* attempting to use that material in BLSS.

Even if you have never used computers before, it is easy to perform nontrivial operations on data with BLSS. The best way to learn BLSS is to set aside several blocks of time, each about one or two hours. For each session, sit down at the computer terminal, work thoroughly through several sections of this book, and *think* about the interaction happening between you and the BLSS system.

If you are learning BLSS as you learn statistics, you will want your textbooks as well as this book at the terminal—you will find that you spend more time thinking about statistics than you spend thinking about BLSS.

1.1 How to Read This Book

This book, like BLSS itself, is intended for people at various levels. Its chapters are reasonably independent and there are many paths through it. We outline a few here.

In any case, read this chapter first.

---

1. BLSS is pronounced *bliss.*

We recommend—especially if you have no previous computer experience—that you read Chapters 1, 2, 3, and the first half of 4 (through Section 4.6) in order. Working through them while at a computer terminal forms a solid introduction to BLSS.

The first half of Chapter 4 is essential. Some people may want to read Section 4.3 (about printing output) or Section 4.4 (about entering data) very early on; some people may want to skip the second half of Chapter 3 (about detailed use of stem-and-leaf diagrams).

Select material beyond Chapter 4 according to your needs—consult the table of contents. Chapters 5, 6, 7, and 8 each concentrate on a specific area of statistics, although they by no means exhaust BLSS's capabilities. Chapters 9, 10, and 11 are each more general, in their own way. Chapter 9 presents several extended examples which combine material from the previous chapters. If you are an experienced computer user impatient to get the flavor of BLSS, a route through this book is: read Chapters 1 and 9; then skim Chapters 8, 10, and 11. However, this route will leave you short on basic mechanical knowledge of the system.

Part One of this book will help you understand many features of BLSS. But it is not a complete explanation of the system, because BLSS has more statistical abilities than can be explained in a narrative text of this size. Part Two of this book is the *BLSS Reference Manual* and contains complete descriptions of all BLSS commands. Chapter 11 explains how to use the manual.

Perhaps the best way to read this book (indeed, the best way to read any computer software book or manual) is to first *skim* the chapters, sections, and manual entries that interest you, to get a general outline of what they contain. Then, at the terminal, reread and study carefully the sections that you need when you need them.

1.2   Getting Started

Before you can use BLSS, you need access to a UNIX computer system which has BLSS installed on it. You need:

1. A *user name* (also called an *account name* or *log name*) which identifies you to the UNIX system.

2. A secret *password* which lets you prove to UNIX that you are allowed to use that user name.

3. Access to a computer terminal which connects to that UNIX system. Large organizations may have many different UNIX systems, so you may need to know the name of the UNIX system to which you have been assigned and where to find terminals attached to that particular system. The person who issues your computer account and password should tell

you if there are any special local instructions for connecting terminals to your UNIX computer system.

Logging in. The next step is to log in to the UNIX computer system. Here is how. When you have connected a terminal to UNIX, it prints the message:

    login:

Type your user name and then press the RETURN key. UNIX then asks for your password by printing the message:

    Password:

Type your password and press the RETURN key again. In order to keep it secret, UNIX prevents your password from appearing on the terminal when you type it. If you mistyped either your account name or password, UNIX will inform you and print the login message so that you can try again.

At some sites, there may be additional steps to the login procedure described above.

When you have logged in, wait for the UNIX prompt symbol. At sites that run BSD (Berkeley) UNIX or its derivatives,[2] this is the '%' character, or a word which ends in the '%' character. At sites that run System V UNIX,[3] this is the '$' character. In either case, when the UNIX prompt symbol appears it means that UNIX is waiting for you to give it a command.

Once you have the UNIX prompt (which in our examples we show as '%'), type the UNIX command:

    % blss

to enter BLSS. Remember that you must finish this command, and every computer command, by striking the RETURN key. The computer will not pay attention to UNIX or BLSS commands until the RETURN key has been struck. Then wait for the BLSS prompt symbol, which is a '.' (dot) character followed by a space. *Note:* Certain class accounts are set up so that you get the BLSS prompt without typing the command 'blss'. In this case there is no need to type 'blss' but you should still wait for the '.' prompt before typing anything. If you have such an account, your instructor should inform you in advance.

1.3   First Session with BLSS

A short example session with UNIX and BLSS appears below. Try it, in order to acquire a feeling for the interaction that happens between you and the BLSS system. In this and all examples, what the computer prints is shown in

---

2. BSD stands for Berkeley Software Distribution. Throughout this book, when we refer to BSD UNIX we include its derivatives, such as SUN-OS, DEC ULTRIX, IBM ACIS, etc.
3. System V is AT&T's standard version of UNIX. As with BSD, when we refer to System V UNIX we include its derivatives.

monospace font. What you type is shown in *slanted monospace* font. Remarks are shown in *italic* font, to the right of the '#' symbols. Do not type them.

Do *not* type the prompt symbols ('%' and '.'). Remember to follow each command with the RETURN key so that the computer will know you are finished typing it. Each time you type a command, look at what the computer displays and wait until it gives a new prompt symbol. Then type the next command.

```
login: username            # type your account name
Password:                   # type your password—it does not appear
                            # some login messages may appear
% blss                      # enter BLSS—some BLSS messages appear
. load finalexam            # get a BLSS dataset
Loaded "finalexam" from the BLSS system data area.
. show finalexam            # show the data on your terminal

    45.00    83.00    75.00    71.00    73.00    40.00    61.00    63.00
    71.00     6.000   54.00    67.00    62.00    52.00    73.00    70.00
    92.00    64.00    19.00    32.00    39.00    68.00    79.00    46.00
    77.00    35.00    46.00    55.00    45.00    82.00     9.000   52.00
    29.00    44.00    83.00    50.00    84.00    12.00    69.00    41.00
    36.00    29.00    80.00    58.00    76.00    66.00    61.00    65.00
    82.00    80.00    65.00    78.00    45.00    33.00    39.00    91.00
    86.00    56.00    57.00    58.00    58.00    59.00

. stat finalexam            # compute some statistics

Statistics: finalexam
Col   N     Mean     SD      Min      25%      50%      75%      Max
 1    62    57.68    20.49   6.000    45.00    60.00    73.00    92.00

. stemleaf finalexam        # make a picture of the data

N = 62,  min = 6,  25% = 45,  50% = 60,  75% = 73,  max = 92
Leaf digit unit (ldu) = 1   (1|2 represents 12.)

 0|69
 1|29
 2|99
 3|235699
 4|01455566
 5|02245678889
 6|11234556789
 7|0113356789
 8|00223346
 9|12

. exit                      # exit from BLSS
%                           # back to UNIX
```

Briefly, what happened was that you entered BLSS and looked at a dataset consisting of the final exam scores of 62 students in a statistics class at UC Berkeley. You looked at summary statistics of the data and made a picture of the data known as a *stem-and-leaf diagram*. You then exited from BLSS and once again received the UNIX prompt symbol.

If you are done with UNIX, log out now. You can log out by typing:

```
% exit                          # UNIX log out
```

If you have the BSD UNIX prompt, '%', you can also log out by typing:

```
% logout                        # BSD UNIX log out
```

In addition to these commands, a single control-D (typed by holding down the CTRL button and typing 'd') usually logs you out of any type of UNIX.

The following chapters explain the BLSS commands in the example, and many others. However, the remaining examples in this book will not tell you when to enter or exit BLSS or when to log in or log out of UNIX—you must decide that for yourself.

1.4   Computer Use

Exiting BLSS and UNIX. Whenever you are done using BLSS, type 'exit' to tell BLSS you are finished. You will once again get the UNIX prompt, as in the example session above. When you are done using UNIX, log out as explained in the previous section.

> **Do not forget to log out of UNIX when you are done.**
> Otherwise, little gremlins deep in the system may eat your account.

Note: If you have one of the special class accounts mentioned previously which automatically gives you BLSS when you log in, you do not need to explicitly log out of UNIX after exiting BLSS—UNIX will automatically log you out. In either case, you will know that you are logged out when the computer prints a new login message.

The period of time between logging in to UNIX and logging out is known as one *UNIX session.* The period of time between entering BLSS and exiting it is known as one *BLSS session.* You can have as many BLSS sessions as you like in the course of one UNIX session. Try to remember these terms, as they are used elsewhere in this book.

Typing. Typing characters to a computer system such as UNIX is almost identical to typing characters on a typewriter. However, you must be careful to type the correct character: computers are not as good as humans at guessing what you meant to type. Note that the characters *1* (the numeral 'one') and *l* (the lowercase letter 'ell') are *different*. *1* and *l* are generated by typing different keys on the keyboard; they look different on your terminal's screen and on paper printouts; and they have different meanings to the computer. Similarly, the characters *0* (the numeral 'zero') and *O* (the uppercase letter 'oh') are different. In general, uppercase letters (*A, B, C, ...*) are distinct from lowercase letters (*a*, *b*, *c*, ...). Finally, observe that there are three distinct types of parentheses and brackets on your keyboard. We refer to them as follows: parentheses (), square brackets [], and curly braces { }. Each has its own special meaning in BLSS.

Fixing mistakes. UNIX provides three special characters which allow you to recover from mistakes. When you type the *erase character*, the last character on the line you are currently typing is erased. You can type more than one erase character, with the obvious effect. (If the effect isn't obvious to you, try it and see what happens.) When you type the *kill character*, the current line you are typing is deleted (killed) and you can start that line over. Finally, the *interrupt character* can be used to interrupt the current command or computation—you will get a new prompt.

On BSD UNIX systems, the erase character is usually either DELETE or else control-H (which is typed by holding down the CTRL button and typing 'h');[4] the kill character is usually control-U; and the interrupt character is usually control-C.[5] On other UNIX systems the erase, kill, and interrupt characters may be different. Find out what they are from the same person who issued your computer account and make a note of them. (Inside the front cover of this book is a convenient place.)

Warning. Older UNIX systems often use '#' as the erase character and '@' as the kill character. However, these characters (in fact, all printing characters) have special reserved meanings in BLSS. Therefore, if your system uses '@' or '#' as control characters, BLSS changes them to control-H and control-U when you enter BLSS and changes them back when you exit. If any such changes are made, BLSS informs you with a message such as:

```
Erase character set to Control-H
Kill character set to Control-U
```

Spacing. To a certain extent, BLSS relies on the spacing you use in a command to determine just precisely what you are saying. In many cases, BLSS needs space characters to separate the words in a command—that is, to determine where one word ends and the next begins. In some other cases, space characters may *not* be typed—for example, in the middle of a word or number. Until you acquire a feel for what sort of spacing is needed, it's best to type the examples just as they appear in this book.

On the other hand, don't be afraid to experiment if you are so inclined. If you get an error message, try to understand what went wrong and then try your command again with corrections. The flexibility to quickly experiment—with both commands and with statistical techniques—is one of the great advantages of an interactive system such as BLSS.

---

4. If your terminal has a BACKSPACE key, it has the same effect as control-H.
5. UNIX allows users (and site administrators) to change these characters according to preference. Sometimes the shorthand notation ^*X* is used to denote a control-*X* character.

1.5 BLSS Datasets, Text Files, and Names

BLSS knows how to manipulate two types of objects. The first type is known as a *BLSS dataset*, or *dataset* for short. A BLSS dataset may contain a single number, a simple list of numbers which are observations on a single statistical variable, or a more complex dataset which consists of measurements on several individuals for each of several different statistical variables. The object *finalexam* in the preceding session was a BLSS dataset which contained measurements on one statistical variable.

The second type of object is known as an *ASCII text file*, or *text file* for short.[6] Text files contain text—for example, the pictures or displays that result from a BLSS command, comments about a dataset, the text representation of a dataset (more about this in Section 4.10), or any other combination of characters. If you already know what a text file is in UNIX, text files in BLSS are just the same.

Note that BLSS datasets and text files are *different*. The types of manipulations that can be performed on them are different. You cannot perform dataset manipulations on text files, nor can you perform text file manipulations on datasets.[7]

Sometimes we will jointly refer to BLSS datasets and text files as *files*.[8]

Every BLSS dataset and text file has a unique name to identify it. The rules about dataset and text file names are:

1. They must begin with a letter.
2. They must consist entirely of letters, digits, or the period '.' or underscore '_' characters.
3. They must be 14 or fewer characters long.
4. The names may not conflict with BLSS function names.

In particular, don't use mathematical function names such as *sin*, *cos*, *log*, or *exp*; also avoid *abs*, *fac*, *gam*, *max* (which denote special functions), *PI*, *E*, and *NA*. All these names have special built-in meanings for BLSS.

However, uppercase and lowercase letters are distinct—for example, the names *xy* and *Xy* refer to two distinct files. Thus, you can safely use the names *pi*, *e*, and *na*, because they are distinct from the names *PI*, *E*, and *NA*.

When you choose names for your datasets and text files, try to choose names that are descriptive and help you remember what data or text they contain.

BLSS has a small library of datasets (known as the *BLSS system data library*) which are often used for examples. Among them are: a completely fake

---

6. *ASCII* (pronounced *ask*-ee) is an acronym for *American Standard Code for Information Interchange*.
7. There are a few exceptions. As we will see, you can **load**, **show**, and **save** both BLSS datasets and text files.
8. The reason is that BLSS datasets are stored as binary data files.

dataset named (appropriately enough) *fake* and four real datasets called *finalexam* (which we saw in Section 1.2), *tractor*, *rivers*, and *plover*. Most examples in this book are written in terms of these library datasets, but the same commands work with any dataset. In Section 4.4 we will see how to enter your own data into a BLSS dataset.

1.6 Typographical Conventions

We use several typographical conventions throughout this book.

Bold type indicates literal text: words which you must type exactly as they appear in the book. Thus, we use bold type for the names of BLSS commands, options, and tags (options and tags are introduced in later chapters).

Italic type indicates prototype text: text for which you may substitute your own names or words. Thus, we use italics for BLSS dataset names, text file names, and other UNIX file names. We also use italics for representing 'generic' parts of BLSS commands—for example, in positions where you may substitute specific command or option names of your choice.

As mentioned earlier, in examples we use `slanted monospace` font for what you type and `monospace` font for what the computer types. Within examples, we show remarks to the right of '#' symbols, in *italics*.

The notation *xxx*(*BRM*) refers to the entry entitled *xxx* in the *BLSS Reference Manual* (Part Two of this book). The notation *xxx*(*BDL*) refers to the entry entitled *xxx* in the *BLSS Data Library* (Part Three of this book), which documents the library datasets.

1.7 Movies

BLSS comes with over a dozen prepackaged 'movies'—demonstration modules that illustrate BLSS commands and features while you sit back and watch. For example, simply type the command:

```
. demo.plot1
```

and BLSS will demonstrate its abilities to make visual displays of univariate datasets. For a list and descriptions of these demonstration modules, see *demo*(*BRM*) or type:

```
. help demo
```

(The **help** command uses the *more* mechanism, which is explained in Section 4.8.) For descriptions of the datasets used in the demonstration modules and examples throughout this book, see the *BLSS Data Library*. Or type:

```
. help data.xxx
```

where *xxx* is the name of the dataset you are interested in.

CHAPTER 2

Numbers

2.1 BLSS as a Desk Calculator

BLSS can be used as a desk calculator. To evaluate an arithmetic expression, just type it. BLSS prints the answer. For example:

```
.  7 + 5

    12.00

.  45 - 10

    35.00
```

The '+' key means add; the '−' key means subtract. To multiply, use the '*' key; to divide use '/'. To exponentiate (raise a number to a power), you can use either '^' or '**'. For example, to square 5, you can type either of the following commands:

```
.  5 ^ 2

    25.00

.  5 ** 2

    25.00
```

These examples all use positive integers, but of course you can work with negative numbers and decimal fractions by using minus signs and decimal points. For example:

```
.  -7.4 / .25

  -29.60
```

Functions. BLSS understands many functions, including: **abs** (absolute value); **sqrt** (square root); **sin, cos,** and **tan** (trigonometric functions); **ln** and **log** (both give the natural logarithm); **log10** (logarithm to the base 10); and **exp** (exponentiation of the mathematical constant *e*). For example, to show the square root of 64:

```
. sqrt(64)
```
```
   8.000
```

Assigning results. The results of a calculation can be stored and used in new calculations. For example:

```
. let x = .5 * 35 * 140
. let y = log(x)
```

These commands mean, quite literally, "Let the BLSS dataset *x* have the value .5 times 35 times 140. Let the BLSS dataset *y* have the value which is the logarithm of *x*." When you *assign* the values of calculations to a dataset, as in these examples, BLSS does not display the results. To see the values, use the **show** command:

```
. show x
```
```
 2450.00
```
```
. show y
```
```
   7.804
```

Abbreviations. The **let** and **show** commands are used so often that BLSS allows you to omit the words **let** and **show** from these two commands. For example, the command sequences given above may be abbreviated to:

```
. x = .5 * 35 * 140
. y = log(x)
. x
```
```
 2450.00
```
```
. y
```
```
   7.804
```

You can **show** several BLSS datasets and arithmetic results with a single command. For example:

```
. show x y 4+5 10-6
```
```
 2450.00
```
```
   7.804
```
```
   9.000
```
```
   4.000
```

Arguments. The expressions 'x', 'y', '4+5', and '10–6' in the previous example are called the *arguments* of the command. Note that the arguments are separated by spaces, but there are no spaces within the arguments. If you want to **show** arithmetic arguments which contain spaces, be sure to enclose them within parentheses. For example, this command works:

```
. show (2 + 8) x-(15 * 12)
```

but this does not:

```
. show  2 + 8   x-15 * 12
```

because BLSS thinks that each of '2', '+', '8', 'x–15', '*', and '12' are separate arguments.

Parentheses and precedence. Parentheses may also be used to provide grouping and change the order in which (that is, the *precedence* with which) arithmetic operations are performed. For example, the answers produced by:

```
. show 5+5*2 10-4+6 1/10*10 10*2^3 -2^2
```

are quite different from those produced by:

```
. show (5+5)*2 10-(4+6) 1/(10*10) (10*2)^3 (-2)^2
```

Try them. As you can see, when no parentheses are provided BLSS performs arithmetic operations in the following order (or *precedence*):

1. Exponentiations (raising a number to a power);
2. Unary negations (taking the negative of a number);
3. Multiplications and divisions, left to right;
4. Additions and subtractions, left to right.

These are the standard rules of algebra, but it never hurts to provide extra parentheses if you're not sure. We discuss BLSS's precedence rules in more detail in Sections 8.2 and 8.8.

2.2   Lists of Numbers

Of course, BLSS datasets can contain not only single numbers, but lists of numbers. You can create your own short list of numbers as follows.[1] Suppose you want to create a dataset called *mylist* which contains 2, 6, 1, 3, and 7. Type:

```
. mylist = 2,6,1,3,7
```

Be sure to put commas between the numbers in the list. To verify what it contains, use the **show** command:

```
. show mylist
```

---

1. In Section 4.4, we introduce methods for creating BLSS datasets which are more convenient when you need to enter large amounts of data.

Any of the arithmetic operations and functions discussed in the previous section may be applied to the entire list. For instance, to add 10 to every number in the list, assign the result to a dataset called *a*, and then see the contents of the new dataset, type the commands:

```
. let a = mylist + 10
. show a
```

To square every number in the list and put the result in a dataset called *b*:

```
. let b = mylist ^ 2
```

To take the square root of every number in the list and assign the result to a dataset called *sqrts*:

```
. let sqrts = sqrt(mylist)
```

Observe that we called the new dataset *sqrts*, not *sqrt*. The name *sqrt* is a function name—if we try to use it as a dataset name, BLSS would give an error message.

Arithmetic operations may also be applied to two lists of numbers, provided that they have the same length. For example:

```
. x = 1,2,3
. y = 4,5,6
. x + y
```
```
   5.000      7.000      9.000
```
```
. x * y
```
```
   4.000    10.00    18.00
```

The arithmetic operation is applied to corresponding pairs of elements in the two lists. The result is a new list (of the same length) which contains the individual answers.

As noted in the previous section, you can see several arithmetic results using a single **show** command:

```
. show mylist^2 sqrt(mylist) x+y x-y
```

The advantage of this method is that it takes less typing. The disadvantage is that BLSS discards the result after showing it, whereas with the '=' sign method, the result of the arithmetic operation is available (in the newly created dataset) for further use.

It is possible that when you took square roots of your data, BLSS printed 'NA' for some of the values. What this means is explained in the next section.

Extending lists of numbers. The ',' (comma) operator, which we used above to create lists of numbers, can also be used to extend already existing lists: either by adding more numbers, or by joining (or *catenating*) lists. For example:

```
. u = x,8
. u
   1.000    2.000    3.000    8.000
. v = -1,-2,y,x,9
. v
  -1.000   -2.000    4.000    5.000    6.000    1.000    2.000    3.000
   9.000
```

Note that the list *v* did not fit in a single line, so BLSS continued printing it on a second line. Section 8.3 (in particular, 8.3.3) discusses the ',' operator and related operators in much more detail.

2.3   More about Data Values

This section contains additional information about the representation of numerical data in BLSS. You may wish to skip or skim it on your first reading. But before long you should return and read this section because you will probably have questions which are answered here.

Missing data. What happens if you try to take the square root of a negative number? Type:

```
. sqrt(-1)
```

BLSS shows:

```
        NA
```

BLSS is unable to take the square root of a negative number. It prints 'NA' instead.[2] The 'NA' means *Not Available*, or *missing data value*. In general, if any data value is either missing or uncomputable (such as the square root or logarithm of a negative number) BLSS denotes the fact with an 'NA'.

Even though you may not plan to take square roots of negative numbers, it can happen in practice when you are analyzing lists of numbers. For example, the BLSS system data library contains a short list of numbers in the dataset called *fake*. In the following example, we use the **load** command to fetch *fake* from the data library. Then we **show** it and various functions of it:

```
. load fake
Loaded "fake" from the BLSS system data area.
. show fake
  -3.000   -1.000    0.000    0.000    1.000    3.000
. show sqrt(fake)
        NA       NA    0.000    0.000    1.000    1.732
```

---

2. On some computers, BLSS may print 'NaN' instead of 'NA'.

. *show log(fake)*

    NA   NA   NA   NA   0.000   1.099

BLSS displays 'NA' for each missing or uncomputable data value. Perhaps this happened to you when you computed the square roots of your own data in Section 2.2.

E-notation.   E-notation is a special notation used by many computer programs, including BLSS, for printing very large or very small numbers. The reason is that conventional notation is hard to read when there are many decimal places. For example, conventional notation represents $1.67 * 10^8$ as 167000000.0 and $2.54 * 10^{-11}$ as 0.0000000000254. Obviously, it's easier to tell the magnitude of the number when you know the exponent of 10 and don't have to count 0's. Type:

. *show (1.67 * 10^8) (-2.54 * 10^-11)*

and BLSS shows these numbers in E-notation:

    1.67e+08

    -2.54e-11

Just remember that the 'e' is shorthand for the phrase 'multiplied by 10 exponentiated to (the number that follows).'

E-notation can also be used when typing in numbers. For example:

. *big = 2e+14, 2.5e+15*
. *show (big * 3.3e-20)*

BLSS shows:

    6.60e-06   8.25e-05

How large and how small an exponent you can use depends on the specific type of computer on which BLSS is installed. On many computers, the exponent of 10 can be as large as 37, or as small as −39. However, if a dataset contains very large or very small numbers, it is preferable to rescale it. If the units of the data are 10^{20}, for example, divide the data by 10^{20}. This helps avoid accidentally creating too large an exponent during analysis. (For example, if you try to square 10^{20}, the result is too large for some computers.)

Precision.   By default, BLSS displays most numbers using only three or four significant digits. This suffices for most statistical purposes. However, BLSS stores data internally and performs calculations to (slightly over) six significant digits of working accuracy.[3] Data may be displayed to full six-place precision with the **show {g}** command. For example, try:

. *show {g} exp(1) 4*atan(1)*

---

3. In technical language, BLSS uses single precision floating point numbers.

It shows:

```
2.71828
3.14159
```

The {g} is known as an *option* to the **show** command—the presence of the {g} changes the behavior of **show**. The {g} may be placed anywhere in the command. For example, the following two commands have the same meaning as that above:

```
. show exp(1) 4*atan(1) {g}
. show exp(1) {g} 4*atan(1)
```

The reason is that, in BLSS commands, the placement of the *options* with respect to the *arguments* does not matter (although the placement of the arguments with respect to each other does).

Another option to **show** is {i}. It causes **show** to display data in an integer format: no decimal points or decimal fractions are shown. For example:

```
. show {i} (5, 10, 17, 23.6)

  5   10   17   24
```

Note that the final number in the list, 23.6, was displayed by rounding it to the nearest integer, 24. The {i} option is primarily used with data which are already integers.

CHAPTER 3

Simple Data Analysis

3.1 Summary Statistics

In this chapter we introduce simple techniques for analyzing a dataset which contains one variable. As an example, let's look at the dataset called *finalexam* in the BLSS system data library. It contains final exam scores for a Statistics 2 class taught at UC Berkeley in 1985. We obtain it with the **load** command:

```
. load finalexam
Loaded "finalexam" from the BLSS system data area.
```

and show its contents with the **show** command:

```
. show finalexam
        45.00    83.00    75.00    71.00    73.00    40.00    61.00    63.00
        71.00    6.000    54.00    67.00    62.00    52.00    73.00    70.00
        92.00    64.00    19.00    32.00    39.00    68.00    79.00    46.00
        77.00    35.00    46.00    55.00    45.00    82.00    9.000    52.00
        29.00    44.00    83.00    50.00    84.00    12.00    69.00    41.00
        36.00    29.00    80.00    58.00    76.00    66.00    61.00    65.00
        82.00    80.00    65.00    78.00    45.00    33.00    39.00    91.00
        86.00    56.00    57.00    58.00    58.00    59.00
```

This dataset is fairly large—we cannot tell much about it by looking at the individual numbers except that they are all in the range 0–100. However, we can summarize the data with a few simple descriptive statistics using the **stat** command:

```
. stat finalexam

Statistics: finalexam
Col   N    Mean      SD      Min     25%      50%      75%      Max
 1   62   57.68    20.49   6.000   45.00    60.00    73.00    92.00
```

The display shows:

N	The number of observations in the data.[1]
Mean	The arithmetic mean of the data.
SD	The standard deviation.
Min	The minimum value of the data.
25%	The 25th percentile (the 1st quartile).
50%	The 50th percentile (the median, or 2nd quartile).
75%	The 75th percentile (the 3rd quartile).
Max	The maximum value.

The abbreviation 'SD', for standard deviation, is used throughout BLSS.

The **stat** command can produce many other statistics, such as the variance, skewness, kurtosis, etc. For more information, see *stat*(*BRM*)—that is, the reference manual entry on the **stat** command.

3.2   Stem-and-Leaf Diagrams

The *stem-and-leaf diagram* (for short, *stemleaf diagram*) is a variation on the familiar histogram. But it contains more information than a histogram—in fact, it contains all the information in the original dataset, sorted into increasing order and displayed in a clever fashion.[2]

The **stemleaf** command makes a stem-and-leaf diagram of the data:

```
. stemleaf finalexam

N = 62,  min = 6,   25% = 45,   50% = 60,   75% = 73,   max = 92
Leaf digit unit (ldu) = 1   (1|2 represents 12.)

 0|69
 1|29
 2|99
 3|235699
 4|01455566
 5|02245678889
 6|11234556789
 7|0113356789
 8|00223346
 9|12
```

---

1. The **stat** command ignores missing values (NA's). If a dataset contains missing values, 'N' is actually the number of non-missing observations.
2. The stem-and-leaf diagram was invented by John W. Tukey—who also, among many other inventions, coined the word *bit* as an abbreviation for *binary digit*.

We now explain this diagram piece by piece. The first line of the display shows the number of observations and the minimum, maximum, and quartiles of *finalexam*. The second line contains information about the scale of the data. We will return to it later.

Each row of the diagram itself represents the data in a certain range. For example, the first row:

       0|69

represents the data values 6 and 9. The second row:

       1|29

represents the data values 12 and 19. The row that starts with '7|':

       7|0113356789

represents the data values 70, 71, 71, 73, 73, 75, 76, 77, 78, and 79. It's just a short way of writing down the numbers.

Each of these rows is called a *branch*. The numbers to the left of the '|'s in the rows are called the *stem labels*. Each stem label represents the most significant digit of a data value.

The digits to the right of a '|' are called *leaves*. Each leaf represents the next most significant digit of a data value.

The *stem* is the column of stem labels together with the vertical line of '|'s:

       0|
       1|
       2|
       3|
       4|
       5|
       6|
       7|
       8|
       9|

Think of the branches as growing off the stem, and the leaves as growing off the branches.[3]

Comparison to histograms. The stem serves the same purpose as the axis on a histogram. Note that every observation in the data is represented by exactly one leaf digit in the stemleaf diagram, and vice versa. As a result, the length of each branch in the diagram is proportional to the number of observations in that range. Thus, the shape of a stemleaf diagram is like the shape of a histogram. But, unlike a histogram, a stemleaf diagram lets you read the first two digits of every observation in the dataset.

---

3. Some authors use the term *stem* to refer to what we call the branches—but we generally think of stems as vertical, not horizontal.

Practice reading the stemleaf diagram by answering the questions: How many students scored 82 on the final exam? How many scored in the 80's? If the instructor sets the 'F' range at 25 and below, how many students score an F? If she sets the 'A' range at 85 and above, how many students score an A? If she changes the A range to 82 and above, how many score an A?

3.2.1   The Scale of a Stem-and-Leaf Diagram

We looked at the *finalexam* data before making its stem-and-leaf diagram, so we already knew its scale—in other words, we knew that the stem-and-leaf pair '1|2' represented the data value 12, not 0.12, or 1.2, or 120, or 1200. But what if we did not already know (or had forgotten) the scale? The second line of the stemleaf display reminds us what the units of the leaf digits are:

```
Leaf digit unit (ldu) = 1   (1|2 represents 12.)
```

In other words, the leaf digit '2' represents the value $1 * 2 = 2$, so '1|2' represents the value 12.

Of course, not all data are in the range 0 to 100. If we make a stemleaf diagram of *finalexam* multiplied by 10, it shows:

```
. stemleaf 10*finalexam

N = 62,  min = 60,  25% = 450,  50% = 600,  75% = 730,  max = 920
Leaf digit unit (ldu) = 10   (1|2 represents 120.)

 0|69
 1|29
 2|99
 3|235699
 4|01455566
 5|02245678889
 6|11234556789
 7|0113356789
 8|00223346
 9|12
```

This diagram is identical to the original one, but now the leaf digit unit is 10 (the leaf digit '2' represents $10 * 2 = 20$), so the branch:

```
 1|29
```

represents the values 120 and 190, and the branch:

```
 7|0113356789
```

represents the values 700, 710, 710, ..., 790. If we make a stemleaf diagram of *finalexam* divided by 100, it shows:

```
. stemleaf finalexam/100

N = 62,  min = 0.06,  25% = 0.45,  50% = 0.6,  75% = 0.73,  max = 0.92
Leaf digit unit (ldu) = .01   (1|2 represents 0.12)
```

```
0|69
1|29
2|99
3|235699
4|01455566
5|02245678889
6|11234556789
7|0113356789
8|00223346
9|12
```

Again, the diagram is identical to the original, but now the leaf digit unit is 0.01 (the leaf digit '2' represents 0.01 * 2 = 0.02), so the branch:

```
1|29
```

represents the values 0.12 and 0.19, and the branch:

```
7|0113356789
```

represents the values 0.70, 0.71, 0.71, . . . , 0.79.

3.2.2   Stem-and-Leaf Diagrams with Split Branches

The **stemleaf** command automatically adjusts the number of branches in the display to best show the shape of the histogram. To do this, in some cases it splits a branch in two, or in five.

For example, the *tractor* dataset in the BLSS system data library contains the time to brake failure (in hours of operation) for Caterpillar tractors. Its stem-leaf diagram has two branches for each stem label value. **Stemleaf** displays the leaf digits 0 through 4 in the first branch for a given label, and the leaf digits 5 through 9 in the second branch:

```
. load tractor
. stemleaf tractor

N = 107,  min = 56,  25% = 1029,  50% = 1795,  75% = 2604,  max = 7739
Leaf digit unit (ldu) = 100  (1|2 represents 1200.)

0|1111234
0|55566777788888999
1|0000111122223333
1|555555666667889
2|000000111222222233344
2|55666778999
3|012224
3|6788
4|0023
4|5
5|1
5|66
6|
6|9
7|
7|7
```

The dataset *rivers* in the BLSS system data library contains the length in miles of 141 'major' rivers in North America. Its stemleaf diagram has five branches for each stem label value:

```
. load rivers
. stemleaf rivers

N = 141,  min = 135,  25% = 310,  50% = 425,  75% = 680,  max = 3710
Leaf digit unit (ldu) = 100  (1|2 represents 1200.)

0|1
0|2222222222233333333333333333333333333333333333333
0|44444444444444444444444444444445555555555555555
0|66666666666677777777777
0|8888999999
1|00011
1|22233
1|55
1|
1|89
2|
2|33
2|5
2|
2|
3|
3|
3|
3|7
```

Each stem with a given label contains (if anything) only the leaf digits 0 and 1, or 2 and 3, or 4 and 5, or 6 and 7, or 8 and 9.

3.2.3   Stemleaf Options

The {**terse**} option omits the two lines of header information. We use it in the examples that follow to save space.

Sometimes we want to know how many data values are in a given interval. To find out, we can count the leaves on the appropriate branch in the stemleaf diagram. Alternatively, the {**count**} option does this for us—the count for each branch precedes its stem label:

```
. stemleaf finalexam {count;terse}

 2   0|69
 2   1|29
 2   2|99
 6   3|235699
 8   4|01455566
11   5|02245678889
11   6|11234556789
10   7|0113356789
 8   8|00223346
 2   9|12
```

We see immediately that 11 people scored in the 50's, and another 11 scored in the 60's.

Sometimes we want to know the total number of data points which are at or below a given branch in the diagram. Again, we can add up the counts—but the {**cum**} option shows these values for us. (The option name {**cum**} is short for *cumulative count*.) The {**cum**} and {**count**} options can be used together:

```
.  stemleaf finalexam {cum;count;terse}
   2    2   0|69
   4    2   1|29
   6    2   2|99
  12    6   3|235699
  20    8   4|01455566
  31   11   5|02245678889
  42   11   6|11234556789
  52   10   7|0113356789
  60    8   8|00223346
  62    2   9|12
```

42 people scored in the 60's or lower.

The {**depth**} option is similar to the {**cum**} option, but it counts in from the nearest end of the diagram. That is, for each branch it shows the total number of data points from that branch (inclusive) to the nearest end of the diagram, unless the branch contains the median. For the branch that contains the median it shows instead the count, enclosed in parentheses. The {**depth**} and {**count**} options can be used together:

```
.  stemleaf finalexam {depth;count;terse}
    2    2   0|69
    4    2   1|29
    6    2   2|99
   12    6   3|235699
   20    8   4|01455566
   31   11   5|02245678889
  (11)  11   6|11234556789
   20   10   7|0113356789
   10    8   8|00223346
    2    2   9|12
```

10 students scored in the 80's or higher; 20 scored in the 70's or higher.

Notice that, as in these examples, we can combine any number of options within a single set of curly braces { } by separating them with semicolons ';'.

The manual entry *stemleaf(BRM)* contains more information on these and other options to the **stemleaf** command.

3.3 Data Transformations

We often want to analyze a dataset whose histogram is approximately symmetric about its center. If a dataset is not symmetric, sometimes a simple function, or *transform*, of it is. In this case, we might analyze the transformed dataset instead of the original.

As shown by their stemleaf diagrams, the *finalexam* dataset is approximately symmetric about its center, but the *tractor* and *rivers* datasets are not. Let's try some transforms on these datasets and see what happens. The transforms to try are those which spread out the lower data values while bringing the higher data values closer together.[4]

If we take the square root of each value of *tractor*, the stemleaf diagram is more nearly symmetric (compare the stemleaf that follows with the original *tractor* stemleaf shown in Section 3.2.2):

```
. stemleaf {terse} sqrt(tractor)

0|79
1|01
1|67
2|11123
2|56666788999
3|001222334444
3|555667889999
4|0000022244
4|55555566666777788889
5|00111222444
5|566669
6|0112334
6|67
7|1
7|55
8|3
8|8
```

The cube root is stronger than the square root—the stemleaf is starting to be pulled toward the high end:

```
. stemleaf {terse} tractor^(1/3)

 3|8
 4|479
 5|
 6|27
 7|577
 8|0257888
 9|113445679
10|002234455668
11|00034445666777
12|0124566667789999
13|0001223355777899
14|022346777
15|1445699
16|135
17|278
18|
19|08
```

---

4. If you're good at math, try to work out for yourself why the transforms we use do this, and hence why they have the effects we observe on the shape of the stemleaf diagram. However, we will not go into this reasoning here.

If we take the natural log of each value of *tractor*, this pull is even more pronounced:

```
. stemleaf {terse} log(tractor)

4|04
4|68
5|
5|57
6|111234
6|555566777888999
7|0000011111222333333344444
7|5556666666666777777777888888899999
8|000001111222233344
8|5668
9|0
```

However, the *rivers* data is so clumped toward the lower values that even the logarithm is not strong enough to make the histogram symmetric:

```
. stemleaf {terse} log(rivers)

4|9
5|
5|333
5|44445555555
5|6666666666677777777777
5|8888888888999999999999
6|0000000001111111111
6|22222333333333
6|4444444445555
6|66666667777
6|88888999
7|001111
7|233
7|55
7|7
7|88
8|
8|2
```

We can use transforms which are stronger still, such as reciprocals and reciprocal square roots, but they have side effects we do not wish to discuss here.[5]

3.4   Sorting Data

The stem-and-leaf diagram shows the data sorted into increasing order. Sometimes we want the sorted data itself, with no diagram. The **sort** command, used as follows, creates a new dataset called *sorted* which contains the sorted final exam data:

---

5. Briefly, the reciprocal transforms 'flip' the data around the center. You must be careful not to confuse this effect with pulling the data toward or beyond the center.

```
. sort finalexam > sorted
. show sorted
    6.000    9.000    12.00    19.00    29.00    29.00    32.00    33.00
   35.00    36.00    39.00    39.00    40.00    41.00    44.00    45.00
   45.00    45.00    46.00    46.00    50.00    52.00    52.00    54.00
   55.00    56.00    57.00    58.00    58.00    58.00    59.00    61.00
   61.00    62.00    63.00    64.00    65.00    65.00    66.00    67.00
   68.00    69.00    70.00    71.00    71.00    73.00    73.00    75.00
   76.00    77.00    78.00    79.00    80.00    80.00    82.00    82.00
   83.00    83.00    84.00    86.00    91.00    92.00
```

The '>' symbol in the **sort** command means: create a new dataset (called an *output dataset*) whose name follows. In this example, we named the new dataset *sorted*. (In Chapters 4 and 5 we return to this idea of output datasets.)

From the sorted data, we can readily answer such questions as: What is the minimum value? What is the maximum? How many people scored in the 80s? In the 90s? Of course, for specific questions such as these, the stemleaf display serves just as well if not better.

The manual entry *sort(BRM)* contains further information about sorting datasets (for example, into decreasing order).

CHAPTER 4

Utility Commands

Most chapters in this book are concerned with *statistics commands*—commands that perform the numerical or statistical tasks whose results you wish to see. This chapter, in contrast, discusses *utility commands*—commands and features that allow you to print results, enter your own data, save your files and datasets for later use, etc.

The information in the first half of this chapter (Sections 4.1–4.6) is essential. Whatever else you do in BLSS, you will need this material. However, you will understand it better if you have already used BLSS for one or two small sessions. Information in the later sections of this chapter will help you make better use of BLSS as you become more experienced with it. The best way to read this chapter is to skim it all, and then return to and read carefully specific sections as you need them.

4.1 General Information

Recall, from Section 1.5, that the term *files* refers to both BLSS datasets and text files.

The work area. When you work in BLSS, it gives you a temporary active work area, known simply as the *work area*. All datasets and text files you create are placed in this work area, and BLSS statistics commands work on datasets which are in it.

The purpose of the **load** command (which we used in Section 1.2, 2.3, and elsewhere) is to copy preexisting BLSS datasets into the work area where you can use them.

The **list** command describes the contents of your work area. For example, if you give the **list** command after working some of the examples in Chapter 2 and Chapter 5, it might respond as follows:

```
. list
Contents of your work area:
a              dataset, dims=(1,7)      (row vector)
b              dataset, dims=(1,7)      (row vector)
mylist         dataset, dims=(1,7)      (row vector)
plover         dataset, dims=(68,5)     (matrix)
vol            dataset, dims=(68,1)     (column vector)
x              dataset, dims=(1,1)      (scalar)
y              dataset, dims=(1,1)      (scalar)
```

It lists all the files in your work area and gives a short description of each. In this case, all your files are datasets. The numbers enclosed in parentheses are called the *dimensions* of the dataset. The first number is the number of rows in the dataset; the second is the number of columns. Datasets which contain only one row and one column—in other words, only a single number—are called *scalars*. Those which contain only a single row are called *row vectors*; those which contain only a single column are called *column vectors*. Those which contain several rows and several columns are called *matrices* (plural for *matrix*). We discuss these terms further in Section 5.1.

You can also use the **list** command to find out about specific files only, instead of all the files. For example, if you want to find out the dimensions of *a* and *b*, but nothing else, type:

```
. list a b
Specified files in your work area:
a              dataset, dims=(1,7)      (row vector)
b              dataset, dims=(1,7)      (row vector)
```

If a specified file does not exist, **list** says 'No such file' to the right of the name in the listing.

Exiting from BLSS. As noted in Chapter 1, you exit from BLSS with the **exit** command.

Note well: **When you exit from BLSS normally (via the 'exit' command), your temporary active work area and its entire contents vanish.**

The reason is that, when you work in BLSS, you create many temporary datasets and text files—it can become quite a mess! When you finish, BLSS cleans up after you automatically: the next time you use BLSS, it gives you a clean (empty) work area. Section 6 of this chapter describes how to save files for use in a later BLSS session.

Recovering an aborted session. If BLSS exits abnormally (for instance, if the computer crashes), then when you next use BLSS it should inform you that you have an old work area and ask whether you wish to use it. The message looks like this:

```
You have an old BLSS work area, "blss/tmp0", dated Wed Sep 30 14:38:58 1987.
Type one of [ulnrq], or type ? for help:
```

If you type a 'u', BLSS will let you use your old work area—your temporary datasets and files will all be there and you can resume work where you left off. (The 'u' means 'use it.')

If you type a '?', BLSS will give you information about all the choices.

On occasion, BLSS may exit abnormally but not inform you about any old work areas when you next use it. This can happen if you are suddenly logged out due to a broken phone connection, broken portselector connection, or broken remote connection. To recover old work areas in this case, enter BLSS with the UNIX command:

```
% blss -a
```

instead of the usual command 'blss'.

4.2   Error Messages

By now, you have probably made a few mistakes when typing BLSS commands and encountered the resulting error messages. Most error messages in BLSS are self-explanatory. Here are some hints for understanding four common error messages whose meanings may not be obvious.

Syntax error. This means that BLSS did not understand your command. Most likely, you made a typing error. Perhaps you used the wrong punctuation, special character, or spacing. Check what you typed against what you should have typed and try again.

Does not exist. You tried to access a dataset or text file that does not exist. Possibly you misspelled its name. If not, perhaps you misremember its name. Or perhaps you forgot to **load** it during the current BLSS session. In either case, use the **list** command to find out the names of your current files.

Cannot find ... Areas searched ... This error message is produced by **load** when it cannot find the file you asked for. Check your spelling or use the **list** command options explained in Section 4.6 to find out what files are available to be loaded. If it's your own file, did you remember to **save** it (see Section 4.6) during a previous BLSS session?

Invalid option. You gave a bad option to a command—you spelled it wrong, or requested a nonexistent option. When BLSS gives this error message, it also gives a list of options recognized by the command. Section 11.4 explains this error message in more detail.

Many other error messages can be caused by simple typing mistakes. When you receive an error message, the first thing to do is check what you typed against what you meant to type.

4.3   Printing Output—the {lpr} Option

To make printed output from regular BLSS commands, use the {**lpr**} option.[1] This option sends the text output of a command to the lineprinter on your system instead of to your terminal. *Note:* the first character in the option name is the lowercase letter *l* ('ell'), not the digit *1* ('one'). The option name {**lpr**} is an abbreviation of the word *lineprinter*.

For example, to make a printed copy of the dataset *plover* (assuming you have already **load**-ed it):

```
. show plover {lpr}
```

To make a printed copy of stem-and-leaf diagrams of *plover*:

```
. stemleaf plover {lpr}
```

Recall, from the end of Chapter 2, that the placement of options with respect to arguments does not matter. So the following commands have the same meaning:

```
. show {lpr} plover
. stemleaf {lpr} plover
```

Note that the {**lpr**} option allows you to make printouts from only one command at a time. If you need printouts from many commands, printing each one separately wastes paper and makes it difficult to organize your printouts. In Section 4.5 we explain how to combine text output from several commands into a single printout.

4.4   Reading in Your Own Data

In Sections 2.1 and 2.2 we saw how you can create your own small BLSS datasets (a single number, or a short list of numbers) with the **let** command. This section explains how to enter larger quantities of data into a new dataset. This operation is called *reading in* data because from the computer's point of view, data are read. Of course, from your point of view the data are written.

Before you read in any data, you must first decide what shape the new dataset will have. Will it be a row vector, a column vector, or a matrix? If you are not sure which to use, or how to arrange data in a matrix, refer to Section 5.1.

Reading in a column vector. Suppose you want to read the numbers 11 through 15 into a column vector named *mycol*. To do so, issue the command:

```
. read > mycol
```

---

1. {**lpr**} is one of several options which are recognized by all regular BLSS commands. Other such options include: {**more**}, discussed in Section 4.8; and {**quiet**}, {**terse**}, and {**long**}, which control the amount of output displayed by a command. A complete discussion of such options is in the manual entry *options*(BRM). **Let** commands and the *internal* commands (discussed in Chapter 10) do not provide these options.

BLSS gives the message:

```
Type your data one row per line; finish with RETURN and CTRL-D.
```

You then type your data, followed by a control-D:

```
11
12
13
14
15
```
control-D

Note that: 1) you type each number on a separate line (press the RETURN key after each number); 2) you will not get the BLSS prompt again until you type the control-D (hold down the CTRL button and type 'd'). After you type the control-D, BLSS says:

```
Read 5 values into "mycol"; dims=(5,1) (column vector).
```

and gives a new BLSS prompt.

Reading in a matrix. Suppose you want to read this table of data into a matrix named *mymat*:

1	2	3	4
5	6	7	8
9	10	11	12

To do so, use the **read** command just as with a column vector. Enter the data one row per line. Spacing within each row does not matter (provided that the numbers are separated by at least one space or tab character). Here is the complete conversation with BLSS:

```
. read > mymat
Type your data one row per line; finish with RETURN and CTRL-D.
1 2 3 4
5 6 7 8
9 10 11 12
```
control-D
```
Read 12 values into "mymat"; dims=(3,4) (matrix).
```

Be careful to type the same number of data values on each line. BLSS computes the number of columns in the matrix from the first line you type—if you type a different number of values on any other line, it will give you a warning message.

Reading in a row vector. To read data into a row vector, you could use the read command as shown above and type all the data on one row. However, this is impractical if the row is too long. To read in long rows of data, use the {**row**} option. For example:

```
. read {row} > myrow
```

BLSS gives the message:

```
Type your data; finish with RETURN and CTRL-D.
```

Note that it did not say 'one row per line.' With the {**row**} option, where you choose to start a new line does not matter. Simply enter the data from the keyboard, separating the values with at least one space or tab character, or else by starting a new line. For example, you might enter:

```
6 3 10 4
20 22 7 19 5
16
Control-D
```

BLSS then says:

```
Read 10 values into "myrow"; dims=(1,10) (row vector).
```

Verifying and correcting your data. After you read in a new dataset, **show** it to verify that you entered it correctly. If you made a few mistakes—but not enough to warrant reentering the entire dataset—you can fix them with the **let** command and double subscripts as follows. Suppose that the number in the 2nd row and 4th column of the matrix *mymat* should be 18 instead of 8. In double subscript notation, this is the [2,4] position. To make this change, type:

```
. mymat[2,4] = 18
```

Subscripts are discussed further in Section 5.1.1.

Alternatively, if you know how to use a text editor you can correct your data using the **editdata** command described in Section 4.10.3.

Saving your dataset. If you have created any valuable datasets with the **read** command, be sure to save copies of them with the **save** command before you exit from BLSS. For example:

```
. save mydata
```

We discuss the **save** command further in Section 4.6.

Dimensioning. By default, **read** sets the row and column dimensions of the new BLSS dataset equal to the number of rows and columns in the data as it is typed. This is known as 'auto-dimensioning', or {**autodims**}. When we created the column vector *mycol* and the matrix *mymat* in the examples above, we relied on this feature.

The **read** command provides several options for specifying dimensions. We already saw the {**row**} option, which forces the new dataset to be a row vector regardless of the arrangement in which the data are typed. Similarly, the {**col**} option forces the new dataset to be a column. The {**dims=**} option forces the new dataset to have any specified set of dimensions. For more information and examples, see *read(BRM)*—that is, the reference manual entry on **read**.

Reading data from text files. The **read** command can read data from a text file as well as from the keyboard, provided that the data in the text file is arranged just as you would type it (taking any dimensioning options into account). If your data is in a text file called *mytext* in your work area, use the command:

```
. read mytext > mydata
```

to read it into a dataset called *mydata*. However, if you created the text file outside of BLSS, it is probably in your *home directory*, not your work area. (Your *home directory* is the default location where you create files when you use UNIX.[2]) To read data from the file *mytext* in your home directory into a BLSS dataset called *mydata*, use the {**home**} option:

```
. read {home} mytext > mydata
```

Note that you *must* use the {**home**} option to read text files from your home directory—otherwise BLSS will not know to look there.[3]

If you need to enter a reasonably large amount of data which is not already on the computer, and if you know how to use a UNIX text editor, it is more convenient to first enter the data into a text file with the editor and then **read** it into a BLSS dataset. Be sure to arrange the numbers in the text file just as you would type them in to the **read** command. Here are the commands to use:

```
. edit mytextfile                    # puts you in the UNIX editor

#  ... enter your data into the text file ...

. read mytextfile > mydataset
```

We will return to the **edit** command in Section 4.9.3.

Output datasets. Note that the **read** command creates a new BLSS dataset. The new dataset is called the *output dataset* of the command. The '>' symbol in the **read** commands above serves to separate the output dataset name from the rest of the command which precedes it, including any options or input arguments. In this context, the '>' symbol is called the *output dataset separator*. We will see this same form of command:

$$command\ \{options\}\ input\text{-}arguments\ >\ output\text{-}datasets$$

throughout BLSS.

---

2. *Directory* is a UNIX concept, not a BLSS concept. We will make occasional reference to directories for people who want to interface BLSS to other UNIX commands, but knowledge of directories is not necessary for BLSS. If you are interested, Appendix A contains an overview of directories.

3. An exception: BLSS understands UNIX pathnames and the ~ convention of the UNIX C shell: ~ is an abbreviation for the user's home directory, and *~joe* is an abbreviation for the home directory of the user *joe*. For example, the commands:

```
. read ~/mydata
. read ~joe/hisdata
. read /tmp/jane/herdata
```

all work (assuming that those files exist).

4.5   Creating and Printing Text Files

As first mentioned in Section 1.4, text files can contain any text material. Examples are: written comments, documentation, the text representation of a dataset, or—of particular interest in this section—the *output text*, or *output display*, from BLSS commands which is normally displayed on your terminal.

In Section 4.3, we saw that the {**lpr**} option to a command causes the output text from that command only to be sent to the lineprinter. In order to combine output text from several commands into a single printout, we accumulate the output text from the commands in a text file, and then print the text file using the **show** {**lpr**} command. Here is how. To send the output text from any regular BLSS command to a text file instead of the terminal, add the following to the end of the command:

$$>> \text{ } \textit{filename}$$

where *filename* is the name of the text file. For example, to send the output text from the command:

```
. scat {x=1;y=2} plover
```

to the text file named *lab2*, type:

```
. scat {x=1;y=2} plover >> lab2
```

Nothing appears on your terminal except for a new BLSS prompt (and error messages, if anything goes wrong) because the output text is sent to the text file *lab2* instead.[4]

Output from additional commands may be *appended* to (that is, added to the end of) the text file using the same notation and same text file name. For example:

```
. regress {x=1;y=2} plover > fit res >> lab2
. scat plover {x=1} res >> lab2
```

adds the output text from these two commands to the text file *lab2*—*lab2* now contains the output text from all three commands.

The **show** command can show text files as well as BLSS datasets—so you can make a printout of the text file using **show** with the {**lpr**} option:

```
. show {lpr} lab2
```

Multiple text files. If you are saving the output display from many commands, you may prefer to store the output text from each command, or group of related commands, in different text files. To do this, you might type '>> lab2.1' after the first group of commands, '>> lab2.2' after the second group of commands, and so forth. The text files *lab2.1, lab2.2, . . .* may be printed

---

4. The **scat** and **regress** commands, which we use in this section simply as examples of commands from which to save or print output displays, are discussed in detail in Chapter 5.

together or separately, whichever is most convenient. It is up to you to decide how to group your printouts. For example:

```
. stat rivers >> lab2.1
. stemleaf rivers >> lab2.1
. let sqrivers = sqrt(rivers)
. stat sqrivers >> lab2.2
. stemleaf sqrivers >> lab2.2
. let logrivers = log(rivers)
. stat logrivers >> lab2.3
. stemleaf logrivers >> lab2.3
. show {lpr} lab2.1 lab2.2 lab2.3
```

The '>>' symbol used here is called an output text *redirection symbol*. We return to redirection symbols in Section 4.10.1.

The **show** {**lpr**} command in the preceding example sends three separate text files to the printer. However, they will be printed one after another on the same page of printout. To force **show** to print files on separate pages, use the {**ff**} option together with the {**lpr**} option. (The {**ff**} stands for *formfeed*—a formfeed makes the lineprinter advance to a new *form*, or page of printer paper.) For example:

```
. show {lpr} {ff} lab2.1 lab2.2 lab2.3
```

The manual entry *print(BRM)* contains more information about printing files.

When two or more options are given, as in the previous command, they may be combined within a single set of curly braces:

```
. show {lpr;ff} lab2.1 lab2.2 lab2.3
```

Section 11.3 explains, in detail, this and other general rules about BLSS commands and options.

The list command and text files. If you give the **list** command after creating text files as above, the output looks like this:

```
. list
Contents of your work area:
fit           dataset, dims=(42,1)    (column vector)
lab2          text file, 2231 bytes,  58 lines,  widest is 78 chars
lab2.1        text file,  491 bytes,  26 lines,  widest is 79 chars
lab2.2        text file,  533 bytes,  33 lines,  widest is 79 chars
lab2.3        text file,  485 bytes,  25 lines,  widest is 79 chars
logrivers     dataset, dims=(141,1)   (column vector)
plov          dataset, dims=(42,5)    (matrix)
res           dataset, dims=(42,1)    (column vector)
rivers        dataset, dims=(141,1)   (column vector)
sqrivers      dataset, dims=(141,1)   (column vector)
```

For each text file, **list** tells: its size in bytes,[5] how many lines it contains, and how many characters are in its widest line.[6]

---

5. A *byte* is the unit of storage required to store one character (including non-printing characters such as newlines, tab characters, etc).

4.6 Saving Files

When you work in BLSS, your files are stored in a temporary *work area* which disappears (datasets, text files, and all) at the end of your BLSS session. BLSS provides two permanent storage areas in which you may save files between BLSS sessions. One is for BLSS datasets and is called the *data area*. The other is for text files and is called the *text area*.[7] To save datasets and text files in these areas, use the **save** command. For example, to save the files *a*, *b*, and *c*, type:

```
. save a b c
```

before you exit from BLSS. This is a good idea if you have spent much time creating the files, whether by entering data from the keyboard or as a result of a lengthy series of commands. Note that **save** only copies the files, so the saved files are still available in your work area until you exit from BLSS.

Thereafter, whenever you use BLSS you can use the **load** command to obtain a copy of the permanently saved dataset or text file. This is just like **load**-ing data from the BLSS system data area—a copy of the saved file is placed in your work area. For example:

```
. load a b c
```

The **save** command automatically determines whether to use the data area or the text area for each file. The **load** command searches first the data area, and then the text area, for files to load.

To see what is saved. The **list** command normally makes a list of files in your work area. To make a list of datasets stored in your permanent data area, use the {**data**} option:

```
. list {data}
```

To make a list of files stored in your permanent text area, use the {**text**} option:

```
. list {text}
```

If you have many files saved in your permanent storage areas, you can use the {**more**} option (see Section 4.8) to display the list one screenful at a time. Alternatively, if you want to find out about specific files only, you can specify them as arguments to **list**. For example, the command:

```
. list {data;text} a b c
```

makes a list of files named *a*, *b*, and *c* in your data and text areas.

---

6. The line width takes into account any possible tabs and backspaces.
7. The BLSS data and text areas are the UNIX *blss/data* and *blss/text* directories. More information about BLSS areas and UNIX directories is in Section 4.12.

4.7 Removing, Renaming, and Copying Files

Removing files. To remove datasets or text files from your work area, use the **remove** command. For example, to remove *d, e* and *f*, type:

```
. remove d e f
```

To remove datasets from your permanent data area, use the {**data**} option. For example, to remove *g, h, i* and *j* from your data area, type:

```
. remove {data} g h i j
```

To remove text files from your permanent text area, use the {**text**} option. For example:

```
. remove {text} k l
```

Occasionally you may need to access files in your home directory. All BLSS utility commands recognize the {**home**} option introduced at the end of Section 4.4. The {**home**} option allows you to **save** files into your home directory, **load** files from it, **list** files in it, **remove** files from it, and so forth.

The **save**, **load**, and **remove** commands all have an {**ask**} option. This option causes these commands to ask you to confirm, for each file, whether you want to save, load, or remove it. If you give one of these commands the {**ask**} option, but no file arguments, then the command asks you about all possible files in the specified (or default) area. For example, the command **remove** {**ask**} asks you whether you wish remove each file, in turn, in your work area. The conversation with BLSS looks like this:

```
. remove {ask}
Remove "fit" in your work area? Yes
Remove "lab2" in your work area? yes
Remove "lab2.1" in your work area? y
Remove "lab2.2" in your work area? y
Remove "lab2.3" in your work area? y
Remove "logrivers" in your work area? y
Remove "plov" in your work area? no
Remove "res" in your work area? y
Remove "rivers" in your work area? n
Remove "sqrivers" in your work area?
7 files removed.
```

Any reply which begins with a 'Y' or 'y' means 'yes', and BLSS removes the file. Any other reply (such as 'no' or 'n' or just the RETURN key) means 'no', and BLSS does not remove the file. At the end, BLSS tells you how many files it removed.

Renaming files. Occasionally, you may want to rename a dataset or text file. To do so, use the **rename** command. For example, suppose you want to change the name of a file from *mybingo* to *artichoke*. Just type:

```
. rename mybingo artichoke
```

The {**data**} and {**text**} options may be used with the **rename** command too—
they cause files in the data area or text area to be renamed.

Copying files. There is no explicit copy command in BLSS, because you can
copy files using other features of BLSS. For example, to make a copy of the
BLSS dataset *x* and call it *y*, type:

```
. y = x
```

To make a copy of the text file *a* and call it *b*, type:

```
. show a >! b
```

The '>!' text redirection symbol, explained in Section 4.10.1, is similar to the
'>>' text redirection symbol. ('>!' and '>>' have the same effect if *b* does not
already exist.)

When you copy files, be careful not to accidentally overwrite an existing file.

4.8   The {more} Option

As noted previously, the **show** command can show both text files and BLSS
datasets. Thus, you can use **show** to preview a text file before printing it. For
example, to show the text file *lab2:*

```
. show lab2
```

If *lab2* has more than about 23 lines of text (as it will if you created it following the example in Section 4.5) and if you are using a video terminal, then
some of the text it contains may disappear off the top of your terminal's
screen before you have a chance to read it. The same problem arises when
you try to **show** the *plover* dataset on the screen, and with any other command
that produces a long text output. For this situation, BLSS provides another
useful option—the {**more**} option. To see both *lab2* and *plover* using this
option, type:

```
. show {more} lab2 plover
```

At the bottom of each screenful of data or text, the computer types the message:

```
--More--[Press space to continue, 'q' to quit.]
```

This means: to continue and see the next screenful, press the space bar. To
skip the rest of the data or text (that is, to quit the **show** command and get the
BLSS prompt again), hit the 'q' key. Several other responses are also possible:
the 'd' key shows the next half-screenful; the RETURN key shows one more
line. The 'h' key shows a complete menu of responses available when using
{**more**}.

The {**more**} option is available for every regular BLSS command. For example, the **regress** command with the {**long**} option produces a display which is
too long to fit on most terminal screens. The {**more**} option can be used to

view it one screenful at a time. The *more* mechanism is used in other contexts as well. You may already have encountered it when using the **help** or **man** commands.

The *more* mechanism is not a feature of BLSS itself, but rather a feature of UNIX which BLSS uses. Different versions of UNIX may provide different versions of *more*. For example, on some versions of UNIX the message which appears at the bottom of each screenful is:

```
--More--[Hit space to continue, Rubout to abort]
```

although the 'q' key, 'd' key, and other responses work just as described above. Some versions of UNIX may provide no *more* mechanism, or an alternative to the *more* mechanism.

4.9   Adding Remarks to Text Files

In order to know what you were doing when you look at your printed output later on, it can be useful to enter remarks into your output text files. BLSS provides several ways to do so.

4.9.1   The 'echo' Command

The **echo** command allows you add remarks to a text file one line at a time.[8] Type:

```
. echo This is a remark.
This is a remark.
```

As you can see, the **echo** command simply echoes its arguments, so to speak. To put remarks into an output file, you could proceed as follows:

```
. echo Plot of Egg Widths against Lengths >> lab2
. scat plover {x=1;y=2} >> lab2
. echo Regression of Egg Widths on Lengths >> lab2
. regress plover {x=1;y=2} >> lab2
```

Warning. In BLSS, certain characters have special meanings. We have already seen that '>' is one such character. Another special character is '#'. It introduces a comment. That is, everything typed on a line after a word that begins with '#' is ignored.[9] To remove the special meaning of characters such as '>' and '#' in BLSS, either surround them by double-quote characters " " or precede them by a backslash character '\'. For example, either of these commands:

---

8. The BLSS **echo** command and '#' comment character work just like the **echo** command and '#' comment character of the UNIX shells. Comments are discussed further in Sections 10.8.1 and 11.3.2.

9. The symbols '#*', '#^', and '#,' in **let** commands are exceptions, because they denote matrix operations. These symbols are discussed in Chapter 8.

```
.  echo ">> Assignment #3 <<"
.  echo \>\> Assignment \#3 <<
```

echoes the (overly decorative) remark '>> Assignment #3 <<'.[10]

4.9.2 The 'addtext' Command

Unlike the **echo** command, the **addtext** command allows you to add as many lines of text as you like to the end of a text file using a single command. Moreover, characters such as '>' and '#' have no special meaning when using **addtext**, so there is no need to surround them by double-quotes "" or precede them by backslashes '\'.

To add to a text file called *mytext*, give the command:

```
.  addtext mytext
```

BLSS gives the message:

```
Adding to text file "mytext" in your work area.
Type your text; finish with RETURN and CTRL-D.
```

Type your text. You can fix mistakes using your UNIX erase and kill characters. Finish by typing a control-D at the beginning of a new line. For example:

```
        >> Assignment #3 <<
Analysis of the "plover" dataset
using scatterplots and regression.
control-D
```

After you type the control-D, BLSS gives a new prompt.

The **addtext** command can add text to a new text file as well as an already existing one.

4.9.3 Editing Text Files

If you already know how to use the UNIX text editor on your system, you may wish to edit your output text files in order to add remarks, fix mistakes, etc. To edit the text file *xx,* type the BLSS command:

```
.  edit xx
```

The default editor when you use the **edit** command is (at most sites) **vi**. To choose a different editor, use the {**editor**=} option. For example, to edit the text file *xx* with the editor **ex**, give the BLSS command:

```
.  edit {editor=ex} xx
```

If you do not know how to use a text editor, it is not worth learning simply to edit your output text files from BLSS. However, as mentioned in Section 4.4, if you need to enter large amounts of data, knowledge of a text editor can be very helpful. Appendix B contains a brief introduction to the **vi** editor.

---

10. Note that '\' and '"' are themselves special characters in BLSS.

If you accidentally entered the **ex** or **vi** editor by typing the commands above and do not know how to exit, type ':q' and then the RETURN key. You will return to BLSS.

4.10   More about Text Files and Datasets

This section contains information about text files and datasets that will become useful as you gain more experience with BLSS.

4.10.1   Redirection Symbols

The '>>' symbol, introduced in Section 4.5, is called an output text *redirection symbol*, because it redirects a command's output display to a text file instead of the terminal.

Here is the full meaning of the '>>' symbol: If the text file whose name follows the '>>' symbol does not already exist, create it and send the output text there. If it does already exist, append the output text to the existing contents of the text file.[11]

BLSS provides other redirection symbols. The redirection symbol '>!' (used in place of '>>') overwrites the text file (if it exists) with the new output text from the current command. Any previous contents of the text file disappear. Thus, if you give the command:

    . stat rivers >! xx

then the text file *xx* contains the output display from that command and nothing else. If you next give the command:

    . stemleaf rivers >> xx

then *xx* consists of the output display of both commands. If, instead, you give the command:

    . stemleaf rivers >! xx

then *xx* contains only the output text of the 'stemleaf rivers' command—the output text from the previous command, 'stat rivers', is lost.

The redirection symbols '>!' and '>>' allow error messages to go to the terminal, which is what you usually want. The redirection symbols '>&' and '>>&' are analogous to '>!' and '>>', but redirect both the output text and any possible error messages from a command to a text file.[12] Section 11.3 and the manual entry *output(BRM)* both summarize this information.

Note. The '>' output dataset separator, introduced at the end of Section 4.4,

---

11. If a BLSS *dataset* with the same name exists, BLSS gives an error message and refuses to create the text file.
12. The BLSS redirection symbols '>!', '>>', '>&', and '>>&' have identical meanings to those of the UNIX C shell.

is not an output text redirection symbol.[13] In commands that use both output datasets and output text redirection, put the output redirection symbol and text file last:

command > output-datasets >> output-text-file

as we did in an example in Section 4.5.

Be careful to distinguish between *output datasets* produced by BLSS statistics commands, which are BLSS datasets, and the *output text* (or *output display*) of a command, which is normally shown on the terminal. Note that some commands produce output datasets but not output text displays; some commands produce output text but not datasets; and some commands produce both.

4.10.2 Text Representations of Datasets

Another important distinction is between a text representation of a dataset (such as you might create using a text editor) and the corresponding BLSS dataset itself.

There is only one internal BLSS dataset representation for any given dataset (on any given computer type). But there are many possible text representations of a dataset, simply because the appearance of the text can be different. For example, different notation (E-notation or regular notation) might be used for very large or small numbers, different spacing might be used between numbers, different amounts of data might be on each line, etc.

We already saw that the **read** command converts a text representation of data to a BLSS dataset. To convert a BLSS dataset to a text representation, use the **show** command. For example:

```
. show plover >! plover.text
```

The **show** command has various options which affect the appearance of a dataset's text representation. The {**g**}, {**f**}, {**i**}, and other format options control the spacing between and appearance of individual numbers: the number of decimal places shown, whether E-notation is used, etc. If the dataset has more columns than can conveniently fit across a page, the {**col**=} option controls the number of columns printed in a group, and the {**width**=} option controls the overall text width. For more information and examples, see the manual entry *show(BRM)*.

Converting a BLSS dataset to a text representation can be useful not only for making a printout of the data, but also for transferring data between different computers. (It is often possible to transfer the binary BLSS datasets, but not always. Different types of computers may store binary data differently, and some transfer methods—such as electronic mail—may not permit binary data transfer.) Another reason is to transfer data between statistical systems. Most statistical systems store data in a form analogous to BLSS datasets, but each one has its own specific internal format.

---

13. Despite the fact that, in the UNIX shells, '>' *is* used as an output text redirection symbol.

4.10.3 Editing Datasets

On occasion, you may wish to edit the contents of a dataset. Although you cannot edit datasets directly (because they are stored in a binary form), you can edit their text representation. The **editdata** command automates this procedure. The command:

```
. editdata a > b
```

does the following:

1. Using the **show** command, it creates a temporary text file which is a text representation of the dataset *a*.

2. It puts you into the editor for that text file. (As with the **edit** command, you can specify an alternate editor with the {**editor=**} option.)

3. When you exit the editor after making any changes to the text file, it **read**s the text file (using the {**autodims**} option) into the output dataset *b*; the dataset *a* remains unchanged. If no output dataset is specified, **editdata** creates a backup copy of *a*, named *a.bak*, and uses *a* as the output dataset.

Whatever changes you make to the text representation of the dataset must be acceptable to the **read** command. For example, you can change the values of any elements of the dataset; you can add or delete entire rows or columns; and you can change the spacing within rows. But if, for example, you add characters which cannot be interpreted as numbers, or you add an extra element to some but not all rows, the **read** command will give an error message and the new dataset *b* will not be created properly.

4.11 On-Line Help

The **help** command provides on-line information about BLSS. Three types of information are available. The first is the manual entries which comprise Part Two of this book. You can see any BLSS manual entry on-line by typing:

```
. help entry-name
```

where *entry-name* is the name of the manual entry. Some manual entries discuss particular topics. For example, the manual entry *options(BRM)* describes options common to every BLSS command, such as the {**more**} option. In addition, every BLSS command has a manual entry which describes it in detail (some groups of closely related commands share a single manual entry). The manual entry name for a command is simply the name of the command—in other words, you can type:

```
. help command-name
```

for any BLSS command. To save space, the BLSS manual entries follow a common format and observe certain conventions. These are explained in Chapter 11.

The second type of on-line information is the library dataset descriptions which comprise Part Three of this book. You can see any such description by typing:

. *help data.xxx*

where *xxx* is the name of the dataset. The command:

. *help data*

gives brief descriptions of all datasets in the library.

The third type of on-line 'help' information is the help files which contain general information about BLSS, such as information about consulting, news about recent changes, etc. For example, type:

. *help consult*

to find out about consulting for BLSS at your site; type:

. *help news*

to read the news about BLSS.[14] Your local administrator or the instructor for your class (if you are a student) may have installed additional help files with information specific to your site or your class. These can be accessed on-line via the **help** command just like the others.

The **help** command automatically uses the *more* mechanism (described in Section 4.8) when displaying long manual entries and help files.

Keyword search for manual entries. You may not always know the name of the manual entry or command you need. In this case, use the {**key**} option of the **help** command to make a keyword search. For example, the command:

. *help {key} regress*

displays the names and one-line descriptions of all manual entries related to regression (that is, whose one-line descriptions contain the keyword 'regress').

Your own help files. On occasion, you may wish to create your own on-line help files—perhaps to document your own datasets, or to provide your own reminders about BLSS or particular statistical techniques. To make your own help file named *xxx*, simply create a text file *xxx* using either the **addtext** or **edit** commands with the {**help**} option. The {**help**} option places the text file in your *help area*. Thereafter, whenever you type:

. *help xxx*

the file will be displayed.

A help file is simply a text file which is saved in a help area. If you already have a text file which you want to place in your help area, **save** it using the {**help**} option.

---

14. The command **news** is equivalent to the command **help news**.

The man command. A related command, **man**, also shows manual entries and help files. The difference between **man** and **help** is what they show in case of a name conflict: **help** prefers to show a help file; **man** prefers to show a manual entry. For example, suppose you create your own help file with notes on regression and call it *regress*. Then the command:

     . help regress

displays your help file; whereas the command:

     . man regress

displays the manual entry *regress(BRM)*.

4.12   Summary of Utility Commands and Area Options

Table 4.1 summarizes the BLSS utility commands, the purpose of each, and the sections in this book where each is discussed.

Table 4.1.   Summary of utility commands.

Command	Purpose	Section
addtext	Add text to the end of a text file.	4.9.2
edit	Edit text files using the UNIX text editor.	4.9.3
editdata	Edit a BLSS dataset.	4.10.3
help	Display on-line help files and manual entries.	4.11
list	List and describe datasets and text files.	4.1, 4.5, 5.1
load	Load (copy) files from permanent storage areas into the work area.	2.3, 4.6
man	Display on-line manual entries and help files.	4.11
read	Enter your own data into a BLSS dataset.	4.4
remove	Remove files.	4.7
rename	Rename a file.	4.7
save	Save (copy) files into permanent storage areas.	4.6
show	Show BLSS datasets and text files.	2.1, 2.3, 4.5

The utility commands have certain capabilities in common. First, most of them allow you to specify as many file name arguments as you like. For example, the command:

     . load a b c d e f

loads the six named files. We have already seen examples of specifying multiple arguments with the **list**, **save**, and **remove** commands. (The exceptions are **read** and **rename**—you can **read** or **rename** only one file at a time.)

Second, all utility commands (except **editdata**) recognize the area options listed in Table 4.2. These options override the default areas where the commands look for files (in the case of **list**, **load**, **help**, **man**, **read**, and **show**), act upon files (in the case of **addtext**, **edit**, **editdata**, **remove**, and **rename**), or save files (in the case of **save**).

Table 4.2 shows which area each option refers to, what it is used for, and where in this book it is first discussed. (Note that the *bin area* is not introduced until Chapter 10.)[15]

Table 4.2. Summary of area options.

Area	Option	Purpose	Section
bin area	{bin}	Permanent storage of command files.	10.6
data area	{data}	Permanent storage of datasets.	4.6
help area	{help}	Permanent storage of help files.	4.11
home directory	{home}	Where you first login to UNIX.	4.4
text area	{text}	Permanent storage of text files (other than command or help files).	4.6
work area	(none)	Temporary active work area.	4.1

We have seen several examples of using these options in this chapter. Another useful application of them is with the **show** command. For example, the command:

    . show {data} a

shows the contents of the saved BLSS dataset *a*; the command:

    . show {text} t

shows the contents of the saved text file *t*. This is convenient, because it saves you the trouble of **load**-ing saved files merely to see what they contain.

The manual entry *area(BRM)* gives a complete list of area options recognized by the utility commands with a more technical discussion. The manual entries for the individual commands explain more precisely how the options affect each command.

As mentioned earlier, the manual entry *options(BRM)* gives a list of options recognized by *all* BLSS utility and statistics commands.

---

15. BLSS *areas* correspond to UNIX directories. Your BLSS bin, data, help, and text areas are your UNIX *blss/bin*, *blss/data*, *blss/help*, and *blss/text* directories; your BLSS work area is also a subdirectory of your *blss* directory. Note that BLSS normally uses the *blss* tree (that is, directory hierarchy) in your home directory, regardless of the directory from which BLSS is invoked. However, the **–d** option allows you to specify that the *blss* tree is in a different directory. For example, the UNIX command:

    % blss -d project1

causes BLSS to use *project1/blss* as the BLSS tree instead. This option is useful if you are using BLSS for several large, unrelated projects. See the manual entry *blssf(BRM)* for full information on how BLSS uses UNIX directories.

CHAPTER 5

Analyzing Two
or More Variables

5.1 Data Arrays

Statistical data is often arranged in tables. When such tables are stored on a computer, they are called *arrays* or *matrices*.[1] The usual arrangement is that each column of a statistical data array contains observations for a particular statistical variable, and each row contains observations on one individual. The observations (rows) are also known as *cases*.

The dataset called *plover* in the BLSS system data library contains measurements on 5 statistical variables (columns) for each of 68 snowy plover eggs (cases, or rows) and on the chicks that hatched from the eggs:

Snowy Plover Egg/Chick Data

column	statistical variable
1	Egg length, in millimeters
2	Egg breadth, in millimeters
3	Egg weight, in grams
4	Chick weight, in grams
5	Did the chick survive to fledgling? (0 = no; 1 = yes)

The snowy plover is a species of shorebird. These data are part of a much larger dataset collected on free-living snowy plovers in the Monterey Bay area by the Point Reyes Bird Observatory of Stinson Beach, California.

---

1. The word *matrices* is plural for the word *matrix*.

Now let's examine this dataset. Type:

```
. load plover
. list plover
```

The **list** command tells you that *plover* is a BLSS dataset in the form of a matrix—it says:

```
Specified files in your work area:
plover          dataset, dims=(68,5)    (matrix)
```

The first number, 68, is the number of rows. The second number, 5, is the number of columns. These numbers—(68,5)—are called the *dimensions* of the dataset. You can find out the dimensions of any BLSS dataset with the **list** command.

To see the data, you might type:

```
. show plover
```

However, if you are using a video terminal with a display less than 68 lines long (the number of rows in the dataset) the top rows will probably run off the screen before you have a chance to read them. One remedy is to look at the data one column at a time, as described below. Another remedy is the {**more**} option, described in Section 4.8.

5.1.1 Referring to Portions of Datasets

Columns. Individual columns of datasets are referred to using *single subscripts* such as [1], [2], etc. For example, the individual columns of *plover* are referred to as *plover*[1], *plover*[2], etc. The number inside the square brackets is the number of the column. To see the third column (egg weight) of *plover*, type:

```
. show plover[3]
```

Note that the entire word *plover*[3] is typed without any spaces. When you **show** individual columns this way, you read the numbers going across, not down—just as you did earlier with the datasets *finalexam*, *fake*, and *rivers*.[2] Note that some of the egg weight observations are missing—they are printed as NA's.[3] Check the result of the last **show** command to see that the weight of the second egg is 7.4 grams and the weight of the second last egg is 9.0 grams.

Elements. Individual elements of a dataset are referred to using *double subscripts*. Here are two examples. To reference the two elements we just checked above—2nd row, 3rd column and 67th row, 3rd column—use the

---

2. The {**shape**} option to the **show** command makes it show column vectors in their true shape—as columns. See *show(BRM)* for more information.

3. Missing values (NA's) were introduced in Section 2.3. The reason some egg weights are missing is that those eggs were not discovered until they were several days old. The researchers were interested in weights of newly laid eggs only, because eggs lose weight as the embryo develops.

forms *plover*[2,3] and *plover*[67,3]. For example, to **show** just those elements, type:

```
. show plover[2,3] plover[67,3]
```

The general form for referring to individual elements of arrays is:

$$arrayname[rownumber, colnumber]$$

Rows. Individual rows of a dataset are referred to as follows: *plover*[1,] is row 1 of *plover*; *plover*[2,] is row 2 of *plover*; etc. This notation is an extension of double subscript notation: the row number and the comma are typed just as in a double subscript, but the column number is omitted. When the column number is omitted from a double subscript, every element in the row—in other words, the entire row—is referenced.

The reverse is also true: when the row number is omitted from a double subscript, every element in the column—in other words, the entire column—is referenced. Thus, individual columns may be referred to using the forms *plover*[,1], *plover*[,2], etc. However, because individual statistical variables (columns) are more often needed than individual cases (rows), BLSS provides the simpler single subscript notation described above for referring to columns.

5.1.2   Vectors

On occasion, you may want to work extensively with a single column or row from a BLSS data matrix. If so, it is convenient to create a new dataset which contains the column or row of interest. You do this using the subscript methods described above. For example, to create a dataset called *survived* which contains column 5 (the survival code) of *plover*, type:

```
. survived = plover[5]
```

To create a dataset called *case27* which contains the data for case 27 (that is, row 27) only, type:

```
. case27 = plover[27,]
```

Here is how the **list** command describes the new datasets:

```
. list survived case27

Specified files in your work area:
case27         dataset, dims=(1,5)      (row vector)
survived       dataset, dims=(68,1)     (column vector)
```

Datasets that have only one row are called *row vectors*. Datasets that have only one column are called *column vectors*.

In Section 8.3 we return to the subject of matrix subscripting in much greater detail.

5.1.3   Descriptive Statistics for Data Arrays

The **stat** command produces descriptive statistics for all five columns (statistical variables) of *plover*. Type:

```
. stat plover
```

The result is:

```
Statistics: plover
Col   N     Mean      SD       Min      25%      50%      75%      Max
 1    68    31.42    1.173    28.07    30.66    31.47    32.15    34.41
 2    68    22.84    0.5372   21.49    22.51    22.88    23.20    23.99
 3    52     8.638   0.5018    7.400    8.300    8.650    9.000    9.900
 4    57     6.144   0.3946    5.200    5.900    6.200    6.400    7.100
 5    65     0.4462  0.5010    0.000    0.000    0.000    1.000    1.000
```

The first heading, 'Col'—which we have ignored until now—gives the column number. The second heading, 'N', is the number of non-missing observations in the column (recall that *plover* contained 68 cases). As you can see from the display, only columns 1 and 2 are free of missing values. The remaining headings we discussed in Section 3.1.

Similarly, the **stemleaf** command makes stem-and-leaf diagrams for all five columns:

```
. stemleaf plover
```

If you typed the preceding command on a video terminal, all but the last one or two diagrams probably flew off the top of your screen. There are several remedies. You can use the {**more**} option described in Section 4.8 to show the display one screenful at a time. Alternatively, you can make stemleaf diagrams for just one column at a time. There are two methods. First, you can refer to individual columns using the forms *plover*[1], etc., as described above. Type:

```
. stemleaf plover[1]
```

Second, the **stemleaf** command provides an option that allows you to directly specify the column it is to work on. Type:

```
. stemleaf {x=1} plover
```

As mentioned earlier, *options* change the behavior of a command. In this case, the option {**x=1**} tells **stemleaf** to make a diagram of column 1 only. Instead of {**x=1**} you can use {**x=2**} to make a stemleaf diagram of only column 2, or {**x=3**} for column 3, etc.

The same two methods for processing one column at a time also work with the **stat** command. Try either of the following:[4]

```
. stat {x=2} plover
. stat plover[2]
```

---

4. The two commands are not quite identical. The output labeling is slightly different and—more important—the {x=} method executes somewhat more efficiently.

The ability to specify individual statistical variables in a dataset—whether via options or subscripts—is used extensively in the following sections, when we look at the relationships between statistical variables.

5.2 Scatterplots

The **scat** command makes scatterplots from data stored in BLSS datasets. To make a scatterplot for *plover* with egg length on the horizontal axis (X-axis) and egg weight on the vertical axis (Y-axis), type:

```
. scat plover {x=1;y=3}
```

In this command, we enclosed two options within curly braces { }, and we separated them with a ';' symbol. The option {x=**1**} tells **scat** to use column 1 as the X-variable. The option {y=**3**} tells it to use column 3 as the Y-variable. Another example: to plot column 3 as the X-variable and column 4 as the Y-variable, type:

```
. scat plover {x=3;y=4}
```

The scatterplot display. Figure 5.1 shows the scatterplot produced by the first example.[5] Each star '*' in the scatterplot represents one observation. The '2' characters indicate two observations at the same point. In general, if 2 through 9 observations must be plotted at the same location, then instead of printing a star at that location, **scat** prints the number of observations which go there. If 10 through 19 observations are to be plotted at one location, **scat** puts a '%' symbol at that location; if 20 or more observations, a '#' symbol.

More about scat. It is possible to plot two vectors against each other provided that they have the same length. Here is how. For this example, we create a new column vector. It contains a variable which is roughly equal to the egg volumes: $\pi/6$ times the length times the breadth squared.[6] Call it *vol*:

```
. vol = PI/6 * plover[1] * plover[2]^2
```

(In BLSS, the mathematical constant π is denoted by **PI**. Section 8.2 contains a short discussion.) To plot the variables *vol* and *plover*[3] (egg weight) against each other, type:

```
. scat vol plover[3]
```

The first vector (in this case, *vol*) is used as the X-variable. The second vector (in this case, *plover*[3]) is used as the Y-variable. Can you explain to yourself why these vectors have the same length?

---

5. In truth, the output in Figure 5.1 (and in all **scat** displays in this book) was generated using **scat** with the {**big**} option. The {**big**} option generates a larger display, better suited for printouts. The default size is appropriate for video display terminals.

6. The ratio of the volume of an egg to the volume of the circumscribing rectangular solid is approximately equal to the ratio of the volume of a sphere to the volume of the circumscribing cube, which is $\pi/6$.

Figure 5.1

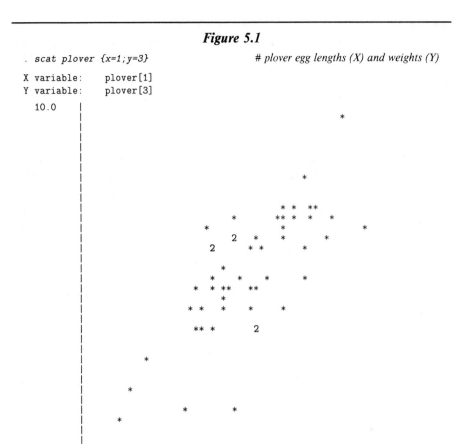

```
. scat plover {x=1;y=3}                    # plover egg lengths (X) and weights (Y)

X variable:    plover[1]
Y variable:    plover[3]
```

It is also possible to make scatterplots of columns from two different datasets—again, provided that they have the same length. There are two ways to do it. The first uses single subscripts and the second uses the {x=} and {y=} options. Suppose that you have two datasets of the same length called *u* and *v* and that you want to make a scatterplot with column 2 of *u* as the X-variable and column 4 of *v* as the Y-variable. To use the single subscript method, you would type:

```
. scat u[2] v[4]
```

To use the options, you would type:

```
. scat u {x=2} v {y=4}
```

(Note that this example is hypothetical. Just typing this command will produce an error message, because, unless you create them yourself, there are no datasets named *u* or *v* in your work area.)

Simultaneous scatterplots. It is possible to make scatterplots of two or more different Y-variables against one X-variable. For example, to plot both columns 1 and 2 (egg length and egg breadth) as the Y-variables against column 3 (egg weight) as the X-variable, type:

```
. scat plover {y=1,2;x=3}
```

The scatterplot from this command is shown in Figure 5.2.

Each *a* character in the plot denotes a single point which corresponds to the first Y-variable (in this case, column 1, or egg length). Each *b* character in the plot denotes a single point which corresponds to the second Y-variable (in this case, column 2, or egg breadth). The *A* characters denote two or more points from the first Y-variable (that is, more than one *a*) at the same location; the *B* characters denote two or more points from the second Y-variable. If there

---

Figure 5.2

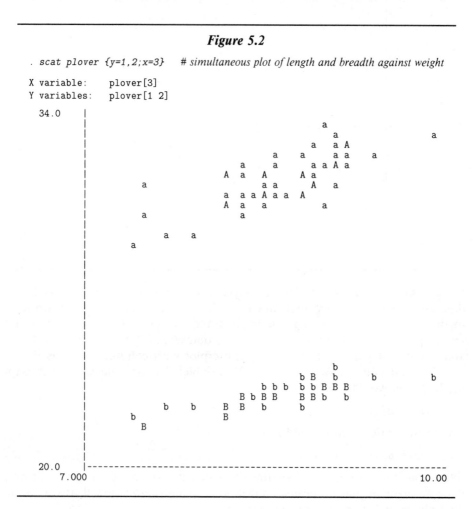

were a third Y-variable, it would be plotted with c's and C's; a fourth Y-variable with d's and D's; and so forth.

In Figure 5.2, the a's and the b's form two separate clusters, because the lengths are all greater than the breadths. However, in general, points corresponding to different Y-variables might overlap. If observations from more than one Y-variable must be plotted at the same location, that location is marked with a '$' symbol. An example of this is in Figure 5.3.

Simultaneous scatterplots can also be made of columns from two different datasets. Continuing the hypothetical example with u and v, you would type:

. scat u {x=2} v {y=3,4,5}

to simultaneously plot columns 3, 4, and 5 of v against column 2 of u. If you were to omit the {y=} option in this case, **scat** would make a simultaneous scatterplot of all columns of v against the second column of u.

The **scat** command has several more options which control how large a plot to make, the minimum and maximum X and Y values to plot, etc. For more information and examples, see *scat(BRM)*—that is, the reference manual entry on **scat**.

5.3   Correlation and Covariance

The **stat** command has an option, {**cor**}, which makes it display correlations in addition to the usual descriptive statistics. Type:

. stat {cor} plover

On occasion, you may want the correlations only and not the usual **stat** output. If so, use the {**coronly**} option:

. stat {coronly} plover

In either case, **stat** produces the correlations for all possible pairs of statistical variables (columns) in the dataset. The output is in the form of a *lower triangular correlation matrix*:

```
Correlation Matrix: plover          (N = 42)
        1         2         3         4         5
1     1.000
2     0.405     1.000
3     0.793     0.839     1.000
4     0.678     0.734     0.847     1.000
5     0.127     0.099     0.167     0.229     1.000
```

This is a table which contains the various correlations. For example, the correlation of variable 3 with variable 1 can be found by reading off the number in row 3, column 1: 0.793. Because the correlation of variable i with variable j equals the correlation of variable j with variable i, the correlation matrix is said to be *symmetric*. Therefore, only the lower triangular portion of the correlation matrix is actually shown—it contains all the information. This arrangement is just like the tables of mileages between cities sometimes found on road maps: because the distance from city *A* to city *B* equals the distance from city *B* to city *A*, only one distance is actually shown.

Notice that the correlation of each variable with itself is 1.

The notation '(N = 42)' in the display indicates that the correlations were computed from 42 cases of *plover* which were free of missing data. We discuss this point further in the next section.

The **stat** command has two options that produce the covariance matrix: {**cov**} and {**covonly**}. The {**cov**} option produces the covariances along with the other statistics; the {**covonly**} option produces the covariances only. For example:

```
. stat {covonly} plover

Covariance Matrix: plover          (N = 42)
        1         2        3        4        5
1    1.221
2    0.1996    0.1987
3    0.4254    0.1816   0.2359
4    0.3129    0.1368   0.1719   0.1745
5    0.07077   0.02233  0.04077  0.04826  0.2538
```

The covariance of variable 1 with variable 3 is 0.4254. The covariance of variable 1 with itself—in other words, its variance—is 1.221.

The variance and other univariate moment statistics can also be shown using the {**mom**} option. For more information about this and other options to the **stat** command, see the reference manual entry *stat(BRM)*. Moreover, all statistics computed by **stat** can be saved as *output datasets*, which are discussed in the next section.

5.4 Missing Data

Some datasets are 'complete', in the sense that they are not missing any data. But real data often contains missing observations. Such is the case with our *plover* dataset. How does BLSS handle missing values?

Some BLSS commands do the obvious: they simply ignore missing values. For example, when the **stat** command computes means, SDs, and percentiles for each column, it ignores elements in the column which are missing. When the **scat** command makes scatterplots, it does not plot any point for which either coordinate is missing. (Although *plover* has 68 cases, the scatterplot in Figure 5.1 has only 52 points.)[7]

---

7. In contrast, it is not obvious how to compute correlations and covariances when a dataset contains missing values. The **stat** command provides two options: {**casex**}, which causes

In contrast, some statistical techniques make no sense when applied to datasets which contain missing values. The corresponding BLSS commands refuse to work on datasets with missing values. One such command, which we use in the next section, is **regress**. In this situation, the simplest solution—although not the only one, and not necessarily the best—is to create a new dataset composed of the completely non-missing cases of the original dataset.

We do this using the **select** command—it 'selects' the non-missing cases. Type:

```
. select plover > plov
```

We first saw this command form in Section 4.4, but let's review it now. The word 'select', of course, is the name of the command, and the word 'plover' is its *argument* or *input dataset*—the name of the dataset on which the command operates. The '> plov' means: put the selected data (the cases which are free of missing values) into a new BLSS dataset (called the *output dataset*) which has the name *plov*. The '>' symbol is called the *output dataset separator*: it serves to separate the input dataset from the output dataset.

Of course, you can choose any name you like for your output dataset. But it is best to choose names which are suggestive of what the dataset contains. In this case, we chose the name *plov* simply because it is short for *plover*. As the **list** command shows:

```
. list plover plov

Specified files in your work area:
plov            dataset, dims=(42,5)     (matrix)
plover          dataset, dims=(68,5)     (matrix)
```

the new dataset contains only 42 cases. The other 16 cases were excluded because of missing data elements.

We discuss missing values again, in more detail, in Sections 8.5 and 8.7.

5.5   Regression

The **regress** command makes least squares regressions.

5.5.1   Simple Regression

Suppose you want to use the *plov* dataset (as created in the previous section) to regress the weight of the plover eggs (column 3) on their lengths (column 1). You can obtain the regression display only, or you can also save the numerical results as output datasets.

---

casewise exclusion of missing data (if a case contains *any* missing values, it is excluded from the calculations for *all* the statistics); and {**pairx**}, which causes *pairwise exclusion* of missing data. By default, **stat** uses {**casex**} whenever correlations or covariances are computed. See *stat(BRM)* for full information.

The regression display. Type the command:

```
. regress plov {x=1;y=3}
```

The option {**x**=**1**} tells **regress** to use column 1 as the independent variable. The option {**y**=**3**} tells **regress** to use column 3 as the dependent variable. Of course, you can use different column numbers to specify different independent or dependent variables. The output display from this command is:

```
Dependent variable:      plov[3]
Independent variable:    plov[1]
Observations  42         Parameters  2

Parameter    Estimate      SE         t-Ratio      P-Value
intercept    -2.3165     1.3296      -1.7423       0.0891
  coef 1      0.34851    0.042365     8.2263       0.0000

Residual SD  0.29970    Residual Variance   0.089823
Multiple R   0.79278    Multiple R-squared  0.62850
```

The first three lines tell you: the dependent variable was *plov*[3]; the independent variable was *plov*[1]; the regression was based on 42 observations; and it estimated 2 parameters. The rest of the display consists of regression statistics. For each parameter, including the intercept and the regression coefficient ('coef 1'), it shows: the estimated value of the parameter ('Estimate'); the standard error of the estimate ('SE'); the t-ratio (also called the t-statistic: it is the estimated value divided by its SE); and the P-value (two-sided) of the t-ratio. At the bottom the display shows: the residual SD (that is, the RMS error); the residual variance (that is, the MS error); the multiple correlation *R*; and the multiple R^2.

The regression equation can be formed from the estimated value of the intercept and the regression coefficient. In this example, the regression equation for the fitted values is:

fitted value for weight = −2.3165 + 0.34851 ∗ *length*

Some textbooks use the term 'estimated value' or 'predicted value' instead of the term 'fitted value'.

Regression output datasets. The following command displays the same statistics as above. It also saves the fitted values and residuals from the regression in new datasets:

```
. regress plov {x=1;y=2} > fit res
```

The '> *fit res*' means: store the fitted values and the residuals in new datasets called, respectively, *fit* and *res*. Here is a detailed explanation. As we saw in Section 5.4, the '>' symbol means: store the numerical results of a command in the BLSS datasets whose names follow—the output datasets. In this case, there are two output datasets. The first output dataset contains the fitted values, so we call it *fit*. The second output dataset contains the regression residuals, so we call it *res*.

In place of the names *fit* and *res*, you can use any dataset names. We chose *fit* and *res* simply to help remind you what is stored in these particular datasets.

Regression plots. You can see the regression line by making a scatterplot of the fitted values (Y-axis) against the independent variable (X-axis):

```
. scat plov[1] fit
```

However, it is more useful to put the original data and the fitted values on the same plot. Type:

```
. scat plov[1] (plov[3],fit)
```

This produces a simultaneous scatterplot, which is shown in Figure 5.3. The data are plotted as *a*'s (for one data point at a location) and *A*'s (for several data points at a location). The fitted values (which fall on the regression line)

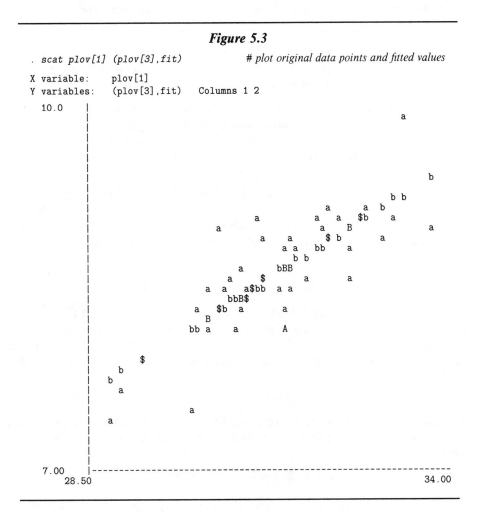

Figure 5.3

are plotted as *b*'s and *B*'s. Where both data values and fitted values fall at the same location, they are plotted as a *$* character. Thus, the regression line is the line that passes through the *b*, *B*, and *$* characters.

It may be helpful to compare the scatterplot in Figure 5.3 with that in Figure 5.1—the *a*'s correspond to the *'s.[8]

In the **scat** command for Figure 5.3, we used a comma ',' to create a new dataset whose first column is the original dependent variable and whose second column is the fitted value. We discuss this ability further in Section 5.6.3. The important point to remember for now is that you can make simultaneous scatterplots this way, even if the columns come from different datasets. Just be sure the columns have the same lengths.

The residual plot is a plot of the regression residuals (Y-axis) against the independent variable (X-axis). To make it, use the **scat** command:

```
. scat plov[1] res
```

The resulting residual plot is shown in Figure 5.4. The X-axis runs through the center of the plot. Note that when a point to be plotted falls on one of the axes, the '*' symbol replaces the corresponding axis symbol.

Warning. Output datasets created by a command overwrite any preexisting datasets or text files with the same name. If you run several different regression commands and you want to save the fitted values and residuals from each command, use different names for the output datasets from each command.

As an exercise, use the *plov* dataset to regress chick weights on egg weights. Make scatterplots which show the fitted regression line and which plot the residuals against the independent variable. Create the output datasets in such a way that the output datasets *fit* and *res* from the previous regression are not overwritten.

5.5.2 Multiple Regression

To perform a multiple regression, use the {x=} option to list all the columns that are to be the independent variables. Use commas to separate the column numbers within the option. Everything else works just the same as with simple regression. For example, to regress egg weight on both length and breadth, type:

```
. regress plov {x=1,2; y=3}
```

A second example: to regress chick weight (column 4) on the first three variables, and to save the fitted values and residuals in datasets called *wt.fit* and *wt.res*, type:

```
. regress plov {x=1,2,3; y=4} > wt.fit wt.res
```

---

8. Can you explain why some *'s in Figure 5.1 do not have corresponding *a*'s in Figure 5.3?

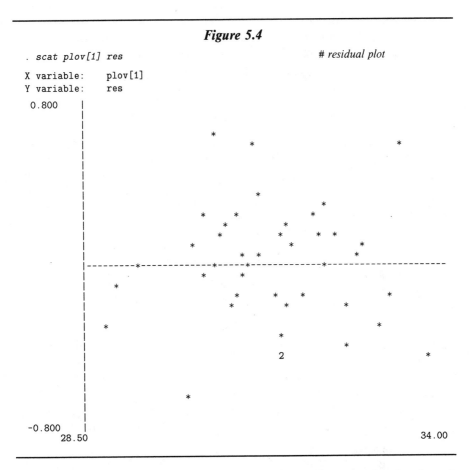

Figure 5.4

To plot the regression residuals (on the Y-axis) against the fitted values (on the X-axis), simply plot those vectors against each other:

```
. scat wt.fit wt.res
```

5.5.3   Regression Statistics

Several options cause the **regress** command to produce additional statistics. The {**rstat**} option displays three different types of R^2 statistics, the autocorrelation of the residuals, and the Durbin-Watson statistic.[9] The {**anova**} option displays the analysis of variance table for the regression. The {**bcov**} and {**bcor**} options display, respectively, the estimated covariance and correlation of the regression coefficients. The {**long**} option displays all the

---

9. The Durbin-Watson statistic is (asymptotically) equal to 2 * (1 – residual autocorrelation) and thus provides equivalent information. But many users of the Durbin-Watson statistic are unaware of this; hence {**rstat**} shows both statistics. Of course, the residual autocorrelation and Durbin-Watson statistics make sense only when the data have a time dependence—which the *plover* data do not.

Figure 5.5

```
. regress plov {x=1,2;y=3} {long}

Dependent variable:      plov[3]
Independent variables:   plov[1 2]
Observations  42         Parameters  3

Parameter   Estimate      SE       t-Ratio    P-Value
intercept   -14.264     0.91567   -15.5779    0.0000
 coef 1      0.23824    0.017456   13.6480    0.0000
 coef 2      0.67424    0.043260   15.5859    0.0000

Residual SD  0.11289    Residual Variance   0.012744
Multiple R   0.97397    Multiple R-squared  0.94861    Centered
Multiple R   0.97261    Multiple R-squared  0.94597    Centered, adjusted
Multiple R   0.99992    Multiple R-squared  0.99984    Uncentered
Res Autocor -0.00676    Durbin-Watson       1.98501

Anova Table
 Source       df       SS        MS          F                    P-Value
Fit           2     9.1744    4.5872    359.936 (df=2,39)        0.0000
Residual     39     0.49703   0.012744
Total        41     9.6714
Mean          1     3116.6
Grand Total  42     3126.3

Estimated Covariance of Coefficients:
           intercept     1         2
intercept  0.8385
    1      -0.002564   0.000305
    2      -0.03316   -0.000306   0.001871

Estimated Correlation of Coefficients:
           intercept     1         2
intercept  1.000
    1      -0.160      1.000
    2      -0.837     -0.405      1.000
```

aforementioned statistics. Figure 5.5 shows output from the first multiple regression example, invoked with the {**long**} option.

See the manual entry *regress(BRM)* for explanations of these statistics and information on how to save them in new datasets. See Section 11.3.1 for general information about saving only one output dataset when many are available.

5.6   Adding Variables to a Regression

Sometimes we want to create new variables for a regression model, and perhaps add them to the dataset on which we are making regressions.

5.6.1   Simple Regression with Transformed Variables

In the examples of the previous section, we regressed egg weight on length alone, or on length and breadth. Common sense tells us that egg weight is closely related to egg length and breadth not directly, but rather through volume—we expect the weight to be roughly proportional to the volume, and the volume, in turn, to be proportional to the length times the breadth squared. Suppose that we want to regress egg weight on egg volume. First, we create a new dataset called *volume* which is roughly equal to the volume:

```
. volume = PI/6 * plov[1] * plov[2]^2
```

Then we use the new dataset in the regression:

```
. regress volume plov {y=3}
```

Note that we specified the independent and dependent variables as separate datasets. The **regress** command allows this, just as does the **scat** command. The output from this command is:

```
Dependent variable:      plov[3]
Independent variable:    volume
Observations  42         Parameters  2

Parameter    Estimate      SE          t-Ratio    P-Value
intercept    0.92331       0.27924     3.3065     0.0020
 coef 1      0.0008962     3.248e-05   27.5931    0.0000

Residual SD  0.10986      Residual Variance   0.012068
Multiple R   0.97472      Multiple R-squared  0.95009
```

If we were planning to use the weight variable extensively, or if we wanted the regression output to state that the dependent variable is 'weight' instead of 'plov[3]', we could explicitly create a separate *weight* vector and use it:

```
. weight = plov[3]
. regress volume weight
```

```
Dependent variable:      weight
Independent variable:    volume
```
(same regression output as above)

5.6.2 Transformed Variables in a Multiple Regression

In the previous section, we used the new dataset *volume* as an independent variable in a simple regression. We could also use it as the dependent variable in either a simple or multiple regression, because, as we saw above, the **regress** command allows the dependent variable to come from a different dataset than the independent variables.

However, the independent variables in a regression must all come from the same dataset. In order to add independent variables to a regression model, we create a new dataset which contains the original data and also the new variables. For example, suppose we want to use *volume* and another variable, *density* (equal to *weight* divided by *volume*), as independent variables along with egg length, breadth, and weight in a regression model for the chick weights. (This is not a particularly good regression model. We make it for the purpose of illustration.)

There are several ways to create the new dataset. All involve using the comma ',' notation we saw in Section 5.5.1. First, we can create *density* as an intermediate dataset, just as we created *volume* and *weight* above:

```
. density = weight / volume
```

and then create the new dataset using commas:

```
. nplov = plov, volume, density
```

Alternatively, we can create the new dataset *nplov* directly, without using the intermediate datasets *density* or *weight:*

```
. nplov = plov, volume, plov[3]/volume
```

In either case, the new dataset *nplov* consists of the columns of the first dataset listed, followed by the columns of the second, followed by the columns of the third. Thus, columns 1 through 5 of *nplov* contain same data as *plov*, column 6 contains the volume, and column 7 contains the density. This new dataset can be used for regressions or any other purpose. For our original purpose—to regress column 4 (chick weights) on columns 1, 2, 3, 6 and 7— type:

```
. regress nplov {x=1,2,3,6,7; y=4}
```

Section 9.4 contains several more advanced examples of regression.

5.6.3   Catenating Datasets

When the ',' is used to create new datasets as in the examples above, it is known as the *columnwise catenation operator*, because it joins together (or *catenates*) the columns of the listed datasets into a single new dataset. Any number of datasets may be catenated together, and they may contain any number of columns. The only restriction is that the datasets to be catenated together must contain the same number of rows (that is, the columns must have equal lengths).

If the new dataset is needed for only one command, it does not need to be explicitly created with a **let** command. For example, when we made the simultaneous scatterplot of the data and the regression residuals in Section 5.5.1, we catenated the vector containing the dependent variable and the vector containing the residual into a temporary dataset, and made a simultaneous scatterplot of the two Y-variables in the temporary dataset without explicitly storing it. Similarly, in Section 3.3 we used temporary datasets, such as *log(tractor)*, without explicitly creating or storing them.

We first saw the ',' operator in Section 2.2, where we used it to create lists of numbers. That was a special case: the datasets being catenated were individual numbers (one row and one column each) or row vectors, and the resulting list was a row vector.

Rows can also be catenated together: the rowwise catenation operator is the double-comma ',,'. These operators, and others for manipulating and rearranging arrays, are discussed in more detail in Section 8.3 (8.3.3 in particular).

CHAPTER 6

Confidence Intervals and Hypothesis Tests

Although confidence intervals and hypothesis tests have different goals—one estimates the value of a population parameter whereas the other tests a hypothesis about the value of a population parameter—we present them together because they make the same assumptions about a population and the sample drawn from it, and because the theory and calculations behind them (which we do not cover) are nearly identical. For these reasons, the syntaxes of **ci**, the BLSS command for constructing confidence intervals, and that of **ht**, the command for performing hypothesis tests, are nearly identical.[1]

This chapter explains how to use the **ci** and **ht** commands to calculate confidence intervals and hypothesis tests for a population mean, probability of success, or proportion of successes, and the difference between two population means, probabilities of success, or proportions of successes. It shows many examples of typical uses of the commands, but does not give complete information on the command inputs and options, nor any information on the outputs. For this information see the manual entries, *ci(BRM)* and *ht(BRM)*.

Populations and parameters. In this chapter, two types of data are considered. The first type is an independently drawn random sample from a Normal (or approximately Normal) population with unknown mean. For this type of data, the parameter of interest is the population mean.

---

1. As the manual entries *ci(BRM)* and *ht(BRM)* show, the only real difference in usage is the default order of the output datasets.

The other type of data is *Bernoulli* data, which is discussed in Section 6.2. Bernoulli observations arise from two kinds of problems. The first kind is repeated experiments: the occurrence ('success') or nonoccurrence ('failure') of a specific event for a fixed number of independent identical experiments is a Bernoulli observation. In this case, the parameter of interest is the *probability* of success. The second kind is drawing a simple random sample from a population: the presence or absence of a specific characteristic in a subject is a Bernoulli observation. In this case, the parameter of interest is the *proportion* of a population in which a specific characteristic is present.

6.1   Confidence Intervals for Means

A *confidence interval* is an interval constructed from a sample which, with a prespecified probability, contains the parameter of interest. The **ci** command computes confidence intervals for the types of parameters listed in the previous section. Its general form is:

ci x [y] {**alpha**=;**b**;**level**=;**nx**=;**ny**=;**p**;**pool**;**sdx**=;**sdy**=;**x**=;**y**=;**z**;...} > *outputs*

where: x is the sample; y, which is optional, contains the second sample if any; and the items enclosed in curly braces are options. **Ci** creates a confidence interval for every column of x (and y, if it is present). Specific columns of x and y can be selected using the {**x**=} and {**y**=} options.

The probability that a confidence interval contains the parameter of interest is called its *confidence level*. A level p confidence interval is also called a ($p*100$)% confidence interval: for example, a 95% confidence interval has confidence level .95. Some authors call the confidence level the *confidence coefficient*.

The confidence level is specified by either the {**level**=} or {**alpha**=} option: *level* = 1 − α. Several values of the level (or of α) may be given; **ci** constructs an interval for each. The *default values* (that is, the values used when no option is given) are {**level**= .5, .95, .99} or, equivalently, {**alpha**= .5, .05, .01}.[2]

Assumptions. Classical confidence interval estimation for means makes two assumptions:

1. The observations are independent.

2. The underlying population from which the data are sampled has a Normal, or approximately Normal, distribution.

If these assumptions are met, then the computed test statistic has the presumed distribution (standard Normal or Student's t) and the actual

2. No single confidence level is best under all circumstances. For this reason, BLSS provides not one but three default confidence levels. If you want one single level, you must choose it yourself.

confidence level of the computed interval is approximately equal to the specified level. Otherwise, the actual confidence level may be different.

Independence is the more important assumption: for most population distributions, if the observations are independent, then when the sample size is large, the confidence interval for the mean based on the t distribution has approximately the requested confidence level. (The first example in Section 9.5.2 explores this issue further.)

BLSS, of course, does not check these assumptions. It is the user's responsibility to do so.

6.1.1   One Sample—for a Population Mean

This section tells how to use **ci** to construct confidence intervals for a population mean. These are called *one-sample* confidence intervals because they are based on a single sample.

One sample, population SD known. If the population SD is known, specify it with the {**sdx=**} option. **Ci** constructs the confidence interval based on the Normal distribution for the sample mean.

The BLSS library dataset *plover* mentioned in Chapter 5 contains measurements on snowy plover eggs and chicks. Suppose we know that the SD for snowy plover egg weights is .5 grams. To construct a confidence interval for the mean of the egg weights (column 3):

```
. load plover
. ci plover {x=3} {sdx=.5}

Confidence Interval for the mean.  Data: plover.

Col   N     sample     known       level    z     confidence interval
            mean       SE                          lower end   upper end
 3    52    8.6385     0.069338    .5000   0.67    8.5917      8.6852
                                   .9500   1.96    8.5026      8.7744
                                   .9900   2.58    8.4599      8.8171
```

Ci displays the statistics and the Normal quantile[3] (z) used to construct the confidence intervals along with the lower and upper endpoints of the confidence intervals for each level. For example, the 95% confidence interval for the mean is [8.5026, 8.7744]; the Normal quantile used to construct the interval is 1.96.

One sample, SD unknown. If the population SD is unknown, don't specify it. When the SD is not specified, **ci** automatically estimates it from the data and constructs the confidence interval using Student's t distribution.

For example, we do not in fact know the population SD of the egg weights. To construct the t-based confidence interval:

---

3. That is, the value z such that the probability of a standard Normal random variable not exceeding z is $1-\alpha/2$. Quantiles are discussed in detail in Section 7.4.2.

```
. ci plover {x=3}

Confidence Interval for the mean.  Data: plover.
```

Col	N	sample mean	sample SE	df	level	t	confidence interval lower end	upper end
3	52	8.6385	0.069590	51.0	.5000	0.68	8.5912	8.6857
					.9500	2.01	8.4988	8.7782
					.9900	2.68	8.4523	8.8247

Instead of a Normal quantile, **ci** displays a Student's t quantile (t) and its degrees of freedom (df). In this case the 95% confidence interval for the mean is [8.4988, 8.7782].

The {z} option. Under the {z} option, **ci** uses the standard Normal distribution to compute the confidence interval instead of Student's t distribution even when the SD is estimated from the data. If the sample size is large, the resulting confidence interval is approximately the same as that calculated from the t distribution because as the sample size (and hence the degrees of freedom) gets large, the t distribution converges to the standard Normal distribution. But if the sample size is small, a confidence interval computed using the Normal distribution is smaller than if it had been computed using the t distribution.

The {z} option is offered for pedagogical reasons and for comparison with the t distribution. It is not recommended in practice because the probability that the resulting smaller confidence interval contains the mean is always less than $1-\alpha$ (unless the SD is in fact known—which it seldom, if ever, is).

As an example of a confidence interval using the {z} option, we construct a confidence interval for mean egg weight using the Normal distribution instead of Student's t:

```
. ci {z} plover {x=3}

Confidence Interval for the mean.  Data: plover.
```

Col	N	sample mean	sample SE	level	z	confidence interval lower end	upper end
3	52	8.6385	0.069590	.5000	0.67	8.5915	8.6854
				.9500	1.96	8.5021	8.7749
				.9900	2.58	8.4592	8.8177

Because the sample size is fairly large, the confidence intervals for mean egg weight computed here are just a bit smaller than those computed using Student's t distribution. For smaller sample sizes, the difference between the t-based and Normal-based confidence intervals is much more pronounced.

6.1.2   Two Samples—for the Difference Between Population Means

This section tells how to use **ci** to construct two-sample confidence intervals for the difference between two population means—that is, an interval which, with a prespecified probability, contains the true difference between the population means.

Let us construct confidence intervals for the difference between the mean weight of the snowy plover chicks that survived to fledgling and the mean weight of those that did not. We construct two datasets: *plov0*, which contains the chicks that did not survive to fledgling (those for which *plov*[5] = 0); and *plov1*, which contains those that did (for which *plov*[5] = 1):

```
. select plover (plover[5]==0) {log} > plov0
. select plover (plover[5]==1) {log} > plov1
```

The **select** command is discussed further in Section 8.7 and, of course, in the manual entry *select(BRM)*.

Two samples, SDs known. If both population SDs are known, specify them using the {**sdx=**} and {**sdy=**} options. **Ci** constructs the confidence interval using the Normal distribution.

Suppose, for example, we know that the population SD of chick weight for *plov1* is .42 grams, and the population SD of chick weight for *plov0* is .35 grams:

```
. ci plov0 {x=4} plov1 {y=4} {sdx=.35; sdy=.42}
```

Confidence Interval for difference of means. Data: plov0 and plov1.

Col	N	sample mean diff	known SE	level	z	confidence interval lower end	upper end
4	26,29	-0.22215	0.10390	.5000	0.67	-0.29222	-0.15207
				.9500	1.96	-0.42578	-0.018517
				.9900	2.58	-0.48977	0.045469

SDs unknown. If at least one of the population SDs is unknown, **ci** constructs the confidence interval using Student's t distribution.

Two samples, SDs unknown but equal. If the SDs are unknown but assumed to be equal, we can *pool* together the samples in order to estimate the value of the unknown SD. The {**pool**} option does this. For example, to create a confidence interval for the difference between mean chick weights from *plov0* and *plov1* using a pooled estimate of the SD:

```
. ci plov0 {x=4} plov1 {y=4} {pool}   # pool the samples
```

Confidence Interval for difference of means. Data: plov0 and plov1.

Col	N	sample mean diff	sample SE (pooled)	df	level	t	confidence interval lower end	upper end
4	26,29	-0.22215	0.10465	53.0	.5000	0.68	-0.29322	-0.15108
					.9500	2.01	-0.43205	-0.012249
					.9900	2.67	-0.50175	0.057455

Two samples, SDs unknown and not known to be equal. In the two preceding examples, we made unrealistic assumptions: in the first, that we knew the population SDs and in the second, that even though the SDs were unknown, they were equal. With this data—as with most—these assumptions are unjustified. To drop the assumptions, drop the corresponding options. By default, **ci** estimates the SDs (separately) and uses an approximation procedure—a t distribution with degrees of freedom estimated using Welch's approximation[4] to construct the confidence interval.

For example, to compute a confidence interval for the difference between mean chick weights from *plov0* and *plov1*:

```
. ci plov0 {x=4} plov1 {y=4}

Confidence Interval for difference of means.  Data: plov0 and plov1.

Col    N      sample     sample   df   level    t      confidence interval
              mean diff    SE                          lower end    upper end
 4   26,29   -0.22215    0.10356  52.7  .5000   0.68   -0.29248    -0.15181
                                        .9500   2.01   -0.42989    -0.014408
                                        .9900   2.67   -0.49890     0.054599
```

In this example, the confidence intervals are nearly equal to those with the {**pool**} option because although the t quantile is larger, the sample SE is smaller. Note that Welch's approximation usually yields noninteger degrees of freedom. (The t distribution is defined for such degrees of freedom.)

*The {**z**} option.* If you want to use the Normal distribution instead of Student's t distribution to calculate a confidence interval even though the SDs are unknown, use the {**z**} option. For example:

```
. ci {z} plov0 {x=4} plov1 {y=4}
```

Of course, as discussed in Section 6.1.1, the actual confidence level of the resulting interval is smaller than the specified level unless the sample sizes are large.

6.1.3   Confidence Intervals from Sample Means

Either the x or y input to the **ci** command may contain the sample mean instead of the data itself. If x is a sample mean, specify the size of the sample from which it was computed by {**nx=**} and the SD, either sample or known, by {**sdx=**}. If y is a sample mean, use the {**ny=**} and {**sdy=**} options. If the SDs are population SDs, also give the {**z**} option so that **ci** uses the Normal distribution to construct the confidence interval; otherwise it uses Student's t distribution. (If both x and y are sample means, then either both SDs must be estimated, or both SDs must be known.)

Example: One sample. Suppose we have only the following summary statistics on *plover*, and have lost the data itself:

```
Statistics: plover
Col   N    Mean     SD      Min     25%     50%     75%     Max
 3   52   8.638   0.5018   7.400   8.300   8.650   9.000   9.900
```

We use these statistics in **ci**: we specify the SD with the {**sdx=**} option and the sample size with the {**nx=**} option.

---

4. B. L. Welch (1938). *The significance of the difference between two means when the population variances are unequal.* **Biometrika**, **29**, 350-62. The formula is given in *ci(BRM)*.

```
. ci 8.638 {sdx=.5018; nx=52}

Confidence Interval for the mean.  Data: 8.638.

Col    N      sample      sample    df   level    t      confidence interval
              mean          SE                           lower end    upper end
 1    52     8.6380     0.069587  51.0  .5000   0.68    8.5907      8.6853
                                        .9500   2.01    8.4983      8.7777
                                        .9900   2.68    8.4518      8.8242
```

Example: Two samples. Again, suppose we have only the summary statistics
from *plov0* and *plov1*, and not the data itself:

```
Statistics: plov0
Col    N     Mean      SD      Min      25%      50%      75%      Max
 4    26    6.019   0.3487   5.200    5.800    6.100    6.200    6.500

Statistics: plov1
Col    N     Mean      SD      Min      25%      50%      75%      Max
 4    29    6.241   0.4188   5.300    6.000    6.300    6.500    7.100
```

To construct a confidence interval for the difference of the two means, we
specify the SD and sample size for *plov0*[4] with the {**sdx=**} and {**nx=**}
options, and the SD and sample size for *plov1*[4] with the {**sdy=**} and {**ny=**}
options:

```
. ci 6.019 {sdx=.3487; nx=26} 6.241 {sdy=.4188; ny=29}

Confidence Interval for difference of means.  Data: 6.019 and 6.241.

Col    N      sample      sample    df   level    t      confidence interval
             mean diff      SE                           lower end    upper end
 1   26,29   -0.22200   0.10356   52.7  .5000   0.68   -0.29234     -0.15167
                                        .9500   2.01   -0.42974     -0.014261
                                        .9900   2.67   -0.49875      0.054746
```

The {z} option. If the {**z**} option is specified when the input data are sample
means, **ci** assumes that the SDs specified by {**sdx=**} and {**sdy=**} are known
population SDs and uses the Normal distribution to construct the confidence
intervals. Otherwise, **ci** assumes that the SDs are estimated from the data and
uses the t distribution as in the preceding examples.

6.2   Confidence Intervals with Bernoulli (0-1) Data

Bernoulli data consists of 1's and 0's which denote the outcomes ('successes'
or 'failures') of identically performed random events. The total number of
successes and the total number of observations are usually recorded instead of
the individual 1's and 0's. If your data are actual 1's and 0's, use the **count**
command (described in Section 8.7 and *count(BRM)*) or the **sum** command
(described in *vecop(BRM)*) to obtain the number of 1's observed.

For Bernoulli data, the parameter of interest is either the probability of success for a single experiment, or the proportion of a population that possesses a
given characteristic. This probability or proportion is usually denoted as *p*.
Use the {**b**} option if *x* is the observed *number* of successes—that is, a

binomial random variable (the sum of Bernoulli random variables is one binomial random variable). Use the {**p**} option if *x* is the observed *proportion* of successes.

Ci constructs confidence intervals for *p* in the one-sample case, or the difference between the values of *p* for two populations in the two-sample case. Except under the {**exact**} option—which, in the one-sample case, computes an exact confidence interval based on the binomial distribution—all confidence intervals are based on an approximation using the Normal distribution.

6.2.1   One Sample

For the one-sample case, **ci** constructs confidence intervals for *p*.

One sample, from the number of successes. To form a confidence interval for *p* from the observed number of successes (or 1's), use the {**b**} option. Put the number of successes in the *x* input dataset and specify the sample size using the {**nx=**} option.

To construct confidence intervals for the proportion of snowy plover chicks that reach fledgling, for example, we count the number of such chicks in *plover*[5] and use the result as the input dataset. To keep our results from being biased downward, we first remove the missing data from *plover*[5].

```
. select plover {x=5} > plov5        # remove missing data
. count (plov5==1)

29/65 = 44.62% of the cases meet the condition.
```

Our result is 29 'successes' from 65 observations. Thus the command to construct the confidence interval is:

```
. ci {b} 29 {nx=65}

Confidence Interval for the proportion.  Data: 29.
```

Col	N	estimated prob	sample SE	level	z	confidence interval lower end	upper end
1	65	0.44615	0.061657	.5000	0.67	0.40457	0.48774
				.9500	1.96	0.32531	0.56700
				.9900	2.58	0.28734	0.60497

One sample, from the proportion of successes. If your data is the observed *proportion* of successes (or 1's) in a sample, use the {**p**} option. Specify the proportion as the *x* input dataset and specify the sample size with the {**nx=**} option.

If we use the observed proportion of plovers that survived to fledgling in the example above instead of the number of plovers, the command is:

```
. ci {p} .44615 {nx=65}
```

In this case, the output display is the same as before.

Exact confidence intervals. For Bernoulli data, an exact one-sample confidence interval can be constructed based on the binomial distribution. The {**exact**} option of the **ci** command does this.[5] For example, to find an exact confidence interval for the proportion of plovers that survived to fledgling:

```
. ci {p; exact} .44615 {nx=65}

Confidence Interval for the proportion (exact).  Data: .44615.

Col   N    estimated    sample      level        confidence interval
             prob         SE                    lower end   upper end
 1    65    0.44615    0.062017      .5000        0.41287     0.49571
                                     .9500        0.33702     0.57468
                                     .9900        0.30255     0.61163
```

Note that the output display includes '(exact)' to indicate that the confidence interval is exact. Exact confidence intervals are slightly shorter than those based on the Normal distribution. Unlike approximate confidence intervals using Bernoulli data, exact confidence intervals do not use an intermediate z value; hence none is shown in the output.

6.2.2 Two Samples

For the two-sample case, **ci** constructs confidence intervals for the difference between two values of p. The two input datasets must both contain the observed number of successes (or 1's), or must both contain the observed proportion.

Two samples, from the number of successes. Use the {**b**} option. Give the observed number of successes in the first sample as the x input dataset, and the observed number of successes in the second sample as the y input dataset. Specify the size of the first sample with the {**nx=**} option and the size of the second sample with the {**ny=**} option.

As an example, let's toss two coins with unknown probabilities of heads and construct a 95% confidence interval for the difference between the two probabilities. To make this example more realistic, we use two coins whose probabilities of heads are unknown by specifying the {**mystery=**} option with two different arguments. See *coin*(*BRM*) for more information.

```
. coin {ntoss=100;mystery=1x}

Tossing a mystery coin ...

Cumulative results of 100 tosses:
  Observed number of heads:  44      Observed proportion of heads: 0.4400000
  Expected number of heads:  ???     Probability of heads:            ???
  Estimated standard error:  4.96    Estimated standard error:     0.0496387

. coin {ntoss=100;mystery=2x}

Tossing a mystery coin ...

Cumulative results of 100 tosses:
  Observed number of heads:  54      Observed proportion of heads: 0.5400000
  Expected number of heads:  ???     Probability of heads:            ???
  Estimated standard error:  4.98    Estimated standard error:     0.0498397
```

---

5. The formula is given in *ci*(*BRM*).

```
. ci {b} 44 54 {nx=100; ny=100}
Confidence Interval for difference of proportions.  Data: 44 and 54.
Col   N      estimated    sample      level     z      confidence interval
             prob diff      SE                         lower end   upper end
 1  100,100  -0.10000    0.070342     .5000   0.67   -0.14744    -0.052555
                                      .9500   1.96   -0.23787     0.037868
                                      .9900   2.58   -0.28119     0.081189
```

Two samples, from the proportion of successes. Use the {p} option. Give the observed proportion of successes in the first sample as the *x* input dataset, and the observed proportion of successes in the second sample as the *y* input dataset. Specify the size of the first sample with the {nx=} option, and the size of the second sample with the {ny=} option.

When constructing our confidence interval for the difference in probability of heads between mystery coins **1x** and **2x**, we could have used the observed proportions of heads as the data instead of the numbers of heads:

```
. ci {p} .44 .54 {nx=100; ny=100} {alpha=.05}
```

The output display is the same as that of the previous example.

6.3   Hypothesis Tests

A statistical hypothesis test can be viewed as a procedure for making a decision about a given hypothesis for a population parameter based on data from that population. The given hypothesis is called the *null hypothesis*. There are two possible decisions: to *accept* the null hypothesis, or to *reject* it in favor of an *alternative hypothesis*.[6] For our purposes, the alternative hypothesis is simply that the null hypothesis is false.

The **ht** command performs hypothesis tests for the value of a population mean (or proportion) or the difference between two population means (or proportions) with prespecified probability of error α. The general form of the command is:

ht *x* [*y c*] {**alpha**=;**b**;**ht**=;**nx**=;**ny**=;**p**;**pool**;**sdx**=;**sdy**=;**x**=;**y**=;**z**;...} > *outputs*

The {**ht**=} option specifies the form of the null and alternative hypotheses. Possible forms are described in Sections 6.3.1, 6.3.2, 6.4.1, and 6.4.2. The {**alpha**=} option specifies the *significance level*[7] of the test; that is, the probability of rejecting the null hypothesis when it is true. More than one value of α can be specified; the default values are {**alpha**= .05, .01, .001}.[8]

---

6. Throughout this section, we use the terms *accept* and *reject* with their conventional statistical hypothesis testing meanings. How these meanings correspond to their customary English language meanings depends on one's school of thought. For more discussion, see a statistics textbook. Different authors disagree.

7. Note that for hypothesis tests, the significance level refers to α, whereas for confidence intervals, the confidence level refers to $1-\alpha$.

8. As with confidence levels, no single value of α is best for all circumstances. For this reason, BLSS provides three default values of α. If you want a single level, you must choose it yourself.

The x input dataset contains the sample; the y input, which is optional, contains the second sample if any. **Ht** performs a hypothesis test for every column of x (and y, if it is present). Specific columns of x and y can be selected using the {**x=**} and {**y=**} options.

The output display. For each sample, **ht** displays the sample mean (or mean difference), the sample SE, the test statistic, and its *P-value* (that is, the probability, under the null hypothesis, of obtaining a test statistic at least as extreme as the observed value).[9] If the {**terse**} option is specified, **ht** shows only these values. Otherwise, for each value of α, **ht** also displays the corresponding *critical value*. Whether to accept or reject the null hypothesis for a given α is determined by comparing the test statistic to the critical value (or, equivalently, by comparing the P-value to α). If the {**long**} option is specified, **ht** does the comparison and displays 'acc' for *accept* and 'rej' for *reject*. We will see examples of each of the output displays in the following sections.

Assumptions. The P-value depends on the test statistic, the null hypothesis, and the presumed distribution of the test statistic (standard Normal or Student's t). As with confidence intervals, if the observations are independent and the underlying population is approximately Normal, then the actual significance level of the test is approximately equal to α. Otherwise, it may not be. For more detail, see the discussion in Section 6.1.

The remainder of this chapter discusses how to perform hypothesis tests for a population mean, the difference between two population means, the probability (or proportion) of success, and the difference between two probabilities (or proportions) of success.

As mentioned before, the **ht** command is in many respects identical to the **ci** command; hence Sections 3 and 4 of this chapter parallel Sections 1 and 2.

6.3.1   One Sample—for a Population Mean

For one sample, **ht** performs a hypothesis test for the mean μ of a population. The specific forms of the null and alternative hypotheses are determined by the {**ht=**} option, as follows:

Null	*Alternative*	*Type of Test*	*BLSS Option*
$\mu = c$	$\mu \neq c$	two-tail	{**ht=2**} (default)
$\mu \geq c$	$\mu < c$	lower-tail (left-tail)	{**ht=L**}
$\mu \leq c$	$\mu > c$	upper-tail (right-tail)	{**ht=U**} or {**ht=R**}

where μ is the true unknown population mean and c is a constant.[10] The names *two-tail*, *lower-tail*, and *upper-tail* refer to the rejection region for each

---

9. The P-value can also be regarded as the smallest significance level α at which the null hypothesis is rejected.

10. Some authors show the null hypothesis for the lower- and upper-tailed tests as $\mu = c$ rather than $\mu \geq c$ and $\mu \leq c$. The distinction is conceptual: the calculations and acceptance/rejection results are the same.

of these tests; that is, the set of possible test statistic values for which the null hypothesis is rejected.

The default value of *c* is 0. If *c* is not 0, specify its value using the **c@** input tag as in the examples below.[11]

One sample, population SD known (z-test). If the population SD is known, specify it using the {**sdx=**} option. **Ht** computes a z-test (that is, it uses the Normal distribution).

For example, we test the null hypothesis that mean snowy plover chick weight is 8.5 grams against the alternative hypothesis that it is not 8.5 grams. We assume the SD is known and equals .5 grams.

```
. ht plover {x=3; sdx=.5} c@ 8.5

Hypothesis Test: one-sample, two-tail z.  Data: plover.  Null mean: 8.5.

Col   N      null      sample    known         test   P      alpha  crit
             mean      mean      SE            stat   value         value
 3    52     8.5000    8.6385    0.069338      2.00   .0458  .0500  1.96
                                                             .0100  2.58
                                                             .0010  3.29
```

For two-tailed tests, we reject the null hypothesis if the absolute value of the test statistic is greater than or equal to the displayed critical value.[12]

One sample, SD unknown (t-test). If no SD is specified, **ht** automatically estimates it and performs a t-test (that is, it uses Student's t distribution).

For example, we do not in fact know the population SD of the egg weights. Here we test the null hypothesis that mean egg weight does not exceed 8.5 grams against the alternative hypothesis that it is greater than 8.5 grams, without assuming that the SD is known. This alternative hypothesis requires an upper-tailed test so we use the {**ht=U**} option:

```
. ht {ht=U} plover {x=3} c@ 8.5 {long}

Hypothesis Test: one-sample, upper-tail t.  Data: plover.  Null mean: 8.5.

Col   N      null      sample    sample    df    test   P      alpha  crit   ?
             mean      mean      SE              stat   value         value
 3    52     8.5000    8.6385    0.069590  51.0  1.99   .0260  .0500  1.68  rej
                                                               .0100  2.40  acc
                                                               .0010  3.26  acc
```

In this example we used the {**long**} option, which causes **ht** to explicitly state (in the column labeled '?') whether the null hypothesis should be accepted ('acc') or rejected ('rej') at each significance level α. For an upper-tailed test, we reject the null hypothesis if the test statistic is greater than or equal to the critical value.

---

11. Input tags, such as **c@**, tell BLSS to which input argument the name or value following the tag refers. *C* is the third input to **ht** (*y* is the second) so the tag **c@** is necessary to avoid confusion when there is no *y* input. Input and output tags are discussed in detail in Section 11.3.1.

12. Two-tailed tests have two critical values; **ht** displays only the positive value. The other critical value is the negative of the one shown.

***The* {z}** *option.* If you specify the {z} option, **ht** performs a z-test regardless of whether the SDs are known (and specified). As with the **ci** command, the {z} option to **ht** is made available for pedagogical reasons and for comparison of the t-test to the z-test. Although the t- and z-test are nearly identical for large sample sizes, when the sample is small and the true value SD is unknown, the z-test accepts the null hypothesis when it is false with higher probability than α.

Suppose we want **ht** to estimate the SD of egg weight, but perform a z-test. We omit {**sdx=**} but use {**z**}:

```
. ht {z; ht=U} plover {x=4} c@ 6 {terse}

Hypothesis Test: one-sample, upper-tail z.  Data: plover.  Null mean: 6.

Col   N        null      sample    sample        test   P
               mean      mean      SE            stat   value
 4    57      6.0000    6.1439    0.052268       2.75   .0030
```

In this example we used the {**terse**} option, which causes **ht** to suppress α and the critical values. We reject the null hypothesis if α is greater than the P-value, .003. Observe that the P-value is smaller than the one we got when the {**z**} option was not specified.

6.3.2   Two Samples—for the Difference Between Population Means

For two samples, the basic null hypothesis is that the two population means are equal. The types of hypotheses that **ht** tests are:

Null	*Alternative*	*Type of Test*	*BLSS Option*
$\mu_1 = \mu_2 + c$	$\mu_1 \neq \mu_2 + c$	two-tail	{**ht=2**} (default)
$\mu_1 \geq \mu_2 + c$	$\mu_1 < \mu_2 + c$	lower-tail (left-tail)	{**ht=L**}
$\mu_1 \leq \mu_2 + c$	$\mu_1 > \mu_2 + c$	upper-tail (right-tail)	{**ht=U**} or {**ht=R**}

where μ_1 is the true unknown mean of the first population, μ_2 is the true unknown mean of the second population, and c is a constant. The default value of c is 0; in this case the null hypotheses are simply $\mu_1 = \mu_2$, $\mu_1 \geq \mu_2$, and $\mu_1 \leq \mu_2$. If c is not 0, specify its value using the **c@** tag.

Two samples, SDs known (z-test). If both population SDs are known, specify them with the {**sdx=**} and {**sdy=**} options. **Ht** performs a z-test.

For example, we use the datasets *plov0* and *plov1* created in Section 6.1.2 to test the null hypothesis that the egg weights are the same for chicks which survived to fledgling and those which did not against the alternative hypothesis that the weights are unequal:

```
. ht plov0 {x=4} plov1 {y=4} {sdx=.35; sdy=.42} {long}

Hypothesis Test: two-sample, two-tail z.  Data: plov0 and plov1.

Col   N       null       sample     known      test    P      alpha  crit  ?
              mean diff  mean diff  SE         stat    value         value
 4   26,29   0.0000     -0.22215   0.10390    -2.14   .0325  .0500  1.96  rej
                                                             .0100  2.58  acc
                                                             .0010  3.29  acc
```

Because this test statistic is negative, we compare its absolute value, 2.14, to the critical value we choose. In general, we reject the null hypothesis if α exceeds .0325.

Two samples, SDs unknown. For two samples, if either population SD is unknown, a t-test is performed using Student's t distribution.

Two samples, SDs unknown but equal (t-test). If both SDs are unknown and the {**pool**} option is specified, **ht** assumes that the SDs are equal and calculates a *pooled* estimate of the common SD (see Section 6.1.2). Use this option only if you are quite sure that the SDs *are* equal, or the test will be wrong.

For example:

```
. ht {pool} plov0 {x=4} plov1 {y=4}

Hypothesis Test: two-sample, two-tail t.  Data: plov0 and plov1.

Col   N      null     sample   sample SE  df    test    P    alpha  crit
           mean diff mean diff (pooled)        stat  value        value
 4   26,29  0.0000   -0.22215  0.10465   53.0  -2.12 .0385 .0500  2.01
                                                           .0100  2.67
                                                           .0010  3.48
```

Two samples, SDs unknown and not known to be equal. Drop the {**pool**} option and **ht** estimates the SDs (separately) from the data. The test statistic is approximately distributed as Student's t. **Ht** performs a t-test and estimates the degrees of freedom using Welch's approximation.

```
. ht plov0 {x=4} plov1 {y=4}

Hypothesis Test: two-sample, two-tail t.  Data: plov0 and plov1.

Col   N      null     sample   sample   df    test    P    alpha  crit
           mean diff mean diff   SE          stat  value        value
 4   26,29  0.0000   -0.22215  0.10356  52.7  -2.15 .0366 .0500  2.01
                                                          .0100  2.67
                                                          .0010  3.48
```

This is generally the most appropriate use of **ht** when performing two-sample tests.

6.3.3   Paired Comparisons

A *paired comparisons* test is a two-sample hypothesis test in which each observation from a population is paired with an observation from another population. Paired comparisons usually arise when the same subject is observed under two different conditions, or when each subject in a *treatment* group is matched ('paired') with a subject in a *control* group before treatment begins.

The null and alternative hypotheses one might test for paired comparisons are the same as for regular two-sample hypothesis tests described above, but the test is different. To perform a paired comparisons test, create a new dataset by subtracting each observation in the second group from its paired observation in the first group, then perform a one-sample hypothesis test.

The BLSS library dataset *alkaloid*, for example, contains the number of plants in 45 different families found to contain alkaloids at high and low latitudes.

In order to test the null hypothesis that the number of plants containing alkaloids is the same regardless of latitude:

```
. load alkaloid
. ht (alkaloid[1] - alkaloid[3])

Hypothesis Test: one-sample, two-tail t.  Data: (alkaloid[1]-alkaloid[3]).

Col   N       null      sample    sample    df    test    P     alpha   crit
              mean      mean      SE              stat  value           value
 1    45    0.0000     10.333    4.6804    44.0   2.21  .0325  .0500    2.02
                                                               .0100    2.69
                                                               .0010    3.53
```

The sample mean is far from 0, but the sample SE is large enough that although we can reject the null hypothesis at $\alpha = .05$, we cannot at $\alpha = .01$.

6.3.4   Hypothesis Tests from Sample Means

As with **ci**, either the x or y input to **ht** may contain the sample mean instead of the data itself. If x is a sample mean, specify the sample size with {**nx**=} and the SD, either sample or known, with {**sdx**=}. If y is a sample mean, use the {**ny**=} and {**sdy**=} options. If the SDs are population SDs, also give the {**z**} option so that **ht** performs a z-test; otherwise it performs a t-test. (If both x and y are sample means, then either both SDs must be estimated, or both SDs must be known.)

One sample. Suppose, as in Section 6.1.3, that we have lost the *plover* data but still have the summary statistics. We can perform a hypothesis test using the summary statistics instead of the data itself. For example, to test the null hypothesis that mean egg weight does not exceed 8.5 grams against the alternative hypothesis that it is greater than 8.5 grams using the summary statistics:

```
. ht {ht=U} 8.638 {sdx=.5018;nx=52} c@ 8.5 {long}

Hypothesis Test: one-sample, upper-tail t.  Data: 8.638.  Null mean: 8.5.

Col   N       null      sample    sample    df    test    P     alpha   crit  ?
              mean      mean      SE              stat  value           value
 1    52    8.5000     8.6380   0.069587   51.0   1.98  .0264  .0500    1.68 rej
                                                               .0100    2.40 acc
                                                               .0010    3.26 acc
```

The only difference between this output and the corresponding output using *plover*[3] is the labels.

Two samples. From the *plov0* and *plov1* summary statistics in Section 6.1.3 we test the null hypothesis that the mean chick weight for chicks that survive to fledgling is the same as for chicks that do not:

```
. ht 6.019 {sdx=.3487;nx=26} 6.241 {sdy=.4188;ny=29} {terse}

Hypothesis Test: two-sample, two-tail t.  Data: 6.019 and 6.241.

Col    N       null       sample     sample    df    test    P
               mean diff  mean diff  SE              stat  value
 1   26,29   0.0000     -0.22200   0.10356   52.7  -2.14  .0367
```

Of course, the {**pool**} and {**z**} options can also be used when the data are sample means.

6.4   Hypothesis Tests with Bernoulli Data

For Bernoulli data, **ht** tests hypotheses about the probability of success of a single experiment or the proportion of members of a population which possesses a certain characteristic. **Ht** takes Bernoulli data in the same form as **ci** (see Section 6.2). Use the {**b**} option if your data contains the *number* of successes observed; use the {**p**} option if it contains the *proportion* of successes.

The test statistic for a one-sample hypothesis test using Bernoulli data has a (standardized) binomial distribution. However, the hypothesis test itself is based on the Normal distribution. Because a discrete distribution is approximated by a continuous distribution, a *continuity correction* is sometimes applied to the test. The {**cc**} option of the **ht** command does this.

It is also possible to construct an *exact* hypothesis test for one-sample Bernoulli data. The {**exact**} option of **ht** does this.

6.4.1   One Sample

For one sample, **ht** performs a hypothesis test for the probability of success for a single experiment. The null and alternative hypotheses can be one of the following:

Null	*Alternative*	*Type of Test*	*BLSS Option*
$p = c$	$p \neq c$	two-tail	{ht=2} (default)
$p \geq c$	$p < c$	lower-tail (left-tail)	{ht=L}
$p \leq c$	$p > c$	upper-tail (right-tail)	{ht=U} or {ht=R}

where p is the true unknown probability of success and c is a probability specified by the **c@** input dataset (see Section 6.3.1).[13] The default value of c is .5.

One sample, from the number of successes. To perform a hypothesis test about the probability of success, use the {**b**} option. Put the observed number of successes in x and specify the sample size with the {**nx=**} option.

To return to our example in Section 6.2.1, we test the null hypothesis that the proportion of plover chicks that survive to fledgling is .5 against the alternative hypothesis that it is not:

```
. ht {b} 29 {nx=65} c@ .5 {terse}

Hypothesis Test: one-sample, two-tail z.  Data: 29.  Null prob: .5.

Col    N       null     estimated    sample          test    P
               prob       prob        SE             stat   value
 1     65     0.50000    0.44615    0.062017        -0.87  .3853
```

Because .5 is the default value of c, we could have omitted the **c@** input.

---

13. As with tests of population means, some authors show the null hypothesis for the lower-and upper-tailed tests as $p = c$ rather than $p \geq c$ and $p \leq c$. Again, the distinction is conceptual: the calculations and acceptance/rejection results are the same.

One sample, from the proportion of successes. To perform a hypothesis test about the probability of success, use the {**p**} option. Put the observed proportion of successes in x and specify the sample size with the {**nx=**} option. For example, to perform the previous hypothesis test using the proportion of chicks that survived to fledgling:

```
. ht {p} .44615 {nx=65}
```

The output is the same as above.

Continuity correction. To use the continuity correction in our hypothesis test about the proportion of chicks that survived to fledgling, give the {**cc**} option:

```
. ht {b; cc} 29 {nx=65} {terse}
```

Hypothesis Test: one-sample, upper-tail z (cc). Data: 29.

Col	N	null prob	estimated prob	sample SE	test stat	P value
1	65	0.50000	0.44615	0.062017	-0.74	.4568

Note that the output display includes '(cc)' to indicate that the continuity correction was used. Because of the correction, the test statistic is slightly smaller and hence the P-value is somewhat larger.[14]

Exact hypothesis tests. To construct an exact hypothesis test about the proportion of chicks that survived to fledgling, give the {**exact**} option:

```
. ht {b; exact} 29 {nx=65}
```

Hypothesis Test: one-sample, two-tail (exact). Data: 29.

Col	N	null prob	estimated prob	sample SE	P value
1	65	0.50000	0.44615	0.062017	.4570

Note that the output display includes '(exact)' to indicate that the P-value is exact. Unlike the approximate tests, the exact test does not compute a test statistic which can be compared to the Normal distribution. Hence neither it, nor α, nor any critical values are displayed.[15]

6.4.2　Two Samples

For the two sample case, **ht** performs tests for the equality of two probabilities of success. As with **ci**, the two input datasets must both contain the observed number of successes (or 1's), or must both contain the observed proportion. The null and alternative hypotheses can be any of the following:

Null	Alternative	Type of Test	BLSS Option
$p_1 = p_2$	$p_1 \neq p_2$	two-tail	{**ht=2**} (default)
$p_1 \geq p_2$	$p_1 < p_2$	lower-tail (left-tail)	{**ht=L**}
$p_1 \leq p_2$	$p_1 > p_2$	upper-tail (right-tail)	{**ht=U**} or {**ht=R**}

---

14. The precise form of the continuity correction is given in *ht(BRM)*, but see a statistics textbook for an explanation.
15. The formula for the exact P-value is given in *ht(BRM)*.

where p_1 is the true unknown proportion of 1's in the first population and p_2 is the true unknown proportion of 1's in the second population.

Two samples, from the number of successes. Use the {**b**} option. Give the observed number of successes in the first sample as the x input dataset, and the observed number of successes in the second sample as the y input dataset. Specify the size of the first sample with the {**nx=**} option and the size of the second sample with the {**ny=**} option.

For example, let's test the null hypothesis that the probability of heads of mystery coins **1x** and **2x** (tossed in Section 6.2.2) are equal against the alternative hypothesis that they are not:

```
. ht 44 54 {b} {nx=100; ny=100} {terse}

Hypothesis Test: two-sample, two-tail z.  Data: 44 and 54.

Col    N         null      estimated    sample       test    P
                 prob diff prob diff      SE          stat   value
  1  100,100   0.0000     -0.10000    0.070697       -1.41  .1572
```

Two samples, from the proportion of successes. Use the {**p**} option. Give the observed proportion of successes in the first sample as the x input dataset and the observed proportion of successes in the second sample as the y input dataset. Specify the size of the first sample with the {**nx=**} option and the size of the second sample with the {**ny=**} option.

As an example, we perform the preceding hypothesis test using the proportion of observed heads instead of the number of heads.

```
. ht .44 .54 {p} {nx=100; ny=100}
```

The output display is the same as that of the previous example.

CHAPTER 7

Probability and Random Numbers

This chapter introduces BLSS commands that evaluate probability distribution functions (cdf's) and their densities and inverses, generate random numbers from standard distributions, and generate random samples. It contains material appropriate to all levels of courses, from basic introductory statistics through introductory mathematical statistics for advanced undergraduates. Subjects are arranged in order of increasing sophistication. We begin with a section on instructional commands which can be used in the most elementary of courses.

Note that all results of simulation, sampling, and random number commands shown in this chapter are random (in the sense explained in Section 7.3.2). Thus, when you repeat the same commands, your results will be different.

7.1 Probability Demonstrations

This section describes three commands which illustrate basic concepts of probability in an elementary context. The first simulates tossing a coin. The second simulates drawing tickets from a box model (that is, sampling from any finite discrete distribution) and demonstrates the sampling variation of the estimated average and SD. The third illustrates the central limit theorem.

7.1.1 Tossing a Coin

The **coin** command simulates tossing a coin. With no arguments, it asks what it should do. Here is a typical dialogue:

```
. coin
Hello.  I will toss a coin for you and count the number of heads.
Shall I toss a fair coin? no
Enter the probability of heads: .25
How many tosses? 80
Do you want to see the individual tosses? y
Tossing ...

T H T T H T T H T T H H T T T H T T H H H T T T T T T T H H T H T T H T T T H H
T T T T H H H H T T T H T T T T T T H T H T T T T T T T T T H H T T T T T T

Cumulative results of 80 tosses:
    Observed number of heads: 24          Observed proportion of heads: 0.3000000
    Expected number of heads: 20.00       Probability of heads:         0.2500000
    Chance error:             4.00        Relative chance error:        0.0500000

Toss the same coin again? y
How many tosses? 80
Do you want to see the individual tosses? y
Tossing ...

H T T H T H T T T T T T T H T T H H T T H T T T T T T H T T T H H H H H T T T T
H T H H H T T T H T T T T T T T T T H T T T T T T T T T H T T T H T T T T T H H

Results of the last 80 tosses:
    Observed number of heads:  23         Observed proportion of heads: 0.2875000
    Expected number of heads:  20.00      Probability of heads:         0.2500000
    Chance error:              3.00       Relative chance error:        0.0375000

Cumulative results of 160 tosses:
    Observed number of heads:  47         Observed proportion of heads: 0.2937500
    Expected number of heads:  40.00      Probability of heads:         0.2500000
    Chance error:              7.00       Relative chance error:        0.0437500

Toss the same coin again? n
Goodbye.
```

As you can see, the coin need not be fair. BLSS displays the tosses (because we asked it to) and summarizes the results. Note that the responses 'y' and 'n' suffice for 'yes' and 'no'.

If you specify the number of tosses via the {**ntoss**=} option, or the number of samples (groups of *ntoss* tosses) via the {**nsamp**=} option, the **coin** command runs *noninteractively*. That is, it doesn't ask any questions. Another option, {**p**=}, specifies the probability of heads. If it is not given, **coin** uses the default value of .5. For example, the command:

```
. coin {ntoss=80} {nsamp=2} {p=.25}
```

performs the same simulation as above—80 tosses of an unfair coin, repeated twice—without asking any questions. When you try this command, you will notice that it does not show the individual tosses. Use the {**show**} option if you want to see them.

You can save any or all of: the individual tosses (in the **toss@** output dataset);

the results of each group of tosses (in the **group@** output); or the cumulative results at the end of each group (in the **cum@** output). For example, the command:

```
. coin {nsamp=60; ntoss=50} {quiet} > cum@ cumres
```

tosses a fair coin 3000 times—60 samples of 50 tosses each—and saves the cumulative results at the end of each group of tosses in the BLSS dataset *cumres.* The {**quiet**} option (which is recognized by all regular BLSS commands) tells **coin** to be quiet—that is, to print nothing. The '>' symbol (which we discussed in Section 4.4) means that the words which follow it refer to output datasets; the word **cum@** means to save the cumulative results in the output dataset whose name follows—in this case, *cumres.* Words which end in @, such as **toss@**, **group@**, and **cum@**, are called *tags* and are discussed in Section 11.3.1.

The output dataset *cumres* contains all cumulative summary results printed by the **coin** command. Column 1, for example, contains the number of tosses; column 9 contains the relative chance error (that is, the difference between the observed and the expected proportion of heads). Thus, we can plot the relative chance error against the number of tosses:

```
. scat cumres {x=1;y=9}
```

The resulting scatterplot is shown in Figure 7.1. It illustrates the law of large numbers (also called the law of averages): The relative chance error tends to decrease as the number of tosses increases. Of course, every time you try these commands you will get a different plot, because the coin-tossing outcomes are random.

For more information about the **coin** command (including default values, and the complete contents of the output datasets) refer to its manual entry, *coin(BRM).* The **scat** command is discussed in Section 5.2 and in its manual entry, *scat(BRM).*

7.1.2   Box Models—Sample Averages and SDs

Consider a box, filled with tickets. Each ticket has a number on it. The tickets are mixed thoroughly, one is drawn at random in such a way that every ticket has the same chance of being chosen, and its number is observed. This vivid description of a random variable is used extensively by Freedman, Pisani, and Purves in the text *Statistics* (1978, Norton, New York); they call it a *box model.* You can use a box model to model any finite discrete distribution.

The **box** command lets you observe the variation of the sample average and sample SD of tickets drawn from a box model. It draws tickets from the box *with replacement:* that is, after each observation the ticket is replaced in the box, so that successive draws are made under identical circumstances. After some number of draws (which constitute a 'sample'), the average and SD for the sample are computed. This can be repeated for any number of samples.

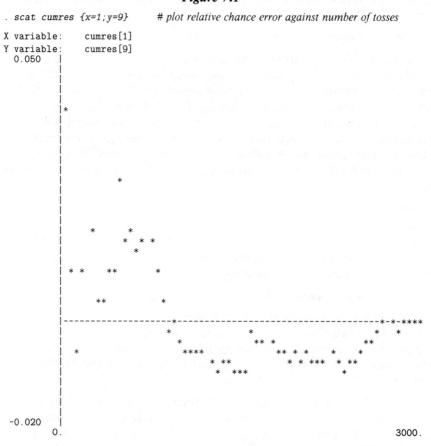

Figure 7.1

If you give the **box** command with no arguments, it asks questions: it allows you to specify any box contents, the number of draws per sample, and the number of samples. In the following example, we enter the ticket values 1, 2, 3, 4, 5, and 6. This box model corresponds to rolling a single fair die. To signify the end of the ticket values, we type a control-D by holding down the CTRL button and typing 'd':

```
. box
Either:  enter the numbers on the tickets in the box;
   Or:  on separate lines, enter each ticket's number and frequency.
Finish with RETURN and CTRL-D.
1 2 3 4 5 6
Control-D                        # hold down CTRL and type 'd'
Box saved in:  box00
The average of the box is 3.5.  The SD is 1.70783.
```

```
How many draws per sample? 10
How many samples? 5

Samples saved in:  samps00

Statistics: samps00
Col  N    Mean     SD      Min     25%     50%     75%     Max
 1   10   3.700   2.111   1.000   2.000   4.000   6.000   6.000
 2   10   2.200   1.476   1.000   1.000   1.500   3.000   5.000
 3   10   4.000   1.414   2.000   3.000   3.500   5.000   6.000
 4   10   3.500   1.958   1.000   1.000   4.000   5.000   6.000
 5   10   3.900   1.663   1.000   3.000   4.000   5.000   6.000
```

Box displays the average and SD of the tickets in the box, draws the samples, and displays summary statistics for each sample.[1]

The **box** command automatically saves the box we entered as a BLSS dataset (in this case, under the name *box00*) and the sample draws (under the name *samps00*). Thus, if we wish, we can reuse the same box in later commands and further analyze the results.

If you specify the box to use as an argument and the number of draws per sample and the number of samples via the {**ndraw**=} and {**nsamp**=} options, the **box** command runs noninteractively. For example, we can repeat the above experiment by typing:

```
. box box00 {ndraw=10} {nsamp=5}
```

No questions are asked.

The **box** command has additional options, including: {**evse**}, which displays the expected value for the sum of the draws and the standard error for the sum and the average; and {**show**}, which shows the individual draws. **Box** can save both the individual draws and the summary statistics as BLSS datasets. See the manual entry *box(BRM)* for more information.

7.1.3   The Central Limit Theorem

As we draw more and more tickets at random with replacement from a box, the probability histogram for the sum of the draws, suitably scaled, approaches that of the standard Normal (also called Gaussian) distribution. This is an elementary version of the *central limit theorem*.

The **demoiv** command lets you see this for yourself.[2] You can give it any box model—either enter the tickets directly, as with the **box** command, or reuse a previously entered box. **Demoiv** displays the probability histogram for a single

---

1. Technical note: The SD of the box is computed with the divide-by-N formula (because it is a population SD), whereas the sample SDs are computed with the divide-by-$(N-1)$ formula. The option {**dn**=0} allows the sample SDs to be computed with the divide-by-N formula if desired.

2. This command in named in honor of Abraham de Moivre (1667-1754), who proved the first version of the central limit theorem—for the number of heads resulting from many tosses of a coin.

draw from the box, for the sum of 2 draws with replacement, the sum of 4, the sum of 8, and so on, doubling each time until you no longer wish to continue. The histograms are scaled if necessary to let you see their shape (and to fit on the terminal screen!).

To run **demoiv** on the box we created previously which modeled the roll of a die, type:

```
. demoiv box00
```

The full dialogue and histograms appear in Figure 7.2. For this box, the probability histogram for the sum of the draws looks much like the Normal distribution after only 4 draws, and very much like it after only 8.

Just like the **box** command, when **demoiv** is given no arguments it asks you to enter the tickets in the box. Similarly, **demoiv** can save output datasets and can run noninteractively in order to display a given number of probability histograms: the {**count**=} option tells it how many histograms to show beyond the first. For example, the command:

```
. demoiv box00 {count=3} >> hists
```

generates the same histograms as before and saves the display in the text file *hists*.

For more information about **demoiv**, see the manual entry *demoiv(BRM)*. For more information about saving output displays in text files, see Section 4.5.

7.2   Random Sampling

The **sample** command allows you to draw a simple random sample from a range of integers, or from the cases (rows) of a dataset.

Sampling with replacement. A *simple random sample, drawn with replacement,* is a sample drawn from a population in such a way that every element in the population has the same chance of being drawn, and successive observations are independent of each other. For example, if you are drawing tickets from a box, then every ticket has the same chance of being drawn; moreover, after you draw each ticket you replace it in the box, so that successive draws are made under identical circumstances.

To draw a random sample of size 20 with replacement from the integers 0 through 10, give the **sample** command as follows:

```
. sample {lo=0;hi=10} {n=20} > a
. show {i} a
    1  7  10  7  9  2  0  0  8  9  7  2  2  1  3  5  6  0  5  3
```

The {**lo**=} and {**hi**=} options determine the range of integers to sample from, by giving the low and high end of the range. The {**n**=} option determines the size of the sample—in this case, 20.

Figure 7.2

```
. demoiv box00
Probability histogram for a single draw:
    1        ****************************
    2        ****************************
    3        ****************************
    4        ****************************
    5        ****************************
    6        ****************************
Do you wish to continue? yes
Probability histogram for the sum of 2 draws:
    2        *****
    3        *********
    4        **************
    5        ******************
    6        ***********************
    7        ****************************
    8        ***********************
    9        ******************
   10        **************
   11        *********
   12        *****
Do you wish to continue? y
Probability histogram for the sum of 4 draws:
    6        **
    7        ****
    8        *******
    9        ***********
   10        ***************
   11        ********************
   12        ***********************
   13        ***************************
   14        ****************************
   15        **************************
   16        *************************
   17        *********************
   18        ***************
   19        ***********
   20        *******
   21        ****
   22        **
Do you wish to continue? y
Probability histogram for the sum of 8 draws:
   16        *
   17        **
   18        ****
   19        *****
   20        ********
   21        **********
   22        *************
   23        ******************
   24        *********************
   25        ***********************
   26        ****************************
   27        ***************************
   28        ****************************
   29        ***************************
   30        *************************
   31        ***********************
   32        *********************
   33        *****************
   34        **************
   35        ***********
   36        ********
   37        *****
   38        ****
   39        **
   40        *
Do you wish to continue? no
```

Note that some of the values are the same—when sampling is done with replacement, a value may be drawn more than once. Of course, when you try this command you will get different results, because they are random.

The {i} option to the **show** command, which we used here, displays data in integer format—it shows no decimal points or fractions. This is useful for more compact display of integer data.

Sampling without replacement. A *simple random sample, drawn without replacement,* differs from a sample drawn with replacement in that, once an element is drawn from the population, it cannot be drawn again. If you are drawing tickets from a box, then after drawing each ticket you leave it out of the box. On any given draw all remaining tickets in the box (or elements in the population) have the same chance of being drawn. Of course, when sampling without replacement the sample size can be no larger than the population size.

To draw a sample without replacement, use **sample** with the {**norep**} option. For example, to draw a random sample of size 5 without replacement from the integers 0 through 10, give the command:

```
. sample {lo=0;hi=10} {n=5} {norep} > b
. show {i} b

  6   2   3   8   0
```

Useful defaults. If the {**lo=**} option is not given, the value 1 is used. If the {**n=**} option is not given, then the sample size is set equal to the population size—in this case, the number of integers in the range *lo . . . hi.* Consider the command:

```
. sample {hi=10;norep} > c
. show {i} c

  5   1   6   9   10   4   2   8   3   7
```

By combining the default values of {**lo=**} and {**n=**} with the {**norep**} option, we obtained a random *permutation* (or shuffle) of the integers 1 through 10.

Sampling from a dataset. Sometimes it is useful to sample the rows (or cases) from a BLSS dataset. The dataset may represent a population from which we wish to sample; or, in the case of a large dataset, it may simply be more convenient to work with a smaller dataset than the original.

For example, the BLSS library dataset *boston* is quite large—it contains 506 cases and 14 variables. To load it from the data library and then take a sample of size 100 without replacement, give the commands:

```
. load boston
. sample boston {norep} {n=100} > b1
```

This command creates a new dataset, *b1*, which contains the sample. As before, the option {**norep**} specifies that the sample is without replacement, and the option {**n=100**} specifies that the sample size is 100.

By default, **sample** samples with replacement and, if the {**n**=} is not given, the sample size is the same size as the input dataset. Thus the command:

```
.  sample boston > b2
```

creates a sample *b2*, of the same size as the original dataset, drawn with replacement.

For more information about the **sample** command, see the manual entry *sample(BRM)*.

7.3 Random Number Generation

Examples. BLSS can generate random numbers from many probability distributions. We start with some examples. The **runi** command generates independent uniform random numbers. For example, the command:

```
.  runi {dims=50} > u
```

generates a row vector *u* of length 50 which contains independent observations from the uniform [0, 1] distribution. (We discuss dimensions and the {**dims**=} option at the end of this section.) The **rgau** command generates independent Gaussian (Normal) random numbers. For example, the command:

```
.  rgau {dims=20,5} > z
```

generates a dataset *z* of dimensions (20,5) which contains independent observations from the standard Gaussian distribution (mean = 0, SD = 1). To specify a mean other than 0, use the {**m**=} option; to specify a SD other than 1, use the {**s**=} option. For example, the command:

```
.  rgau {dims=20,5} {m=50;s=10} > x
```

generates a dataset with the same dimensions as above, but with mean 50 and SD 10. If you prefer to think in terms of variance instead of SD, you can use the {**v**=} option to specify the variance.

Overview. The random number generating commands all have names of the form r*xxx*. The **r** stands for 'random', and the *xxx* is different for each distribution. Thus the commands are named **runi**, **rgau**, **rcauchy**, etc. Distribution parameters are specified via options, as explained below. Successive random numbers are mutually independent.

Table 7.1 shows the full list of distributions from which BLSS can generate random numbers. Column 1 shows the name for each distribution (uniform, Gaussian, ...) and the BLSS name or abbreviation (**uni**, **gau**, ...). Column 2 shows the distribution parameter names, restrictions (if any), and alternates (if any). Column 3 shows the full functional form of the distribution in terms of the cdf (cumulative distribution function). Because different textbooks often use different parameter names and meanings, consult this table to see the exact meanings of the parameters in BLSS.

Table 7.1. BLSS continuous distribution names, parameters, and options.

Distribution / BLSS Name	Parameters and Restrictions	Cumulative Distribution Function, $F(x)$	Parameter Options	Parameter Defaults
Beta / **beta**	$\alpha > 0,\ \beta > 0$	$\dfrac{1}{B(\alpha,\beta)} \displaystyle\int_0^x t^{\alpha-1}(1-t)^{\beta-1}\,dt$	a=; b=	a=1; b=1
Cauchy / **cauchy**	$\alpha,\ \beta > 0$	$\dfrac{1}{\pi}\tan^{-1}[(x-\alpha)/\beta] + .5$	a=; b=	a=0; b=1
χ^2 (Chi-square) / **chisq**	$\nu > 0$	$\dfrac{1}{2^{\nu/2}\Gamma(\nu/2)} \displaystyle\int_0^x t^{\nu/2-1}e^{-t/2}\,dt$	df=	none
Exponential / **exp**	$\lambda > 0$ $\beta = 1/\lambda$	$1 - e^{-\lambda x}$	l= b=	l=1 b=1
Fisher's F / **f**	$\nu_1 > 0,\ \nu_2 > 0$	$\dfrac{\nu_1^{\nu_1/2}\nu_2^{\nu_2/2}}{B(\nu_1/2,\ \nu_2/2)} \displaystyle\int_0^x t^{\frac{\nu_1-2}{2}}(\nu_2+\nu_1 t)^{\frac{-(\nu_1+\nu_2)}{2}}\,dt$	df=	none
Gamma / **gamma**	$\alpha > 0,\ \lambda > 0$ $\beta = 1/\lambda$	$\dfrac{\lambda}{\Gamma(\alpha)} \displaystyle\int_0^x (\lambda t)^{\alpha-1}e^{-\lambda t}\,dt$	a=; l= b=	a=1; l=1 b=1
Gaussian (Normal) / **gau**	$\mu,\ \sigma > 0$	$\dfrac{1}{\sqrt{2\pi}\,\sigma} \displaystyle\int_{-\infty}^x e^{-(t-\mu)^2/2\sigma^2}\,dt$	m=; s= v=(σ^2)	m=0; s=1 v=1

Logistic **logis**	$\alpha>0,\ \beta>0$	$\dfrac{1}{1+e^{-(x-\alpha)/\beta}}$	a=; b=	a=0; b=1
Noncentral t **nt**	$\nu>0,\ \delta$	$1-\displaystyle\sum_{j=0}^{\infty}\left[e^{-\delta^2/2}\dfrac{(\delta^2/2)^j}{2j!}\,I_{\frac{\nu}{\nu+x^2}}\,(\nu/2,j+\tfrac12)\right]$	df=; d=	d=0
Student's t **t**	$\nu>0$	$\dfrac{\Gamma(\frac{\nu+1}{2})}{\sqrt{\pi\nu}\,\Gamma(\nu/2)}\displaystyle\int_{-\infty}^{x}\left[1+\dfrac{t^2}{\nu}\right]^{\frac{-(\nu+1)}{2}}dt$	df=	none
Uniform **uni**	$a<b$	$\dfrac{x-a}{b-a},\quad a\le x\le b$	a=; b=	a=0; b=1

Note: For the F distribution, the {**df=**} option takes two values to specify the two degrees of freedom parameters. All other parameter options listed here take a single value.

Definitions of functions used above:

Gamma	$a>0$	$\Gamma(a)\ =\ \displaystyle\int_{0}^{\infty}t^{a-1}e^{-t}dt$
Beta	$a>0,\ b>0$	$B(a,b)\ =\ \dfrac{\Gamma(a)\Gamma(b)}{\Gamma(a+b)}$
Incomplete beta	$a>0,\ b>0$	$I_x(a,b)\ =\ \dfrac{1}{B(a,b)}\displaystyle\int_{0}^{x}t^{a-1}(1-t)^{b-1}dt$

Column 4 of Table 7.1 shows the BLSS option names which correspond to the distribution parameters. Most often, these are the Roman alphabet equivalents of the Greek letter parameter names: for example, {**m**=} specifies the parameter μ; {**s**=} specifies σ. For the degrees-of-freedom parameters, however, the option is always {**df**=}. In some cases, BLSS allows alternate parameter options. For the Gaussian distribution, for example, you may use either {**s**=} to specify the SD or {**v**=} for the variance; for the exponential distribution you may use either {**l**=} to specify the intensity parameter λ or {**b**=} to specify the scale parameter β.

Column 5 shows the default parameter option values, if any. For example, for the Gaussian distribution the default value of {**m**=} is 0 and the default value of {**s**=} is 1.

Example. We illustrate the use of Table 7.1 with another example. Suppose we want to generate 50 random numbers from the exponential distribution with intensity parameter $\lambda = 10$ and put the result into the dataset *e*. We see from the table that the BLSS abbreviation for the exponential distribution is **exp**. Thus, the command to use is **rexp**. The option {**l**=} specifies λ, so the full command to give is:

    . rexp {dims=50} {l=10} > e

Alternatively, we can specify the scale parameter β instead of the intensity parameter λ. Scale is the reciprocal of intensity, so the corresponding scale parameter is $\beta = 0.1$:

    . rexp {dims=50} {b=0.1} > e

This example illustrates the need to check the BLSS parameter names against the names you may be familiar with from a particular textbook.

Specifying dimensions. All r*xxx* commands recognize the {**dims**=} option, which determines the dimensions of the output dataset—and thus, how many random numbers are generated. If only one number is given after the option, a row vector is created which has that length. For example, {**dims=50**} creates a row vector of length 50. If two numbers are given after the {**dims**=} option, the first number tells how many rows to make and the second tells how many columns. For example, {**dims=20,5**} creates a dataset with 20 rows and 5 columns. Refer to Section 5.1 for more information about dataset dimensions.

Another way to specify the dimensions of a random dataset is to make it be the same size as some other dataset. For example, the command:

    . rcauchy z > z2

creates a dataset *z2* of Cauchy random numbers which has the same dimensions as the dataset *z*. The contents of *z* are ignored; only its dimensions are used.

For more information about the random number generating commands, see the manual entry *random(BRM)*.

7.3.1   Transforming Random Numbers

If you have random numbers generated from one distribution and apply a transformation to them, the result is random numbers from some other distribution. For example, the numbers u above were generated from the uniform [0, 1] distribution. Then the dataset:

. $v = u*10+20$

has the uniform [20, 30] distribution; the dataset:

. $e = -log(1-u)$

has the exponential distribution with $\lambda = 1$; the dataset:

. $c = tan((u-.5)*PI)$

has the Cauchy distribution.

Of course, if you want the result to have a specific distribution, you must apply the proper transformation. The technique used above was transformation by the inverse cumulative distribution function (cdf) and is useful in cases where you know a closed form for it.[3] In Section 9.3 we discuss methods for obtaining multivariate distributions via transformations.

7.3.2   Resetting the Seed

Random numbers generated by a computer are not random in the true sense of the term. Like everything else on the computer, they are generated by a deterministic algorithm; they can be predicted if you know the algorithm. But they do *look* random, in the sense that they pass statistical tests for independence and distribution. Thus, for almost all purposes they can be regarded as if they indeed are random—for this reason they are called *pseudorandom* numbers.[4]

What seems like a disadvantage—predictability if you know the algorithm—is actually an advantage. You can restart the pseudorandom sequence of numbers at a given point. This is useful for exactly repeating a 'random' experiment when you want to replicate results. The **seed** command allows you to do this. For example, we gave the command:

. `seed 11.7`

just before giving the **sample** commands in Section 7.2. If you give the **seed**

---

3. Many distributions, however, do not have a closed-form inverse cdf. This is why the r*xxx* and q*xxx* commands (introduced in the next section) are particularly useful.

4. For a description of the algorithm, see *rans(BML)*; that is, the entry *rans* in the *BLSS Mathematical Library*. It is available on-line within BLSS by typing:

. `man rans`

command with the same argument and then repeat the **sample** commands in Section 7.2, you will get the same results we did.[5]

In general, whenever you give the **seed** command with a given argument, you will obtain the identical sequence of pseudorandom numbers.[6] Different values of the first element of the **seed** argument result in different pseudorandom sequences. (If the **seed** argument is a dataset, only its first element is used.) There is no discernible relationship between the value of the seed and the random numbers which result.

7.4   Probability Functions

This section introduces the BLSS commands for evaluating cumulative distribution functions (cdf's), their inverses, and their densities. It also shows how to use these commands for evaluating tail probabilities, P-values, and critical values (for hypothesis tests).

7.4.1   CDF's

Every probability distribution has a corresponding *cumulative distribution function* (or *cdf*), usually denoted as $F(x)$. This function gives the probability that a random variable X with that distribution is less than or equal to the value x:

$$F(x) = P(X \leq x)$$

See Figures 7.3a–b for a graph of $F(x)$ and its relationship to the probability density function $f(x)$ in the specific case of the standard Gaussian (Normal) distribution.

BLSS commands for evaluating cdf's have names of the form **p***xxx*. The names begin with **p** as a reminder that the result is a probability—that is, a number between 0 and 1. The *xxx* portion of the name indicates the specific distribution, just as with the **r***xxx* commands introduced in the previous section. For example, **pgau** is the cdf for the Gaussian distribution, **pgamma** is the cdf for the gamma distribution, etc.

The **p***xxx* commands have the same parameter options and defaults as the **r***xxx* commands.

Examples. To find the probability that a random variable with the standard Gaussian distribution is less than 1.2, type:

```
. pgau 1.2

    0.884930
```

---

5. Provided that you are using a sufficiently similar type of computer.
6. With the same proviso as before. Refer to the *seed(BRM)* manual entry for further discussion.

To specify a different mean and SD for the Gaussian distribution, use the {**m**=} and {**s**=} options, just as with the **rgau** command. To save the result in an output dataset, give the '>' symbol followed by the name of the dataset. For example:

```
. pgau 1.2 {m=2.1;s=3.5} > p
. show {g} p

    0.398534
```

The **p***xxx* commands are not limited to evaluating the probability $P(X \leq x)$ for a single value of x. They can operate on entire datasets at once—the result is an ouput dataset in which each element contains the probability $F(x)$ corresponding to the x value in the input dataset. In the following example, we create a dataset x which contains the values 0, 1, 2, 3, and 4 and then evaluate and print the probability that an exponential random variable is less than or equal to each of these values:

```
. x = 0:4
. pexp x > p
. show {g} x p

    0.00000     1.00000     2.00000     3.00000     4.00000

    0.00000     0.632121    0.864665    0.950213    0.981684
```

The sequence operator (':'), which we used here to create the sequence of integers 0 through 4, is discussed in Section 8.3.1.

7.4.2   Inverse CDF's

Sometimes we want to ask the opposite question. For example, what value x is such that a standard Gaussian random variable has probability .5 of being less than x? (The answer, of course, is $x = 0$.) In general, suppose you have a random variable X with a known probability distribution. Given a probability value p, what is the value x such that $P(X \leq x) = F(x) = p$? This amounts to solving the equation for x:

$$x = F^{-1}(p)$$

This function, $F^{-1}(p)$, is called the *inverse cumulative distribution function* (or *inverse cdf*) of the probability distribution. Figure 7.4 shows, graphically, the relationship between the cdf and the inverse cdf.

The BLSS commands for evaluating inverse cdf's have names of the form **q***xxx* where, as above, the *xxx* portion of the name indicates the specific distribution. The names begin with **q** as a reminder that the result is a *quantile*, a term explained below. The **q***xxx* commands have the exact same inputs, parameter options, and defaults as the **p***xxx* commands.

Examples. Earlier, we found the probability that a standard Gaussian random variable is less than 1.2: that probability is .884930. To work this example backwards:

```
.  qgau  .884930
     1.20000
```

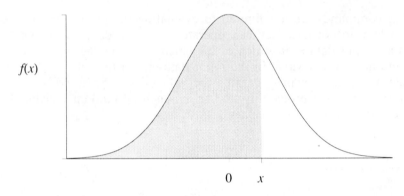

Figure 7.3a. Probability density function $f(x)$ for the standard Gaussian distribution. Area represents probability. The total area under the curve is 1. The area under the curve to the left of the value x (that is, the shaded area) is $P(X \leq x)$, the probability that the random variable X is less than or equal to the value x.

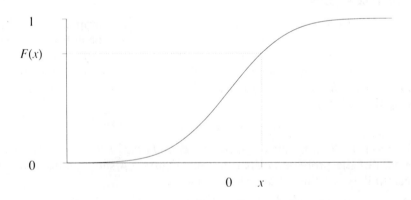

Figure 7.3b. Cumulative distribution function $F(x) = \int_{-\infty}^{x} f(u)\,du = P(X \leq x)$ for the standard Gaussian distribution. Height represents probability. The value of the function $F(x)$ at the value x equals the shaded area in Figure 7.3a and equals $P(X \leq x)$. $F(x)$ approaches 0 as x approaches $-\infty$; it approaches 1 as x approaches ∞.

To compute the 95th percentile of Fisher's F distribution with (2,10) degrees of freedom (that is, the value x such that $P(X \le x) = .95$):

```
. qf {df=2,10} .95

  4.10282
```

The *median* of a distribution is its midpoint: the value x such that $P(X \le x) = .5$. To rephrase our earlier question: What is the median of the standard Gaussian distribution? The answer given by **qgau**:

```
. qgau .5

  0.00000
```

agrees with what we know.

The *quartiles* of a distribution are its quarter-way points; that is, its 25th, 50th, and 75th percentiles. To compute the three quartiles of the χ^2 distribution with 8 degrees of freedom:

```
. qchisq {df=8} .25,.5,.75

  5.07064    7.34412    10.2189
```

The *deciles* of a distribution are its tenth-points; that is, its 10th, 20th, ..., 90th percentiles. To compute, save, and display the deciles of the standard Gaussian and Cauchy distributions:

```
. p = .1:.9:.1              # p is the sequence .1, .2, ..., .9
. qgau p > qg
. qcauchy p > qc
```

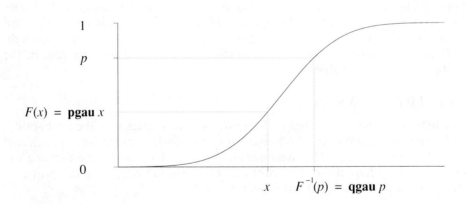

Figure 7.4. The relationship between $F(x)$, $F^{-1}(p)$, and the values **pgau** x and **qgau** p. The same relationships hold for all **p***xxx* and corresponding **q***xxx* commands.

```
. show {g} p,,qg,,qc
```

0.100000	0.200000	0.300000	0.400000	0.500000	0.600000
-1.28155	-0.841621	-0.524400	-0.253347	0.00000	0.253347
-3.07768	-1.37638	-0.726542	-0.324920	0.00000	0.324920

0.700000	0.800000	0.900000
0.524400	0.841621	1.28155
0.726542	1.37638	3.07768

The columnwise-catenation operator ',' used above was introduced in Sections 2.2 and 5.6. It and the rowwise-catenation operator ',,' are explained in detail in Section 8.3.3.

Quantiles. The term *quantile* is a generalization of the terms *quartile, decile,* and *percentile.* The p-th quantile of a distribution is simply the point x such that $P(X \le x) = p$; that is, the $(100p)$-th percentile. The q*xxx* commands let you find *any* quantile of a distribution.

See the manual entry *cdf(BRM)* for more information about the p*xxx* and q*xxx* commands.

7.4.3   Finding Tail Probabilities

As mentioned earlier, probability is represented by the area under the curve of a probability density function. A *tail probability* is the probability that a random variable is less than, or greater than, a given value x. The name *tail probability* refers to the fact that the area which represents it includes one of the extreme ends, or 'tails,' of the distribution.

For left-hand tails, the tail probability corresponding to a value x is $P(X \le x)$. In other words, it is just the cdf, $F(x)$, for the random variable X.

For right-hand tails, the tail probability corresponding to a value x is $P(X \ge x)$. This equals $1 - P(X \le x)$;[7] that is, $1 - F(x)$. Thus, to evaluate a right tail probability for a value x, use the appropriate p*xxx* command and then subtract the result from 1. We show examples of this in the next section, in the context of finding P-values.

7.4.4   Finding P-Values

In a hypothesis test, the *P-value* of an observed test statistic is the probability, under the null hypothesis, of obtaining a test statistic value at least as extreme as the observed value. The meaning of 'extreme' depends on the formulation of the alternate hypothesis, but most often the P-value is either a tail probability or the sum of two equal tail probabilities.

Example: t-test. Suppose you are performing a one-tailed t-test: your observed t-statistic is 1.89 with 18 degrees of freedom and you want to check this against the right tail. To compute the P-value, you have a choice. You can

---

7. Whether the inequality is strict or not is immaterial for continuous random variables and is ignored throughout this section.

either evaluate the right tail probability directly, by subtracting the cdf value from 1:

```
. pt {df=18} 1.89 > pleft
. show {g} 1-pleft

    0.0374860
```

or you can exploit the symmetry of the t distribution in order to take a shortcut: the tail area to the right of 1.89 equals the tail area to the left of −1.89:

```
. pt {df=18} -1.89

    0.0374860
```

If you are performing a two-tailed t-test, the P-value is the sum of the left and right tail probabilities. Because these are equal, simply multiply the left tail probability by 2:

```
. pt {df=18} -1.89 > pleft
. show {g} 2*pleft

    0.0749721
```

Example: F-test. Suppose you are performing an F-test: your observed F statistic is 5.04 with (4,18) degrees of freedom, and you want to check this against the right tail. To compute the P-value (the right tail probability), evaluate the cdf at 5.04 and subtract the result from 1:

```
. pf {df=4,18} 5.04 > pleft
. show {g} 1-pleft

    0.00665277
```

Because the F distribution is not symmetric, you cannot use the shortcut we used for the t distribution.

7.4.5 Finding Critical Values

For a hypothesis test, you may want to know the value x which corresponds to a prespecified P-value, such as .05 or .01. The prespecified P-value is called the *significance level* (or just *level*) and is usually denoted by α. The x value which corresponds to the desired significance level α is called the *critical value*. Critical values can refer to either one-tailed or two-tailed P-values. To find a critical value, you find the x value which gives the appropriate tail probability using the appropriate q*xxx* command.

Example: t-test. Suppose you are performing a one-tailed t-test and the observed t-statistic has 16 degrees of freedom. To find the critical value for a left-tailed test with level $\alpha = .01$, use the **qt** command directly:

```
. qt {df=16} .01

    -2.58349
```

The critical value is –2.58. The critical value for a right-tailed test with level $\alpha = .01$ is 2.58 because the t distribution is symmetric about 0.

To find the critical value for a two-tailed t-test with level $\alpha = .01$, you must find the value for which the left and right tails both have tail probability .005 so that their sum is .01. If we use 16 degrees of freedom as before:

```
. qt {df=16} .005

    -2.92078
```

The critical values are –2.92 on the left side, and 2.92 on the right side.

Example: F-test. Suppose you are performing an F-test and the observed F-statistic has (3,16) degrees of freedom. To find the critical value for a right-tailed test (the common F-test) with level $\alpha = .01$, find the value which gives a right tail probability of .01—in other words, a left tail probability of .99:

```
. qf {df=3,16} .99

    5.29222
```

The critical value is 5.29.

For further discussion of hypothesis tests in BLSS, see Sections 6.3 and 6.4 (which cover the common z- and t-tests) and Sections 9.4.3 and 9.5.2 (which contain examples of F-tests).

7.4.6   Densities

BLSS commands for evaluating probability density functions have names of the form **d***xxx* where, as before, *xxx* indicates the name of the distribution. The **d***xxx* commands have the identical inputs, parameter options, and defaults as the corresponding **p***xxx* commands. The input specifies the value(s) at which the density is evaluated.

Examples. To compute the density of the gamma distribution with parameters $\alpha = 1$ and $\beta = 1$ (the defaults) at the values $x = 0$, 1, and 2:

```
. dgamma 0,1,2

   1.00000      0.367879      0.135335
```

To compute the density of the beta distribution with parameters $\alpha = 5$ and $\beta = 2$ at the values $x = 0, .1, .2, \ldots, .8, .9, 1.0$:

```
. dbeta {a=5;b=2} 0:1:.1

   0.00000   0.00270000   0.0384000   0.170100   0.460800   0.937500
   1.55520   2.16090      2.45760     1.96830    0.00000
```

In this example the sequence operator (described in Section 8.3.1) was used to create a vector of values between 0 and 1 in increments of .1 as the input dataset to **dbeta**.

CHAPTER 8

Mathematical and Matrix Operations

This chapter covers algebraic, matrix, and logical operations in BLSS in detail. The principal command for performing these operations is **let**, and this chapter serves as a reference guide for **let**. As such, it is best read by skimming it all and then rereading carefully specific sections as you need them.

You will understand this chapter more easily if you have already read Chapter 2 and the material on matrix subscripts in Section 5.1, but the essential material from those chapters is repeated here. Chapter 9 contains additional examples which illustrate how individual operations are pieced together to solve larger problems.

This chapter also introduces commands which perform specialized matrix operations. Perhaps the most important are those which perform array and vector operations, documented in *arrop(BRM)* and *vecop(BRM)*, and the commands described in *redim(BRM)* and *select(BRM)*.

8.1   Invoking the Let Command

The **let** command may be invoked *explicitly* in order to assign the value of an expression to a dataset, or *implicitly* in order to pass the value of an expression directly to another command.

8.1.1   Explicit Let Commands

When the purpose of evaluating an arithmetic or mathematical expression is
to assign its value to a BLSS dataset, **let** is invoked *explicitly* as a BLSS command. For example, the BLSS commands:

```
. let x = 10
. let y = x + 2
```

mean: let the value of the dataset x be 10, and let the value of the dataset y
be the value of x plus 2. Any previous values of x and y are lost.

In this context, the word **let** is optional. The commands above can also be
entered as:

```
. x = 10
. y = x + 2
```

Note that the '=' sign in this context has a different meaning from its usual
mathematical meaning. In mathematics, the statement $y = x + 2$ is a statement about the relationship between x and y and can be interpreted as a statement about either variable. Equivalent statements are $y - 2 = x$ and $x = y - 2$.
In contrast, in BLSS (as in most computer languages) the command:

```
. y = x + 2
```

is an *assignment:* it assigns the value of the expression on the right side of the
'=' sign to the dataset whose name is on the left side of the '=' sign, regardless
of any previous value which that dataset may have had. The command:

```
. x = y - 2
```

has an entirely different meaning. (The meanings are clear when you recall
that the word 'let' can appear at the beginning of the command.)

Multiple let commands. More than one explicit **let** command can appear in a
single BLSS command line, provided that: 1) the **let** commands are separated
by semicolons; and 2) other types of BLSS commands do not appear in the
same command line as the **let** commands. The general form is:

> [**let**] *expression* ; [**let**] *expression* ; [**let**] *expression* ; . . .

where the square brackets indicate that the word **let** is optional and where
each *expression* is any valid arithmetic, functional, matrix, or logical expression which obeys the **let** syntax introduced in this chapter. The expressions
are evaluated in order, from left to right.

Due to the implementation of BLSS, multiple **let** commands given on the
same BLSS command line execute more efficiently than do the corresponding
individual **let** commands given on separate BLSS command lines.

Spacing within explicit **let** commands does not matter.

8.1.2 Implicit Let and Let-Arguments

When the purpose of evaluating an expression is to pass its value on to a regular BLSS command, **let** may be invoked *implicitly* within the BLSS command, instead of explicitly in a preceding **let** command. In this example:

```
. show x+1 y-2 (x*y)/10
```

let is called three times, once for each expression. The BLSS command (in this case, **show**) operates on the resulting value(s). No assignment is made and the result(s) cannot be reused without being recalculated.

BLSS invokes **let** implicitly whenever an argument to a BLSS command contains one or more of the following list of characters:

$$- + * / ^ ' , : < [(= \& | ! ?$$

Such an argument is called a *let-argument*, and the characters listed above are called *let-characters*. In particular, any expression enclosed in parentheses is a let-argument, although parentheses are not always necessary. **Let** evaluates each let-argument. The results become input arguments (datasets) to the BLSS command.

All space characters within let-arguments must occur inside parentheses or square brackets. The reason is that, normally, spaces separate arguments to BLSS commands. For example, the following commands all evaluate correctly:

```
. scat plover[1 3]
. stat -a/b
. stat sqrt(c^2 + d^2)
. stat (c^2 + d^2)
. stat c^2+d^2
```

However, this command fails:

```
. stat c^2 + d^2
```

because the spaces surrounding the '+' character do not occur within parentheses. BLSS thinks that '*c^2*', '+', and '*d^2*' are three separate arguments to the **stat** command.

8.2 Arithmetic and Functions

Let provides the standard arithmetic operators and functions. The syntax is similar to that of most computer languages. Some example commands:

```
. y = a*x^2 + b*x + c
. z = sqrt(abs(y))
. f = 1 / (PI * (1 + x^2))
. t = log(p / (1 - p))
. u = exp(-x^2)
. big = 10!
```

Arithmetic operators. Here are the arithmetic operators and their meanings:

+	Addition.
–	Subtraction and unary negation (taking the negative of an expression).
*	Multiplication.
/	Division.
^	Exponentiation (raising an expression to a power).
**	Another symbol for exponentiation.
!	Factorial.

When evaluating these operators, **let** observes the following precedences: first it computes any factorials; then it performs exponentiations right to left; then unary negations; then multiplications and divisions left to right; and then additions and subtractions left to right. These and all operator precedences are summarized in the table in Section 8.8. Parentheses may be used to change the order of evaluation, as in the examples above.

Elementwise operators. The arithmetic operators listed above are *elementwise operators.* They may be applied to scalars[1] or to more complex datasets— vectors and matrices. If both operands[2] of an elementwise operator are scalars, the result is a scalar. If one operand is a scalar and the other is a larger dataset, then the operation is applied *elementwise.* In other words, it is applied, in turn, to the scalar and each element of the dataset. For example, if the dataset *a* contains a 2-by-3 matrix:

```
1   2   3
4   5   6
```

then the result of the command:

```
. b = a * 2
```

is the dataset *b* of the same shape, with every element multiplied by 2:

```
2    4    6
8   10   12
```

If both operands of an elementwise operator are datasets with the same dimensions (number of rows and columns), then the operation is applied elementwise in the sense that it is applied to the corresponding elements of the datasets. The result is a dataset whose elements contain the corresponding individual results. For example, if the datasets *a* and *b* shown above are multiplied together:

```
. c = a * b
```

---

1. A *scalar* (as opposed to a *vector* or *matrix*) is a single data element. We also use the term *scalar* to refer to a BLSS dataset which contains a scalar.

2. An *operand* is an object on which an operator operates. In the expression *a* + *b* the operands are *a* and *b*.

then the resulting dataset *c* contains:

```
    2     8    18
   32    50    72
```

It has the same shape and its individual elements are the products of the corresponding elements from *a* and *b*. (Note that this operation is *not* matrix multiplication.)

We discuss what happens when an elementwise operator is applied to two arrays of different shapes in Section 8.3.5.

Functions which operate on scalars and return scalars, such as **abs**, **exp**, **log**, and **sqrt**, are applied elementwise to BLSS datasets. The result is a dataset whose elements are the corresponding individual results. A complete list of scalar functions is in *let*(*BRM*); that is, the entry on **let** in the *BLSS Reference Manual*.

Special names. The following special names are recognized. They represent scalars:

E	The mathematical constant $e = 2.71828....$
PI	The mathematical constant $\pi = 3.14159....$
NA	The missing value.

As noted in Chapter 1, special names and function names must not be used as BLSS dataset names.

Precision. In BLSS, data values are stored and expressions are evaluated in *single precision*. On most computers, single precision provides slightly more than six decimal places of working accuracy.

8.3   Vector and Matrix Operations

A BLSS dataset may contain a scalar, a vector, or a matrix. By common convention a matrix with *m* rows and *n* columns is said to have dimensions *m*-by-*n*, sometimes written (*m*,*n*). BLSS follows this convention—the row dimension is always given before the column dimension.

In BLSS, vectors are special cases of matrices. Thus, every vector is either a row vector (a matrix with dimensions 1-by-*n*) or a column vector (a matrix with dimensions *n*-by-1). Beware that, although BLSS allows you to add a row vector to a column vector, the result may not be what you expect. This operation is discussed in Section 8.3.5.

A scalar in BLSS is simply a 1-by-1 matrix.

The **list** command informs you of the shape of a BLSS dataset: scalar, row vector, column vector, or matrix. See Section 4.1 for an example. In order to conserve space, however, the **show** command normally displays column vectors as row vectors. The {**shape**} option causes **show** to display column vectors in their true shape—that is, as columns. We will see examples of this in the following sections.[3]

8.3.1   Creating Matrices

There are several ways to create new vectors and matrices in BLSS.

The read command, introduced in Section 4.4, allows individual data elements to be entered from the keyboard or a text file. It is appropriate when the data are observations or follow no mathematical pattern. Refer to the manual entry *read(BRM)* for detailed information and examples.

The sequence operator (':') generates row vectors which are sequences of numbers:

> $a:b$ is the row vector of numbers a through b, incremented by 1.
> $a:b:c$ is the row vector of numbers a through b, incremented by c.

If b is less than a, the sequence is decremented, not incremented. If $b - a$ is not an integer (or, in the case $a:b:c$, an exact multiple of c), the sequence stops just before exceeding b. Here are some examples:

```
. a = -2:3
```

creates the row vector a whose elements are –2, –1, 0, 1, 2, 3;

```
. b = 4:2:0.5
```

creates the row vector b whose elements are 4, 3.5, 3, 2.5, 2;

```
. c = 0:50:10
. d = 0:55:10
. e = 0:59.95:10
```

all create row vectors whose elements are 0, 10, 20, 30, 40, 50.

The catenation operators (',' and ',,') can be used to create small datasets, as well as to join together larger ones. We saw examples of this in Sections 2.2 and 5.6.2 and will see more examples in Sections 8.3.3, 8.3.5, and throughout Chapter 9.

Constant vectors and matrices are vectors and matrices whose elements all equal the same constant value. They are created by the **const** command. For example, a row vector v of length n whose elements all equal the value c is created by the command:

```
. const c {dims=n} > v
```

---

3. The command:
> ```
> . set SHAPE
> ```
sets the SHAPE environment string and is equivalent to including the {**shape**} option in all subsequent **show** commands until the command:
> ```
> . unset SHAPE
> ```
is given. Section 10.3 discusses the **set** and **unset** commands; Section 10.4 discusses environment strings.

A matrix x of dimension *m*-by-*n* whose elements all equal the value c is created by the command:

```
. const c {dims=m,n} > x
```

The {**dims**=} option of the **const** command is just like that of the **read** and the r*xxx* commands discussed in Sections 4.4 and 7.3. Instead of specifying the dimensions with the {**dims**=} option, you can also create a constant dataset whose dimensions are the same as some other dataset *d*:

```
. const c d > x
```

Special matrices. The following commands create special matrices. With all of them, the dimensions of the new matrix can be specified using the {**dims**=} option, or by giving an argument whose dimensions will be used.

idn	Identity matrix.
ltri	Lower triangular matrix of 1's.
utri	Upper triangular matrix of 1's.
colindex	Matrix whose elements contain their column index.
rowindex	Matrix whose elements contain their row index.
sheetindex	Matrix whose elements contain their sheet index (sheets are discussed in Section 8.4).

For more information on these commands, the **const** command, and others, see the manual entry *arrop*(*BRM*).

8.3.2   Matrix Subscripts and Subsets

In Section 5.1.1, we saw that a single element of a matrix is referenced by a double subscript—that is, by appending its row and column numbers within square brackets to the name of the matrix. For example, $x[2,3]$ denotes the element in the second row and third column of x. The numbers 2 and 3 are said to be the *indices* (plural for *index*) of the element. As with dimensions, the row index is always given before the column index.

Submatrices—that is, rectangular subsets of a matrix—are denoted by a generalization of this notation. Suppose that R and C are lists of row and column indices. Then:

$$x[R,C]$$

denotes the submatrix of x whose rows and columns are those listed in R and C. The index lists R and C may take on any of the following forms:

*	The entire dimension—that is, all the rows or all the columns.
(empty)	An empty index list also selects the entire dimension.
i:j	Specifies the indices i through j, inclusive. Both $i < j$ and $i > j$ are acceptable.
i:j:k	Specifies the indices i through j, incremented by k (or decremented by k, if $i > j$). If $j - i$ is not a multiple of k, the sequence of indices stops just before exceeding j.

  i j ... *k* Several items can be placed in the index list, separated by spaces. The items *i* can be single indices, or sequences of indices specified with the colon notation. Any index can be repeated.

  x[*C*] Equivalent to *x*[*,C*] or *x*[,*C*].

The indices must be given as explicit integers, not as datasets. Rows and columns are selected in the same order in which the indices are specified.

Note that the rules provide several alternate ways to denote rows or columns. For example, suppose that *x* is a 20-by-5 matrix. The expressions:

  x[2,*]; x[2,]

both denote row 2 of *x*. The expressions:

  x[*,4]; x[,4]; x[4]

all denote column 4 of *x*. Thus, single subscripts denote columns of a matrix, or individual elements of a row vector. Individual elements of a column vector must be denoted using double subscripts.

To continue the example, the expression:

  x[6:10,]

denotes rows 6 through 10 of matrix *x* (or elements 6 through 10 of a column vector *x*);

  x[,2 5]

denotes columns 2 and 5;

  x[6:10,2 5]

denotes the intersection of rows 6 through 10 with columns 2 and 5;

  x[1 6:10 2,]

denotes rows 1, 6 through 10, and 2 (in that order);

  x[20:1,]

is *x* with its rows in reverse order.

Submatrices may appear on the left side of an assignment command as well as on the right. Thus, the command:

  . x[5:10,] = x[10:5,]

causes rows 5 through 10 of *x* to be replaced by rows 10 through 5. In other words, the positions of rows 5 through 10 are reversed while the remaining rows are unchanged. Similarly, the command:

  . x[,3 1] = x[,1 3]

interchanges columns 1 and 3 of *x* but does not affect the other columns.

As noted above, subscripts must be given as explicit integers, not as datasets.[4] In Sections 8.3.4 and 10.8.2 we show methods for specifying matrix subscripts which are datasets or let-expressions.

8.3.3   Matrix Functions and Operators

Section 8.2 discussed arithmetic operators and functions which are performed elementwise on vectors and matrices. In contrast, the functions and operators in the following list operate on matrices in their entirety. Functions are evaluated first; then operators are evaluated in the order listed below. (As noted before, operator precedences are summarized in Section 8.8, and parentheses may be used to change the order of evaluation.)

det(a)   Determinant of square matrix a.

diag(a)   If a is a matrix, the main diagonal of a as a row vector; if a is a vector, the diagonal matrix constructed from a.

inv(a)   Inverse of square matrix a.

norm(a)   Square root of sum of squares of the elements. (If a is a vector, this is its 2-norm.)

trace(a)   Trace of matrix a.

$'a$ The *row-unravel* operator: the row vector which contains all the elements of a 'unraveled' in *row-major order* (that is, going along the rows).

$''a$ The *column-unravel* operator: the column vector which contains all the elements of a 'unraveled' in column-major order (that is, going down the columns).[5]

a' Transpose of matrix a.

a #^ n Matrix a raised to the power n (that is, multiplied by itself n times). The exponent n may be negative,[6] but it must be a scalar (which is truncated to an integer). The matrix a must be square.

a #* b Matrix multiplication. The number of columns in matrix a must equal the number of rows in matrix b.

a,b The *columnwise matrix catenation* operator: the columns of matrix a followed by the columns of matrix b. The matrices a and b must have the same number of rows.

$a,,b$ The *rowwise matrix catenation* operator: the rows of matrix a followed by the rows of matrix b. The matrices a and b must have the same number of columns.

Most of these functions and operations are in standard mathematical usage. The unravel and catenation operations, however, are particular to BLSS.[7]

---

4. This restriction may be removed in a future release of BLSS.
5. Note that the column-unravel operator consists of two single-quote characters ' together, *not* a double-quote character ".
6. For example, the operation a#^–3 means $(a$#^–1$)$#^3.
7. Of course, the operator symbols #* and #^ are also particular to BLSS.

Examples. These commands use the columnwise matrix catenation operator
',' and the rowwise matrix catenation operator ',,' to create small vectors *a*
and *b*:

```
. a = 1,2,3,4
. show a
     1.000     2.000     3.000       4.000
. b = 1,,5
. show b {shape}
     1.000
     5.000
```

The {**shape**} option, discussed at the beginning of Section 8.3, shows the
column vector *b* in its true shape.

The following command uses both catenation operators to create a small
matrix *x*. As indicated by the list above, the column catenation operator ',' is
evaluated before the row catenation operator ',,' so no parentheses are necessary:

```
. x = 1,2,3,4,,5,6,7,8
. show x
     1.000     2.000     3.000       4.000
     5.000     6.000     7.000       8.000
```

The following two commands are equivalent. Both set *y* equal to the first
three columns of *x*, in reverse order:

```
. y = x[3 2 1]
. y = x[3],x[2],x[1]
. show y
     3.000     2.000     1.000
     7.000     6.000     5.000
```

The following two commands both interchange the rows of *y*:

```
. z = y[2 1,]
. z = y[2,],,y[1,]
. show z
     7.000     6.000     5.000
     3.000     2.000     1.000
```

Recall from Section 8.3.1 that the sequence operator ':' generates sequences
of numbers in the form of row vectors. To generate a sequence of numbers
as a column vector, use the colon operator together with the transpose operator. For example, to create a column vector which contains the numbers 1
through 5:

```
. a = (1:5)'
```

The parentheses are necessary because, as shown in the table in Section 8.8, the transpose operator has higher precedence than the colon operator.

If R is a matrix and w is a vector with the same number of elements as R has rows, then to multiply each row of R by the corresponding element of w, we can premultiply R by the diagonal matrix constructed from w:

```
. R2 = diag(w) #* R
```

Similarly, if C is a matrix and v is a vector with the same number of elements as C has columns, then to multiply each column of C by the corresponding element of v, we can postmultiply R by the diagonal matrix constructed from v:

```
. C2 = C #* diag(v)
```

These two examples used standard matrix algebra to multiply each row or column of a matrix by a different constant. In the Section 8.3.5 we present a more direct method of doing so which is particular to BLSS.

8.3.4 Vector and Matrix Commands

This section surveys several other BLSS commands which manipulate vectors and matrices.

The redim command redimensions (that is, changes the dimensions of) a dataset. For example, if x is a vector of length 40 (and any shape) you can use the **redim** command to convert it into a matrix y of dimensions 10-by-4:

```
. redim x {dims=10,4} > y
```

Elements of x are entered into y in *row-major order* (that is, going along the rows). Thus, this command is the opposite of the row-unravel operator. **Redim** can also be used to replicate a vector. For example, if x is a vector of length 40 as above, then to create a matrix y that contains 5 rows, each of which are x:

```
. redim x {dims=5,40; c} > y
```

Because the matrix y contains more elements than the vector x, we used the {**c**} option, which specifies that the data elements in x are cycled through repeatedly. Without the {**c**} option, NA's would have been used. For more information, uses, and examples of **redim**, see its manual entry, *redim(BRM)*.

The vecop commands compute a variety of operations (sum, mean, minimum, maximum, sum of squares, variance, standard deviation, etc.) on the vectors which comprise a dataset (whether considered as rows, as columns, or as the entire dataset). For example, the **colsum** command:

```
. colsum x > y
```

computes the sums of the columns of x and puts the result in y. If x is a matrix, then y is a row vector: each element of y contains the sum of the elements in the corresponding column of x. Similarly, the commands **colmean**,

colmin, colmax, etc. compute the column means, column minima, column maxima, etc. The commands **rowsum** and **rowmean** compute the row sums and row means; the commands **sum** and **mean** compute the sum and mean of entire datasets; etc. These are only a few of over fifty such commands. See the manual entry *vecop(BRM)* for more information and a complete list.

The arrop commands. The following commands perform special operations on arrays. They are documented, in detail, in *arrop(BRM)*.

cumsum	Cumulative sum of array elements.
cumprod	Cumulative product of array elements.
ncols	Number of columns in an array.
nrows	Number of rows in an array.
nsheets	Number of sheets in an array (sheets are discussed in Section 8.4).
dims	The three dimensions of an array (number of sheets, rows, and columns).

The submat command allows a submatrix to be chosen from a matrix when the indices are specified as datasets rather than as explicit integers. For example, if x is a matrix, the commands:

```
. i = 1,2
. j = 4,3,5
. submat x i j > y
```

create a submatrix y of x which consists of rows 1 and 2 and columns 4, 3, and 5 (in that order). The command sequence:

```
. i = 2:8
. submat x i > y
```

selects rows 2 through 8 of the matrix x—or elements 2 through 8 if x is a vector (whether row or column). For more information, see the manual entry *submat(BRM)*.

Systems of linear equations. If A and B are known matrices of appropriate dimension (A is n-by-n and B has n rows), then the most obvious way to solve the system of linear equations $A \ \#* \ X = B$ for the unknown matrix X is to (conceptually) premultiply both sides of the equation by the inverse of A. In other words, give the command:

```
. X = inv(A) #* B
```

However, for reasons pertaining to numerical analysis, this is not the best method. It is more accurate and efficient to use the **solve** command as follows:

```
. solve A B > X
```

Solving systems of linear equations and computing inverses and determinants are closely related topics. The **inv** and **det** functions are also available as commands; the command versions provide additional diagnostic information. See the manual entry *solve(BRM)* for complete information about the **solve**, **inv**, and **det** commands.

Matrix decompositions. Four standard matrix decompositions are: for symmetric matrices, the Cholesky decomposition and the eigenvalue-eigenvector decomposition; and for general matrices, the QR decomposition and the singular value decomposition. BLSS provides all four: refer to the manual entries *chol(BRM)*, *eigen(BRM)*, *qr(BRM)*, and *svd(BRM)* for information.

8.3.5 Dimension-Expansion with Elementwise Operators

In Chapter 3 and Section 8.2, we saw that when an elementwise binary operator (such as addition, subtraction, multiplication, or division) is applied to a scalar and an array, the result is a new array of the same shape as the original in which each element is the result of the binary operation applied to the scalar and the corresponding element in the original array. In a sense, the scalar is 'expanded' to form an array of the appropriate size. Except for scalar multiplication, this notion is not defined in standard matrix algebra. But it makes intuitive sense.

Sometimes we want to add a different number to every row (or column) of a matrix, or we want to multiply every row (or column) of a matrix by a different constant. These operations can be performed using standard matrix algebra. For example, in Section 8.3.3 we saw that to multiply every row in a matrix by a different constant we can premultiply it by a diagonal matrix, and that to multiply every column by a different constant we can postmultiply it by a diagonal matrix. However, BLSS lets you do this more directly.

Suppose we have a column vector c and a matrix x as follows:

```
.  c = 100,,200,,300
.  show c {shape}
    100.00
    200.00
    300.00
.  x = 1,2,,3,4,,5,6
.  show x
      1.000      2.000
      3.000      4.000
      5.000      6.000
```

To add the i-th element of c to every element in the i-th row of x, simply add c and x.

```
.  y = c + x
.  show y
    101.00     102.00
    203.00     204.00
    305.00     306.00
```

As a second example, suppose we have a row vector r:

```
. r = 10,-20
. show r

  10.00     -20.00
```

To multiply the j-th column of x by the j-th element of r, simply multiply x and r:

```
. z = x * r
. show z

  10.00     -40.00
  30.00     -80.00
  50.00    -120.00
```

What happened? In the first example, before being added to x, the column vector c was 'expanded' (by copying columns) to form a full matrix of the same shape as x. In the second example, before being multiplied by x, the row vector r was 'expanded' (by copying rows) to form a full matrix of the same shape as x.

The general rule is: For *any* elementwise operation, if one operand has length 1 in a given dimension (whether rows or columns) and the other operand has length n in that same dimension, then the operand with length 1 is expanded in that dimension (by copying) to have length n. It does not matter which operand comes first. This is called *dimension-expansion*.[8]

Dimension-expansion can happen for more than one dimension in a single operation. For example, if we add the row vector (1:5) to the column vector (10:30:10)′, we get:

```
. (1:5) + (10:30:10)'

  11.00      12.00      13.00      14.00      15.00
  21.00      22.00      23.00      24.00      25.00
  31.00      32.00      33.00      34.00      35.00
```

In this case, the row vector was expanded in the column dimension, and the column vector was expanded in the row dimension.

When one operand is a scalar and another is an array of any shape, then as we already know, the scalar is matched with every array element. This is a special case of dimension-expansion: all necessary dimensions of the scalar are expanded.

Sometimes dimension-expansion can be confusing. If you want to see the

---

8. Of course, the 'expansion' is conceptual. The actual value and shape of the operands are unchanged.

results of dimension-expansion, set the environment string SHOWEXPAND:

```
. set SHOWEXPAND
```

This causes BLSS to print a message every time it performs dimension-expansion. For example:

```
. (1:5) + (10:30:10)'
Elementwise operand dimensions (1,1,5) and (1,3,1) were dimension-expanded.
      Result dimensions are (1,3,5).

    11.00     12.00     13.00     14.00     15.00
    21.00     22.00     23.00     24.00     25.00
    31.00     32.00     33.00     34.00     35.00
```

Note that three dimensions are shown. The first dimension, which is 1 here for both the operands and the results, refers to *sheets* and is discussed in the next section.

If you attempt a binary operation on operands whose lengths in any given dimension are unequal and not 1, it causes an error. For example, r and c' are row vectors of length 2 and 3. Trying to add them gives an error message:

```
. r+c'
LET:  Elementwise operand dimensions (1,1,2) and (1,1,3) are incompatible.
```

Example: Bicycle gears. How hard (or easy) is a bicycle gear? This is usually measured by how far a bicycle goes forward in each revolution of the pedals. Higher numbers correspond to higher gears. For bicycles with derailleurs (that is, separate front and rear gear clusters) the measurement for any given combination of gears is:

(diameter of wheel) ∗ *(number of teeth on front gear)* / *(number of teeth on rear gear)*

We call this measurement the *gear-distance* (its units are length).

A certain bicycle has 27-inch wheels, two front gear-rings with 38 and 52 teeth, and seven rear gears, with 13, 15, 17, 19, 22, 26, and 32 teeth. We can find out the gear-distance of every possible gear combination with a single step:

```
. 27 * (38,52) / (13,15,17,19,22,26,32)'
    78.92    108.00
    68.40     93.60
    60.35     82.59
    54.00     73.89
    46.64     63.82
    39.46     54.00
    32.06     43.88
```

Section 8.6 gives another example of dimension-expansion, using logical operators.

8.4   Three-Dimensional Data Arrays

As well as scalars, vectors, and matrices, BLSS datasets may contain three-dimensional, or three-way, data arrays. Components of the third dimension are referred to as *sheets*. This term is analogous to the terms *column* and *row*, which refer to components of the first two dimensions. A sheet of a BLSS dataset may be envisioned as follows. Imagine several sheets of paper with identically dimensioned matrices on each. If these sheets were placed one on top of another with the matrices lined up, you would have a three-dimensional array of numbers with each sheet of paper representing one sheet of the array.

When a sheet dimension or index is specified, it is given first, before the row and column dimensions or indices. For example, a dataset with 3 sheets, 4 rows, and 5 columns is said to have dimensions 3-by-4-by-5, or (3,4,5).

8.4.1   Creating Three-Dimensional Arrays

There are several methods to create a three-dimensional array. First, data may be entered into a three-way array from the keyboard or a text file with the **read** command using the {**dims=**} option to specify the three dimensions of the dataset.

Second, *any* command with a {**dims=**} option can generate three-way arrays. For example, the command:

```
.  idn {dims=3,6,6} > box
```

creates a dataset with dimensions 3-by-6-by-6, for which all three sheets are the 6-by-6 identity matrix. Other such commands are **const**, **redim**, the array operation commands introduced in Section 8.3.1 and documented in *arrop(BRM)*, and the r*xxx* random number generating commands introduced in Section 7.3 and documented in *random(BRM)*. When row-major order (such as is used by **redim**) is applied to three-way arrays, the rows of the first sheet come first, then the rows of the second sheet, etc.

Third, three-way arrays may be created using the *sheetwise array catenation* operator:

$$a\#,b$$

is the multisheet array which consists of the sheets of *a* followed by the sheets of *b*. The arrays *a* and *b* must contain the same number of rows and columns. Sheet catenation has lower precedence than row catenation, which has lower precedence than column catenation, which in turn has lower precedence than the sequence operator (see the precedence table in Section 8.8). Thus, no parentheses are needed in this command:

```
.  zz = 1,2,3,,4:6 #, 7:9,,10:12
```

which creates an array *zz* of dimensions (2,2,3) that contains data as follows:

```
.  show {i} zz
   Sheet 1
      1   2   3
      4   5   6
   Sheet 2
      7   8   9
     10  11  12
```

Recall that the {i} option to the **show** command shows data as integers.

8.4.2 Subarrays in Three Dimensions

A single element of a three-way array is referenced by a triple subscript—that is, by appending its sheet, row, and column indices within square brackets to the name of the array. For example, $x[1,2,3]$ is the element in the first sheet, second row, and third column of *x*.

A general subarray is referenced by specifying the indices of its sheets, rows, and columns in a similar manner. The expression $x[S, R, C]$ is the subarray of *x* whose sheet, row, and column indices are given by the index lists *S*, *R*, and *C*. The sheet index list, *S*, may take any of the forms for the row and column index lists given in Section 8.3.2. If the sheet index list is omitted, the expression $x[R,C]$ denotes the subarray consisting of the specified rows and columns in *all* the sheets of *x*. Note that this is consistent with the usual notation when selecting rows and columns from datasets with only one sheet.

Examples. Suppose that *box* has dimensions (3,4,5). Then:

```
box[2, ,]
```

denotes the second of the three sheets;

```
box[,1,]
```

denotes the first row in all three sheets;

```
box[1 2,  2:4,  5 3]
```

denotes the subarray which consists of rows 2 through 4, columns 5 and 3 (in that order), in sheets 1 and 2 only. This subarray has dimensions (2,3,2).

Finally, the expression:

```
box[,4,4]
```

or simply:

```
box[4,4]
```

denotes the *sheet vector*[9] composed of the elements in the [4,4] position in successive sheets of *box*. In general, a sheet vector of length s is an array of dimensions $(s,1,1)$. That is, it is composed of s sheets, each of which contains only one element. This particular sheet vector has dimensions $(3,1,1)$. Sheet vectors may be converted to row or column vectors using the unary unravel operators ′ and ″ introduced in Section 8.3.3. For example, the commands:

```
. x = 'box[,4,4]
. y = ''box[,4,4]
```

create, respectively, a row vector x and a column vector y which contain the same data as the sheet vector *box*[4,4].

8.4.3 Operators on Three-Dimensional Arrays

Elementwise operators and functions (see Section 8.2) operate elementwise on three-dimensional arrays just as they do on all BLSS datasets. Matrix operators and functions (see Section 8.3.3) operate sheetwise on three-dimensional datasets: each sheet is operated on as a separate matrix. For example, matrix transpose (′) transposes each sheet of a three-way array; matrix multiplication (#*) multiplies (separately) corresponding sheets in three-way arrays; columnwise and rowwise catenation apply sheetwise to three-way arrays.

Dimension-expansion (see Section 8.3.5) also applies to three-way arrays. If one operand of an elementwise binary operator has a single sheet, and the other operand has several sheets, then the single-sheet operand is dimension-expanded appropriately.

8.5 Missing Values (NA's)

Missing values can arise in two distinct ways. 1) Mathematical operations may result in datasets which contain undefined values. 2) A dataset to be analyzed may contain no values in certain instances due to no response, a failed experiment, or simply an unrecorded observation.

In either case, BLSS uses a special value called a *missing value* to represent the undefined or unobserved values. BLSS denotes the missing value as NA.[10] The value NA can be used almost anywhere in BLSS that a numerical value can be used.

Example. In the following example we use the **read** command to enter a dataset which contains a missing value, and then the **let** command to set the

---

9. Unlike most terms introduced in this chapter, the terms *sheet* and *sheet vector* are not in standard use. They were coined for use in BLSS.

10. The notation NA has long been used in statistical tables, survey response forms, etc. to denote data items which are 'not available' or 'not applicable'.

second data value to be missing. The **show** command displays the missing values as NA.

```
. read > mydata
1   3   -2   NA   4
Control-D
. show mydata
      1.000      3.000    -2.000         NA      4.000
. let mydata[2] = NA
. show mydata
      1.000         NA    -2.000         NA      4.000
```

With two exceptions, which we shall see in Section 8.6, the result of any function of an NA, or any arithmetic or logical operation which involves NA's, is an NA. To continue the example:

```
. show log(mydata)
      0.000         NA        NA         NA      1.386
. show (mydata + NA)
         NA         NA        NA         NA         NA
```

In Sections 8.6 and 8.7 we will see logical operations on NA's.

8.6   Logical Expressions

Let provides relational and logical operators. There are no special logical values for 'true' or 'false'. Instead, the logical value 'true' is represented by 1 and the logical value 'false' by 0. In contexts requiring logical operands, a value is 'true' if it is not 0 and 'false' if it is 0.

The operators listed in this section are elementwise operators, in the sense described in Section 8.2. Hence, dimension-expansion (described in Section 8.3.5) applies to them.

Relational operators. These operate on numerical values and yield logical values (0 or 1). They have equal precedence and group left-to-right.

>	Greater than.
>=	Greater than or equal to.
<	Less than.
<=	Less than or equal to.
==	Equal to, or both operands are NA's.
!=	Not equal to, or one operand is NA and the other is not.

Note that the '==' and '!=' operators are exceptions to the rule that an operation on an NA (missing value) results in an NA. Thus, they may be used to test whether or not a value is NA.

Logical operators. These operate on—and yield—logical values.

    &&   And (logical intersection).
    || Or (logical union).
    !   Not (logical negation).

'&&' and '||' have lower precedence than the relational operators. '&&' has precedence over '||'. '!' has higher precedence than relational operators.

Conditional evaluation.

    $a\,?\,b:c$ This expression evaluates to b if a is true and to c if a is false.

The conditional operator has lower precedence than all logical and relational operators. Think about its symbols as meaning: 'Is a true ? If yes, the value is b : If no, the value is c.'

Examples.

```
. show x

    1.000     -2.000     -3.000     4.000     5.000

. y = x > 0
. show y

    1.000      0.000      0.000     1.000     1.000

. show sqrt(x)

    1.000         NA         NA     2.000     2.236

. show (sqrt(x) != NA)

    1.000      0.000      0.000     1.000     1.000

. z = (x <= -1) || (x >= 2)
. show z

    0.000      1.000      1.000     1.000     1.000
```

Beware that the expression $x<y<z$ is legal—but it is not, as you might expect, equivalent to $x<y$ && $y<z$. To continue the example above:

```
. show (x<y && y<z)

    0.000      1.000      1.000     0.000     0.000

. show (x<y<z)

    0.000      0.000      0.000     1.000     1.000
```

The results are quite different. In the expression $x<y<z$, the expression $x<y$ was calculated first. The resulting logical array (of 0's and 1's) was then compared to the array z.

The conditional operator is useful for operations which involve elementwise comparison of datasets, such as elementwise maximum or minimum. For example, to create a dataset c which is the elementwise maximum of datasets a and b:

```
. c = a > b ? a : b
```

It is also useful for replacing values in a dataset according to whether or not the values meet a given condition. For example, to replace the elements of x which are negative with the value 0:

```
. x = x < 0 ? 0 : x
```

To replace the elements of x which are NA's with 0's:

```
. x = x == NA ? 0 : x
```

Example: Replacing missing value codes. On occasion, one encounters datasets for which different statistical variables (that is, different columns of the dataset) have different missing value codes. For example, the missing value code for column 1 might be 9, but for columns 2 and 3 it might be 99. Before further analysis in BLSS, the missing value codes should be replaced by NA's so that the codes are not mistaken for legitimate data values. For this example, the appropriate commands are:

```
. x[1]   = (x[1]==9)   ? NA : x[1]
. x[2 3] = (x[2 3]==99) ? NA : x[2 3]
```

Parentheses and extra spaces are included here for readability. The parentheses are not actually necessary because, by the precedence rules stated above, the == operator is evaluated before the ?: operator. Note that these commands do exactly what is necessary and no more—they avoid replacing any 9's in columns 2 and 3 or any 99's in column 1.

Example: Indicator matrices. If a column vector x contains observations on a categorical variable, and the row vector c contains all the possible category codes, it can be useful to have an *indicator matrix*: a matrix such that element $[i,j]$ is 1 or 0 according to whether or not the i-th observation of x is from the j-th category of c. Indicator matrices are used for purposes such as dummies for categorical variables in regression models, etc. As a small example, we might have c and x as follows:

```
. c = 4:6
. sample {n=5} c' > x
. show {i} c x

  4   5   6

  5
  5
  6
  5
  4
```

To construct an indicator matrix y for the vectors x and c:

```
. y = x==c
. show {i} y
    0   1   0
    0   1   0
    0   0   1
    0   1   0
    1   0   0
```

This command used dimension-expansion (see Section 8.3.5) to create the appropriate matrix from the two vectors.

8.7   Using Let with Count and Select

Two BLSS commands, **count** and **select**, allow you to count the number of cases (rows) in a BLSS dataset which meet a specified condition, or to select only those cases and place them in a new BLSS dataset.

Counting cases.   The **count** command counts the number of cases in a BLSS dataset which meet a given condition. The form of the command is:

count {nm} cond

The argument *cond* is a dataset of logical values which expresses the condition. **Count** reports the number and proportion of elements that meet the condition—in other words, the number and proportion that are non-zero (and non-missing).

Let can be used to create such logical conditions. For example, suppose we want to count the cases of *plover* for which the hatched chick survived to fledgling (the fifth element of the row is 1). To do this, use an implicit let-argument:

```
. count (plover[5]==1)
```

The **count** command reports:

```
29/68 = 42.65% of the cases meet the condition.
```

which means that 29 of the 68 cases in the dataset, or 42.65%, meet the condition.

Of course, more complicated conditions can also be counted. Suppose we want to count the cases for which the hatched chick survived to fledgling *and* the chick's weight at birth was less than the median (the fourth element of the row is less than 6.2, obtained from the **stat** example in Section 5.1.3). The command which gives the desired count is:

```
. count (plover[5] == 1 && plover[4] < 6.2)
12/68 = 17.65% of the cases meet the condition.
```

If the {**nm**} option is set, **count** reports the number of cases in the dataset which are completely non-missing ('nm' is an abbreviation for 'non-missing'). For example, the command:

```
. count {nm} plover
```

reports that:

```
42/68 = 61.76% of the cases are completely non-missing.
```

The reported number and proportion can be saved in an output dataset. See the manual entry *count(BRM)* for this and more information.

Selecting cases. The **select** command selects cases (rows) from a BLSS dataset which meet a specified condition and puts them in a new BLSS dataset. The form of the command is:

$$\textbf{select } data\ [cond]\ \{\textbf{nm}\}\ >\ sel\ [rej]$$

where *data* is the dataset from which cases are selected, and *cond* is a vector of logical values. The square brackets [ ] indicate that *cond* is optional. The output dataset, *sel*, contains the selected rows. The second output dataset, *rej*, is also optional. If *rej* is specified, it contains the rejected cases—that is, the cases which were not selected.

Selecting non-missing cases. If *cond* is not specified, **select** selects those rows of *data* for which all elements are non-missing. For example, in Section 5.4 we created a dataset *plov* which contains those cases of *plover* that are entirely non-missing:

```
. select plover > plov
```

If *cond* is specified, it must be a vector whose length equals the number of rows in *data*. In this case, if {**nm**} is specified **select** selects all rows of *data* for which the corresponding elements of *cond* are non-missing. For example, to create a dataset *known* which contains all rows of *plover* for which the fifth column is non-missing (that is, for which it is known whether the chick from the egg lived or died):

```
. select {nm} plover plover[5] > known
```

In order to create a new dataset that contains those rows of *plover* for which the third and fourth columns are non-missing, we can use an implicit **let** expression to create the *cond* vector:

```
. select {nm} plover (plover[3] && plover[4])
```

This works because, as we saw in Section 8.5, any operation in which one of the elements is the missing value results in a missing value. Hence, an element in the *cond* expression is non-missing, and the corresponding row of *plover* is selected, only if that row has non-missing values in both the third and fourth columns.

Logical selection. If a second input is given to the **select** command, then by default **select** selects all rows for which the corresponding element of *cond* is non-missing *and* non-zero. In other words, **select** selects exactly those cases which the **count** command would count as meeting the logical condition. For example, to create a new BLSS dataset which consists of the rows of *plover* which we counted in a previous example of the **count** command:

```
. select plover (plover[5] == 1 && plover[4] < 6.2) > small
```

Another way to select rows of *plover* for which the fifth column is non-missing is to use a *cond* vector which expresses the condition logically:

```
. select plover (plover[5] == NA)
```

Refer to the manual entry *select(BRM)* for full information about the **select** command.

8.8   Summary of Operators

Table 8.1 is a complete list of **let** operators presented in this chapter, listed from highest to lowest precedence. Operators with equal precedence are listed together. Table 8.1 also indicates the associativity of each operator (left or right) and the section of this chapter where each operator is discussed.

Operator vocabulary summary. Some of the words below have been defined previously, but we define here all words associated with operators.

An *operand* is an object on which an operator operates. For example, in the expression $a+b$ the operands are a and b.

Unary, binary, and *ternary* operators respectively operate on one, two, or three operands. For example, $-a$ is a unary operation; $a+b$ is a binary operation; $a{:}b{:}c$ is a ternary operation.

A *prefix* unary operator is one which appears before its operand: for example, the logical negation operator $!x$.

A *postfix* unary operator is one which appears after its operand: for example, the factorial operator $x!$.

The *precedence* of operators establishes the order in which operations are performed in the absence of parentheses to the contrary: higher precedence operations are performed before lower precedence operations. For example, $*$ has higher precedence than $+$, so $a+b*c$ is evaluated as $a+(b*c)$. Parentheses can always be used to change the default order of evaluation.

Left-associative binary operators of equal precedence are evaluated from left to right in the absence of parentheses to the contrary. For example, $a-b+c$ is evaluated as $(a-b)+c$.

Right-associative binary operators of equal precedence are evaluated from right to left in the absence of parentheses to the contrary. For example, $a^b c$ is evaluated as $a^{(b^c)}$.

Table 8.1. Summary of **let** operators, in decreasing precedence.

Operator	Meaning	Associativity	Section
' "	Row-unravel (prefix '); Column-unravel	–	8.3.3
'	Matrix transposition (postfix ')	–	8.3.3
!	Factorial (postfix !)	–	8.2
^ #^	Exponentiation; Matrix exponentiation	Right	8.2, 8.3.3
– !	Arithmetic negation (unary –); Logical negation (prefix !)	–	8.2, 8.6
* / #*	Multiplication; Division; Matrix multiplication	Left	8.2, 8.3.3
+ –	Addition; Subtraction (binary –)	Left	8.2
: ::	Sequence operators (binary and ternary)	–	8.3.1
,	Columnwise matrix catenation	Left	8.3.3
,,	Rowwise matrix catenation	Left	8.3.3
#,	Sheetwise array catenation	Left	8.4
== != < <= > >=	Equality and order relationships	Left	8.6
&&	Logical and	Left	8.6
||	Logical or	Left	8.6
?:	Logical conditional operator (ternary)	Right	8.6
=	Assignment	Right	8.1.1

CHAPTER 9

Extended Examples

BLSS is a flexible statistics system: it allows you combine commands to perform many other procedures beyond those for which it provides single commands. The point of this chapter is to give examples of extended BLSS command sequences. We tried to choose examples that are both interesting in their own right and also illustrate various commands and techniques.

This chapter assumes only that you have read Chapters 1 and 2. Of course, it uses a wide variety of BLSS commands from other chapters as well without providing detailed command explanations. Instead, where appropriate it refers to detailed explanations elsewhere in this book.

We do not show all intermediate results in our examples. Particularly important in this chapter is that you *try out* the examples, look at the intermediate results, and understand how the commands together achieve the desired result.

9.1 Combinatorics

Permutations. The number of possible permutations (arrangements) of n distinct objects is $n*(n-1)* \cdots *2*1$ and is commonly written as $n!$, pronounced 'n factorial.' BLSS recognizes this notation. For example:

```
. 4!

  24.00
```

By convention, $0! = 1$. To compute $0!, 1!, \ldots, 8!$:

```
. show {i} (0:8)!

   1   1   2   6   24 120 720 5040 40320
```

The sequence operator ':' creates a row vector which is a sequence of numbers—in this case, the sequence of integers 0 through 8. It is discussed in Section 8.3.1. The '!' is an elementwise operator: it is applied to each element of the sequence. Thus, we obtain all the factorials for 0 through 8. The ':' operator has lower precedence than '!', so the parentheses in the example are needed.

Alternatively, we can use the **cumprod** command, which calculates the cumulative product of the elements in an array.

```
. n = 8
. cumprod 1,1:n > fact8
. show {i} fact8

    1    1    2    6   24  120  720 5040 40320
```

We put the maximum number, 8, into the BLSS dataset n to show how it enters into the calculation. Note that this is a *recursive method:* it uses the fact that $0! = 1$ and $k! = k*(k-1)!$ for $k > 0$.

Combinations. The number of possible ways to choose k objects from n distinct objects (for $0 \le k \le n$) is $n!/(k!(n-k)!)$. This is often symbolized by $\binom{n}{k}$, pronounced 'n choose k.'

We can compute a single value of $\binom{n}{k}$ using the factorial operator '!'. For example, $\binom{8}{2}$:

```
. 8!/(2!*6!)

  28.00
```

The values $\binom{n}{k}$ for $k = 0$ to n arise in probability as the binomial coefficients. To compute these values, we can use the same technique as above. For example, to compute $\binom{n}{k}$ for k from 0 to n, with $n = 10$:

```
. show {i} 10!/((0:10)!*(10:0)!)

    1   10   45  120  210  252  210  120   45   10    1
```

The two expressions which are multiplied together are the array $(0:10)!$ and itself with the elements reversed. This gives the correct match of $k!$ and $(10-k)!$ for each value of k.

9.1.1 Large Values of N

As n increases, $n!$ increases very rapidly. The value of $n!$ soon becomes too large for computer hardware to represent—n equal to 34 or 35 results in $n!$ too large for many computers. The value of $n!$ is said to *overflow*, and BLSS represents it by NA. Nonetheless, when we want a final result for which $n!$ is only an intermediate value, there are other methods. We discuss two of them.

Logarithms. For large n, we can work with $n!$ indirectly by working with its logarithm. When we cannot compute $n!$ we cannot compute $\log(n!)$, so BLSS provides a special function, **lfac**, which computes log($n!$) without computing $n!$ first. For example, to show log(3500!):

```
. show lfac(3500)

  25066.8
```

We can use the **lfac** function to compute log($\binom{n}{k}$) for large n. The calculation is analogous to that above—recall that, when working with logarithms, multiplication is replaced by addition and division by subtraction. For example, to obtain log($\binom{n}{k}$) for k from 0 to n and $n = 50$:

```
. logcoef = lfac(50) - lfac(0:50) - lfac(50:0)
```

Because binomial coefficients do not increase as rapidly as factorials, we can see their values by exponentiating their logarithms:

```
. show exp(logcoef[1:25])          # show only the first half of them

    1.000     50.00   1225.01   19600.0   2.30e+05   2.12e+06   1.59e+07   9.99e+07
  5.37e+08  2.51e+09  1.03e+10  3.74e+10   1.21e+11   3.55e+11   9.38e+11   2.25e+12
  4.92e+12  9.85e+12  1.81e+13  3.04e+13   4.71e+13   6.73e+13   8.88e+13   1.08e+14
  1.22e+14
```

Recursive methods: cumulative products. Alternatively, we can use the fact that $\binom{n}{0} = 1$, and $\binom{n}{k} = \binom{n}{k-1}(n+1-k)/k$ for $k = 1$ to n. To evaluate binomial coefficients using this recursion instead of factorials, we use **cumprod** the same way we used it to compute the factorials through n:

```
. n = 50
. cumprod 1,(n:1)/(1:n) > coef
. show coef[1:8]                   # show only the first few

    1.000     50.00   1225.00   19600.0   2.30e+05   2.12e+06   1.59e+07   9.99e+07
```

In order to obtain the term $n+1-k$ for k from 1 to n, we could have used the expression $n+1-(1:n)$, but the simpler expression $n:1$ is equivalent.

Because the sequence operator ':' has lower precedence than arithmetic operators such as '/', we enclose the sequences in parentheses. The catenation operators such as ',' have lower precedence still, so no additional parentheses are needed. A full summary of operator precedences appears in Section 8.8.

9.2   Discrete Probability Functions

The binomial distribution. We can use the techniques of Section 9.1 to calculate all binomial probabilities for a given value of n. The probability that a binomial(n, p) random variable assumes the value k is:

$$P(X = k) = \binom{n}{k}p^k(1-p)^{n-k} \quad \text{for } k = 0 \text{ to } n.$$

To compute all binomial probabilities for $n = 10$ and $p = .5$:

```
. n = 10; p = .5
. binomprob = n!/((0:n)!*(n:0)!) * p^(0:n) * (1-p)^(n:0)
```

```
. show binomprob

  9.77e-04   0.009766    0.04395    0.1172    0.2051    0.2461    0.2051    0.1172
  0.04395    0.009766   9.77e-04
```

The binomial coefficients are computed here just as in Section 9.1. Once again, we utilize expressions that correctly match $k!$, $(n{-}k)!$, p^k, and $(1{-}p)^k$ for each value of k.

For distributions on the integers, the cumulative distribution function (cdf) is:

$$F(x) = P(X \le x) = \sum_{k \le x} P(X = k).$$

The **cumsum** command, which calculates the cumulative sum of the elements in an array, lets us calculate $F(x)$:

```
. cumsum binomprob > binomcdf
. show binomcdf

  9.77e-04    0.01074    0.05469    0.1719    0.3770    0.6230    0.8281    0.9453
  0.9893      0.9990     1.000
```

The Poisson distribution. For the Poisson distribution with parameter λ,

$$P(X = k) = e^{-\lambda}\frac{\lambda^k}{k!} \quad \text{for } k = 0 \text{ to } \infty.$$

To compute Poisson probabilities for $k = 0$ to 15 when $\lambda = 5$, we use expressions similar to those above:

```
. lambda = 5
. poissonprob = exp(-lambda) * lambda^(0:15) / (0:15)!
. poissonprob

  0.006738    0.03369    0.08422    0.1404    0.1755    0.1755     0.1462    0.1044
  0.06528     0.03627    0.01813   0.008242  0.003434  0.001321   4.72e-04  1.57e-04
```

The corresponding cdf $F(x)$:

```
. cumsum poissonprob > poissoncdf
. show poissoncdf

  0.006738    0.04043    0.1247     0.2650    0.4405    0.6160    0.7622    0.8666
  0.9319      0.9682     0.9863     0.9945    0.9980    0.9993    0.9998    0.9999
```

is obtained as with the binomial.

9.2.1   Large Values of N

As noted before, large values of n cause overflow when working with $n!$. To avoid overflow with large values of n, we can use logarithms or recursive methods. We show recursive methods only. They are a bit more complex, but are more efficient for very large n.

Poisson. For the Poisson distribution, $P(X{=}0) = \exp(-\lambda)$ and, for $k > 0$, $P(X{=}k) = P(X{=}k{-}1) * \lambda/k$. For a given value of λ (stored in the dataset *lambda*) and n, we can evaluate the Poisson probabilities for 0 through n by:

```
. cumprod (exp(-lambda),lambda/(1:n)) > poissonprob
```

Binomial. The recursive formula for the binomial probabilities is based on that for $\binom{n}{k}$. It is: $P(X=0) = (1-p)^n$ and $P(X=k) = P(X=k-1) * (n+1-k)/k * p/(1-p)$. For a given value of n and p, we can evaluate it thus:

```
. cumprod ((1-p)^n, (n:1)/(1:n) * p/(1-p)) > binomprob
```

For *very* large values of n—where the value of $P(X = k)$ might *underflow*[1]—we can combine recursive methods with logarithms by cumulatively summing the logarithms of appropriate quantities. The BLSS commands **binomial** and **poisson**, which calculate probabilities and cdf's for these distributions, are macros which use this technique.

9.3   Bivariate and Multivariate Random Numbers

In Chapter 7 we discussed how to generate univariate random numbers from different distributions. In this section, we show how to generate multivariate random numbers as linear combinations of independent random numbers.

The examples in this section all begin with standard Normal (Gaussian) random numbers. As a result, the generated multivariate random numbers are Normal because linear combinations of Normal random variables are also Normal. This is not true for arbitrary random variables: if other distributions were used in place of the Normal, the first two moments (means, correlations, and covariances) of the new multivariate random numbers would be the same as obtained here but the exact distribution would be different.

9.3.1   Bivariate Random Numbers

If Z_1 and Z_2 are independent standard Normal, then $X_1 = Z_1$ and $X_2 = rZ_1 + \sqrt{(1-r^2)}Z_2$ are standard Normal with correlation coefficient r. For example, to generate 100 bivariate standard Normal random numbers with $r = .6$, we use the **rgau** command to generate 100 pairs of standard Normal random numbers and then apply the correct transformations:

```
. rgau {dims=100,2} > z
. r = .6
. x = z[1],(r*z[1] + sqrt(1 - r*r)*z[2])
```

We put the correlation coefficient, .6, in the dataset r to show how it enters into the calculations. The final command creates the desired dataset using the column catenation operator, ',', described in Section 8.3.3.

The BLSS command **rbinorm**, which generates bivariate Normal random numbers with a specified correlation, is a macro which uses the technique described above.

---

1. *Underflow* occurs when the magnitude of a floating point number is too small for computer hardware to represent distinctly from 0. On most computers that BLSS runs on, values that underflow are quietly set to 0.

Bivariate random numbers with specified SDs. If every element in a dataset is multiplied by the same constant, then the SD of the new dataset is the original SD multiplied by that constant.

Suppose we want to give the first column of x (generated in the previous example) the SD = 3, and the second column the SD = 5. Because the old SDs are 1, to do this we simply multiply each column by its new SD. We can use the ',' operator to create the two columns of the new dataset separately:

    . w = x[1]*3, x[2]*5

No parentheses are needed because '*' has higher precedence than ','. Alternatively, we can use the dimension-expansion feature of BLSS matrix arithmetic explained in Section 8.3.4:

    . w = x * (3,5)

Parentheses *are* needed here, for the same reason. In either case, the dataset w has the same means and correlation coefficient as x, but the first column of w has SD = 3 and the second has SD = 5.

Note that the following commands will *not* work:

    . w[1] = x[1]*3; w[2] = x[2]*5

because BLSS cannot refer to column 1 (or 2) of w before w itself exists.

Bivariate random numbers with specified means. If the same constant is added to every element in a dataset, then the mean of the new dataset is the mean of the original dataset plus that constant. Suppose we want to give the first column of w a mean of 10, and the second column a mean of 20. Because the old means are 0, to do this we add 10 to the first column and 20 to the second column. Again, we have two alternatives. We can use the ',' operator to create the columns of the new dataset separately:

    . v = w[1]+10, w[2]+20

Or we can use dimension-expansion:

    . v = w + (10,20)

The final dataset, v, is a 100-by-2 dataset of random Normal numbers with means equal to 10 and 20, SDs equal to 3 and 5, and correlation coefficient .6.

In this section we started with uncorrelated random numbers with mean = 0 and SD = 1; first changed the correlation coefficient, then the SDs, and finally the means. If we had performed the same operations in some other order, the resulting dataset would have different moments—addition and multiplication are not commutative. However, as we shall see in the next section, the steps may be combined into a single command.

9.3.2 Multivariate Random Numbers

This section shows how to create a dataset of multivariate random numbers which have any specified mean vector μ and covariance matrix Σ.[2]

If the random vector $\mathbf{z}$ has mean 0 and covariance matrix $\mathbf{I}$ (the identity matrix), then $\mathbf{x} = \mathbf{zT}$ has mean 0 and covariance matrix $\Sigma = \mathbf{T'T}$. Thus, given Σ and a dataset z with mean 0 and covariance $\mathbf{I}$, to transform z into a dataset with mean 0 and covariance Σ it suffices to find a matrix $\mathbf{T}$ such that $\mathbf{T'T} = \Sigma$.

If $\mathbf{A}$ is a symmetric, positive definite matrix, such as a covariance matrix, the *Cholesky decomposition* of $\mathbf{A}$ is an upper triangular matrix $\mathbf{T}$ which has the property $\mathbf{T'T} = \mathbf{A}$. The **chol** command gives a Cholesky decomposition of a symmetric, positive definite matrix.[3]

Suppose that we have BLSS datasets *mu* (a row vector of length 4) and *sigma* (a positive definite matrix of dimension 4-by-4) and we want to create 100 4-variate Normal random numbers with mean *mu* and covariance *sigma*. We start with standard Normal random numbers and apply transformations according to the theory above:

```
. rgau {dims=100,4} > z
. chol sigma > t
. z = z #* t + mu
```

We changed the covariance structure and means of z in a single command by taking advantage of the precedence of operators in BLSS: we first postmultiplied z by t and then added *mu* to the result—the same order of operations as in the previous section.

The BLSS command **rmultinorm** generates bivariate Normal random numbers with a specified covariance matrix. It is a macro which uses the technique described above.

9.4 Regression

This section presents examples drawn from several topics in regression theory.

9.4.1 Polynomial Regression

The n-th order polynomial regression model for a dependent variable y on an independent variable x is:

$$y_i = b_0 + b_1 x_i + b_2 x_i^2 + b_3 x_i^3 + \cdots + b_n x_i^n + \epsilon_i$$

where the error term ϵ_i is a random variable that has mean 0, constant

---

2. In this section we use the typographical convention that bold lowercase letters denote vectors and bold uppercase letters denote matrices.

3. To find a matrix $\mathbf{T}$ such that $\mathbf{T'T} = \Sigma$, it is also possible to use an eigenvalue-eigenvector decomposition (see the manual entry *eigen(BRM)*) or singular value decomposition (see *svd(BRM)*).

Figure 9.1

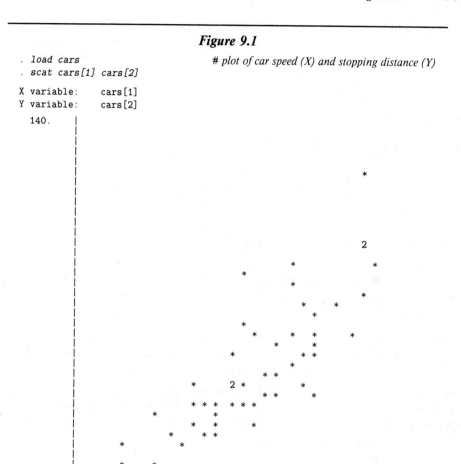

```
. load cars                    # plot of car speed (X) and stopping distance (Y)
. scat cars[1] cars[2]

X variable:    cars[1]
Y variable:    cars[2]
```

variance, and is uncorrelated with other ϵ's.[4] We can use **regress** to fit this model to the pair of datasets x and y by letting the independent variables for the regression be x, x^2, x^3, etc.

The BLSS system library dataset *cars* contains the initial speed in miles per hour (column 1) and stopping distance in feet (column 2) for 50 motorists. To fetch it and plot stopping distance as a function of speed:

```
. load cars
. scat cars {x=1; y=2}
```

The scatterplot appears in Figure 9.1. It suggests that a quadratic regression model for stopping distance in terms of speed may fit the data better than a

---

4. High-order polynomials (order more than two or three) can be bad models. For most data, other methods provide better models.

linear model. Based on physical considerations (when the car speed is 0, the stopping distance must be 0), we exclude the constant term (or *intercept*) from the model. To do so, we use the {**noint**} option. Thus, to fit our model:

```
. x = cars[1],cars[1]^2
. y = cars[2]
. regress x y {noint} > fit res

Dependent variable:      y
Independent variables:   x[1 2]
Observations  50         Parameters  2

Parameter   Estimate      SE        t-Ratio    P-Value
 coef 1     1.2390      0.55997     2.2127     0.0317
 coef 2     0.090139    0.029389    3.0671     0.0035

Residual SD  15.022     Residual Variance   225.65
Multiple R   0.95566    Multiple R-squared  0.91328
```

The plot of the fitted values against the residuals is shown in Figure 9.2. The residual plot suggests that the variance of the error term may be heteroscedastic[5] and that a transformed model (such as $\sqrt{y} = b_1\sqrt{x} + b_2 x$) may be appropriate.

The dataset that contains the independent variables for the regression (in this case, x) is called the *design matrix* for the regression. The first step for all regression examples in this section is to create the design matrix.

In principal, the design matrix also contains a column of 1's which corresponds to the intercept (constant) term, if any, in the regression model. **Regress** includes this term automatically in its internal calculations unless you specify the {**noint**} option.

9.4.2   Confidence Intervals for Regression

This section shows how to construct confidence intervals for regression coefficients and for predicted values from a fitted regression equation. Consider the usual linear regression model:

$$y = X\beta + \epsilon$$

where X is the design matrix (the matrix of values of the independent variables, including the intercept if any); β is the vector of regression coefficients; and ϵ is a vector of random errors with covariance $\sigma^2 I$, where I is the identity matrix. The ordinary least squares regression estimate of β is expressed as $b = (X'X)^{-1} X' y$.[6] From these expressions, it follows that b has mean β and

---

5. The variance of the error term is said to be *heteroscedastic* if its value depends on the values of the independent variables. Otherwise, it is *homoscedastic*. Consult a statistics text for more information.
6. Although this is perhaps the simplest *algebraic* expression of the regression coefficients, it is neither the simplest nor best *numerical* method for computing them. The **regress** command uses a more efficient and accurate algorithm.

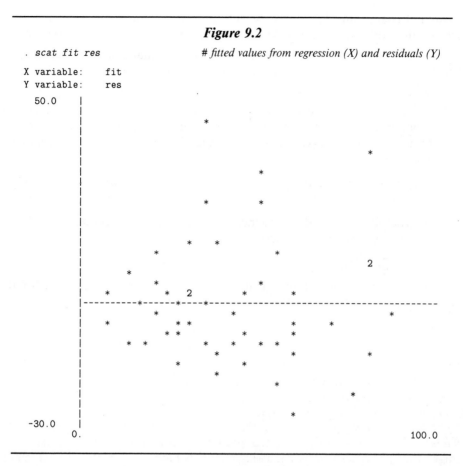

Figure 9.2

```
. scat fit res              # fitted values from regression (X) and residuals (Y)

X variable:    fit
Y variable:    res
   50.0   |
          |
          |                          *
          |
          |                                              *
          |
          |                     *
          |
          |             *              *
          |
          |
          |                 *     *
          |            *                        2
          |         *              *
          |      *
          |          *     2              *       *
          |- - - *- - - - *- - *- - - - - - - - - - - - - - - - - - - - -
          |          *            *                              *
          |      *          * *            *     *
          |             * *       * *          *
          |       *  *       *     *     * *  *
          |                  *          *       *          *
          |             *        *
          |                *
          |                     *
          |                            *
  -30.0   |
          0.                                                   100.0
```

covariance matrix $\sigma^2 C$, where $C = (X'X)^{-1}$. In particular, each individual regression coefficient b_i has mean β_i and standard deviation $\sigma\sqrt{C_{ii}}$.

Confidence intervals for regression coefficients. If the distribution of the random errors ϵ is Normal, then the $(1-\alpha)$ confidence interval for b_i is

$$b_i \;\pm\; t_{m;\,1-\alpha/2}\; s\sqrt{C_{ii}}$$

where s is the estimate of σ; m is the residual degrees of freedom (that is, the number of observations minus the number of parameters); and $t_{m;\,1-\alpha/2}$ denotes the $1-\alpha/2$ quantile of Student's t distribution with m degrees of freedom.[7]

The **regress** command computes all quantities needed to evaluate the confidence interval except the *t*-value: the **b@** output dataset contains the

---

7. The $1-\alpha/2$ quantile results in *individual* level $1-\alpha$ confidence intervals. Level $1-\alpha$ *simultaneous* confidence intervals are computed differently. For example, the Bonferroni method uses α/k in place of α to compute k simultaneous confidence intervals with confidence level $1-\alpha$; that is, it uses the $1-\alpha/(2k)$ quantile. Consult a statistics text for more information.

estimated coefficients b; the **bcov@** output contains the estimated covariance s^2C of the coefficients; the **se@** output contains the standard errors of the coefficients $s\sqrt{C_{ii}}$ (that is, the square roots of the diagonal elements of **bcov@**); and element [2,2] of the **anova@** output contains the residual degrees of freedom.

Let's resume the *cars* example from the previous section and compute confidence intervals for the regression coefficients. First we obtain the necessary outputs from **regress** (the **bcov@** output is not used until the following example):

```
. regress x y {noint;quiet} > b@ b bcov@ bcov se@ se anova@ an
. m = an[2,2]              # residual degrees of freedom
```

To construct $(1-\alpha) = 95\%$ confidence intervals, the $(1-\alpha/2)$ quantile of the t distribution with m degrees of freedom is:

```
. qt .975 {df=m} > t
```

We can now form the column vectors of lower and upper endpoints for the 95% confidence intervals for the column vector of regression coefficients:

```
. lower = b - t * se
. upper = b + t * se
. show lower,upper

  0.1131    2.365
  0.03105   0.1492
```

Confidence intervals for values of the regression function. Let V be a matrix whose rows contain values of the predictor variables of the regression. V might be the original design matrix X from which the regression coefficients are estimated, or it might be some other set of values at which we desire to estimate the regression function—that is, for which we desire the fitted values of the dependent variable.

Under the regression model, the estimated values of the regression function are Vb; they have mean $V\beta$ and covariance matrix $V\text{cov}(b)V' = \sigma^2 V(X'X)^{-1}V'$. If the distribution of the random errors is Normal, then the level $(1-\alpha)$ confidence interval for an individual regression function value $f_i = v_i\beta$ is:

$$v_ib \pm t_{m;1-\alpha/2}\, s\sqrt{[VCV']_{ii}}$$

where v_i is the i-th row of V.

As an example, we construct V so that each column contains values throughout the range of the corresponding column of X:

```
. V = (0:25)'                        # linear term of model . . .
. V = V,V^2                          # followed by quadratic term
```

The commands to obtain $(1-\alpha) = 95\%$ confidence intervals for the vector of regression function values are similar to those for the confidence intervals for the vector of regression coefficients; however, we must compute the SEs from the covariance:

```
. f = V #* b                         # estimated regression function values
. fcov = V #* bcov #* V'             # covariance of estimates
```

The standard errors $s\sqrt{[VCV']_{ii}}$ are the square roots of the diagonal elements of *fcov*. The **diag** function, introduced in Section 8.3.3, extracts the main diagonal of the *fcov* matrix as a row vector. We take its transpose and then the square roots of its elements:

```
. fse = sqrt(diag(fcov)')            # column vector of SEs of estimates
```

We use the same t quantile as above (of course, we could use a different one), and compute the confidence intervals exactly as before:

```
. flower = f - t * fse               # lower c.i. endpoints
. fupper = f + t * fse               # upper c.i. endpoints
```

In this example, the model contains no intercept. If it did, we would instead create *V* as follows, so it contains an initial column of 1's:

```
. V = (0:25)'                        # linear term of model
. const 1 V > V0                     # constant term: vector of 1's
. V = V0,V,V^2                       # constant, linear, and quadratic terms
```

The **const** command, which we used here to create the column vector *V0* of 1's, is described in Section 8.3.1 and in *arrop(BRM)*.

9.4.3   Comparing Two Designs: Are Two Regression Slopes Equal?

Suppose we want to fit a regression line to each of two datasets and test whether the slopes of the two lines are the same. To do this, we run two different regressions—one on the restricted model:

$$y_1 = a_1 + b\,x_1$$
$$y_2 = a_2 + b\,x_2$$

and one on an unrestricted model which allows two different slopes:

$$y_1 = a_1 + b_1 x_1$$
$$y_2 = a_2 + b_2 x_2$$

We then compare the two residual sums of squares using an F-test.

As an example, we use the *plover* data. Suppose we want to compare the chicks which survived to fledgling to those which did not by regressing chick weight (column 4) on egg weight (column 3). We combine both equations of a given model (restricted or unrestricted) into a single design matrix for that model. Equation 1 of each model will refer to the chicks which survived, and equation 2 to those which did not.

First we load the data from the system data library. Because **regress** does not allow missing values, we select the cases which are entirely non-missing:

```
. load plover
. select plover > plov
```

The **select** command is discussed in Section 8.7 as well as in its manual entry, *select*(BRM).

Next we separate the cases into two datasets: *p1*, those for which the chick survived to fledgling (the cases for which *plover*[5] = 1); and *p2*, those for which the chick did not survive (the remaining cases, for which *plover*[5] is not equal to 1). The {**long**} option of **select** states the number of cases in *p1* and *p2*. The option {x=3,4} causes only columns 3 and 4 to be selected: the new datasets, *p1* and *p2*, contain only two columns.

```
. select plov {log} (plov[5]==1) {x=3,4} {long} > p1 p2
Selected cases (23/42 = 54.76%) placed in p1.
Rejected cases (19/42 = 45.24%) placed in p2.
```

For each of the two models we perform a single regression. The data for each model will consist of the cases corresponding to equation 1 (*p1*) followed by the cases corresponding to equation 2 (*p2*). Using rowwise catenation (the ',,' operator), we create a dataset *y* which contains the dependent variable (chick weight) for the *p1* cases followed by that for the *p2* cases:

```
. y = p1[2],,p2[2]    # dependent variable for both regressions
```

Estimating the restricted model. We create column vectors *A1*, *A2*, and *B* which will be the columns of the design matrix *x1* for the restricted model corresponding to the regression coefficients *a1*, *a2*, and *b*:

```
. const 0 {dims=42,1} > A1
. A1[1:23,1] = 1
. A2 = 1 - A1
```

A1 contains 1's in rows corresponding to cases from *p1* and 0's in rows corresponding to cases from *p2*. The length of *A1* and the number of 0's and 1's are provided by the **select** output display. *A2* is the opposite: it contains 0's in rows corresponding to cases from *p1* and 1's in rows corresponding to cases from *p2*. Thus *A2* = 1 − *A1*. The column vector *B* contains the independent variable (egg weight) and is constructed analogously to *y*:

```
. B = p1[1],,p2[1]
```

The complete design matrix for the restricted model consists of the three columns *A1*, *A2*, and *B*:[8]

```
. x1 = A1,A2,B
```

To estimate the model we use the **regress** command with the {**noint**} option, because the intercepts *a1* and *a2* of the two regression lines are already incorporated into the design matrix:

```
. regress x1 y {noint} > res@ r1      # restricted model

Dependent variable:      y
Independent variables:   x1[1 2 3]
Observations  42         Parameters  3

Parameter    Estimate      SE        t-Ratio    P-Value
  coef 1     0.0035100   0.63717     0.0055     0.9956
  coef 2    -0.071661    0.62582    -0.1145     0.9094
  coef 3     0.71581     0.073150    9.7856     0.0000

Residual SD  0.22431     Residual Variance    0.050314
Multiple R   0.99938     Multiple R-squared   0.99876
```

The **res@** output contains the regression residuals. They will be used to construct the F-test.

Estimating the unrestricted model. For the unrestricted model we need vectors *A1*, *A2*, *B1*, and *B2*, which will be the columns of the design matrix *x2* corresponding to the regression coefficients *a1*, *a2*, *b1*, and *b2*.

A1 and *A2* are the same as in the restricted model (although, of course, *a1* and *a2* may well be different). *B1* must contain the *p1* independent variable in the rows corresponding to *p1* and 0's in the rows corresponding to *p2*; *B2* must be just the opposite. We could create *B1* and *B2* using rowwise catenation as we did when creating *y* and *B*, but it is quicker to use elementwise multiplication—*A1* and *A2* already have 1's and 0's in the correct positions:

```
. B1 = A1 * B
. B2 = A2 * B
```

---

8. Here is an alternate, quicker way to construct *A1*, *A2*, *B*, and *y*. It exploits the fact that column 5 of *plov* already contains all the 0's and 1's which constitute *A1*:

```
. A1 = plov[5]
. A2 = 1 - A1
. B = plov[3]
. y = plov[4]
```

Of course, this method does not separate the rows corresponding to the two different groups of plover chicks. That is not necessary, but can be accomplished by first sorting *plov* on its 5th column in reverse (to put the 1's above the 0's):

```
. sort plov {s=5; r} > p
```

and then working with the sorted dataset *p*. See *sort(BRM)* for details. If column 5 contained a categorical variable whose values were not already 0 and 1, it could be converted to 0's and 1's using either arithmetic or the logical operators described in Section 8.6.

The design matrix *x2* is:

```
. x2 = A1,A2,B1,B2
```

Again, we use the {**noint**} option when estimating the model because the intercepts of the two regression lines are already incorporated into the design matrix, and we save the regression residuals:

```
. regress x2 y {noint} > res@ r2       # unrestricted model

Dependent variable:        y
Independent variables:     x2[1 2 3 4]
Observations  42           Parameters  4

Parameter    Estimate      SE        t-Ratio    P-Value
 coef 1      -0.50735    0.81255    -0.6244     0.5361
 coef 2       0.72330    1.0039      0.7205     0.4756
 coef 3       0.77462    0.093382    8.2952     0.0000
 coef 4       0.62258    0.11758     5.2948     0.0000

Residual SD  0.22423    Residual Variance   0.050281
Multiple R   0.99940    Multiple R-squared  0.99880
```

Performing the F-test. The null hypothesis is that $b_1 = b_2$. The test statistic is:

$$F = \frac{(RSS_1 - RSS_2)/(df_1 - df_2)}{RSS_2/df_2}$$

where RSS_1 is the residual sum of squares for the first (restricted) model and df_1 is its degrees of freedom; and RSS_2 is the residual sum of squares for the second (unrestricted) model and df_2 is its degrees of freedom. This type of test is appropriate when the restricted model is a special case of the unrestricted model and under the independence and Normality assumptions stated below. Under the null hypothesis that the slopes are equal, the test statistic has an F distribution with $(df_1 - df_2)$ and df_2 degrees of freedom.

For our models, the number of observations is $n = 42$ and the degrees of freedom are $df_1 = n - 3 = 39$ and $df_2 = n - 4 = 38$.

```
. ss r1 > rss1; ss r2 > rss2           # residual sums of squares
. fstat = (rss1 - rss2) * 38 / (rss2 * (39-38))   # F statistic
. pf fstat {df=1,38} > pleft           # left tail probability
. show 1-pleft                         # P-value

   0.3177
```

The **ss** command calculates the sum of squares of its argument—in this case, the regression residuals. See *vecop(BRM)* for a discussion of it and many similar commands. The P-value is obtained as described in Section 7.4.4. It is .3177; this evidence does not support rejecting the null hypothesis.

It is also possible to obtain the residual sums of squares and the degrees of freedom directly from the **anova@** output of the **regress** command.

Assumptions. Inference based on an F-test requires that the observations be independent, and that the error term in the model be drawn from an (approximately) Normal distribution. Independence is arguable in this case—it depends on what one regards as the population. In Section 9.5.1 we show one method to compare the regression residuals to the Normal distribution.

9.4.4 Partial Correlation

The partial correlation matrix of a set of variables Y ($y_1, y_2, \ldots, y_k$) given the variables X ($x_1, x_2, \ldots, x_m$) is the correlation matrix of the residuals from the multiple linear regression of $y_1, y_2, \ldots, y_k$ on $x_1, x_2, \ldots, x_m$. We denote it as $R_{Y.X}$.

As an example, we calculate the partial correlation matrix of the egg and chick weights (columns 3 and 4) of *plover*, given fixed values of egg length and breadth (columns 1 and 2). The **regress** command allows several simultaneous dependent variables. In this case, its output datasets—such as the fit, residuals, estimated coefficients, etc.—contain one column for each dependent variable. Thus we can compute the partial correlation matrix directly as the correlation matrix of the regression residuals:

```
. select plover > plov
. regress plov {x=1,2} {y=3,4} {quiet} > res@ r
. stat r {quiet} > cor@ Ry.x
. show {f=4} Ry.x
    1.0000    0.2091
    0.2091    1.0000
```

Because **regress** does not allow missing values, we remove those cases which contain missing values using **select**. Then we regress *plov*[3 4] on *plov*[1 2] and save the residuals in *r*. We use the {**quiet**} option of **regress** and **stat** because we are interested only in their output datasets.

A second way to evaluate the partial correlation $R_{Y.X}$ is to evaluate the matrix expression presented for it in many textbooks:

$$R_{Y.X} = D^{-\frac{1}{2}}(\Sigma_Y - \Sigma_{YX}\Sigma_X^{-1}\Sigma_{XY})D^{-\frac{1}{2}}$$

where Σ_Y is the covariance matrix of the variables in Y; Σ_X is the covariance matrix of X; Σ_{YX} is the cross-covariance matrix of the variables in Y with the variables in X; $\Sigma_{XY} = \Sigma'_{YX}$; the joint covariance matrix of Y and X is given by

$$\begin{bmatrix} \Sigma_Y & \Sigma_{YX} \\ \Sigma_{XY} & \Sigma_X \end{bmatrix};$$

and D is the diagonal matrix which consists of the main diagonal of the partial covariance matrix:

$$\Sigma_Y - \Sigma_{YX}\Sigma_X^{-1}\Sigma_{XY}.$$

We evaluate this expression in BLSS as follows:

```
. stat plover {x=1:4} {quiet} > cov@ S    # covariance matrix of X, Y
. Sx = S[1 2, 1 2]
. Sy = S[3 4, 3 4]
. Syx = S[3 4, 1 2]
```

By default, when **stat** computes a covariance matrix, it first deletes missing values on a casewise basis. Thus, in both this method and the previous one, we use only the cases of *plover* which are free of missing values. *Sx*, *Sy*, and *Syx* are the submatrices of *S* that correspond to Σ_X, Σ_Y, and Σ_{YX}. The subscripting methods used to obtain these submatrices, as well as the matrix multiplication operator '#*' and the **inv** and **diag** functions used below, are explained in Section 8.3.

```
. C = Sy - Syx#*inv(Sx)#*Syx'   # partial covariance matrix
. d = diag(1/sqrt(diag(C)))     # square root of main diagonal
. Ry.x = d #* C #* d            # partial correlation matrix of Y given X
```

The dataset *C* is, of course, $\Sigma_Y - \Sigma_{YX}\Sigma_X^{-1}\Sigma_{XY}$; the **inv** function computes the matrix inverse. The **diag** function (which is also available as a command) serves two purposes. First, if the argument to **diag** is a square matrix, it extracts the main diagonal and returns it as a row vector. We take the inverse square root of this result. Second, if the argument to **diag** is a vector, it constructs the corresponding square diagonal matrix. The final result, which we save in the dataset *d*, is $D^{-1/2}$ in the expression for the partial correlation matrix.

This apparently roundabout method for constructing $D^{-1/2}$ lets us avoid taking the square root of any negative numbers, or inverting any zero elements, which may be in the off-main-diagonal positions of *C*.[9]

The partial correlation obtained using this method:

```
. show {f=4} Ry.x

  1.0000    0.2103
  0.2103    1.0000
```

agrees with the first method only to the first two places of precision. The second method is less accurate as a result of the matrix inversion and other extra matrix arithmetic involved.

There is a lesson here. Although complex matrix expressions are useful for theoretical discussion, in practice they have drawbacks. Not only do they require more work to evaluate, but—as a consequence of the extra computations—the accumulated roundoff error is greater.

---

9. Another, quicker, method to obtain a correlation matrix from a covariance matrix is to divide each element by its row and column SD using dimension-expansion rather than matrix multiplication:

```
. s = sqrt(diag(C))
. Ry.x = C / s / s'
```

9.5   Distribution and Random Number Tests

This section presents Q-Q plots, a graphical method for comparing two distributions, and a random number simulation which examines the distribution of two classical test statistics when the underlying distribution is not Normal.

9.5.1   Q-Q Plots

A simple test of whether a dataset comes from a specified distribution is to plot the sample quantiles of the data against the same quantiles of that distribution. A similar test of whether two datasets come from the same distribution is to plot their sample quantiles against each other. Such plots are called Q-Q plots: they plot quantiles against quantiles. If the two distributions are the same, the plot approximates a straight line with slope 1 and intercept $(0, 0)$.

One-sample Q-Q plots. To create a Q-Q plot of the residuals *r1* from the restricted regression model in Section 9.4.3 against Normal (Gaussian) quantiles, we start by sorting the data to obtain its sample quantiles:

```
. sort r1 > r1sort
```

(When used to make a Q-Q plot, a dataset is sometimes first centered or standardized. Because this dataset is composed of regression residuals, it is already centered.) Next we construct the values p at which the sample quantiles are observed:

```
. nrows r1sort > n
. p = (.5:n-.5)/n
```

The **nrows** command gives the number of rows in its argument—in this case, the number of data points n. Thus, p contains n equispaced probability values, $.5/n$, $1.5/n$, ..., $(n-.5)/n$, at which the sample quantiles are observed. These are called the *empirical cumulative distribution function values* or simply the *ecdf values*[10] of the distribution from which the dataset was drawn, and they are the probabilities p at which we evaluate the Normal quantiles *q1gau*:

```
. qgau p > q1gau
```

Qgau and many similar commands are discussed in Section 7.4 and *cdf(BRM)*. To make the Q-Q plot, we plot the presumed distribution's quantiles on the X-axis and the data quantiles on the Y-axis:

```
. scat q1gau r1sort
```

This plot is shown in Figure 9.3. It is fairly straight, except for two largest and two smallest residuals. This indicates that the error distribution is approximately Normal although perhaps not quite so in the tails.

---

10. Some authors use the points $1/n$, $2/n$, ..., $(n-1)/n$, 1 for the ecdf values, but this has drawbacks for Q-Q plots.

Figure 9.3

```
. scat {big} q1gau r1sort            # Normal Q-Q plot of 'r1' regression residuals

X variable:    q1gau
Y variable:    r1sort
  0.600                              |                                    *
                                     |
                                     |
                                     |
                                     |
                                     |
                                     |                            *  *
                                     |                         *
                                     |                       *
                                     |                  ** *
                                     |                  *
                                     |              * *
                                     |          2**
                                     |         **
 ------------------------------------2*-----------------------------------
                                   ***|
                               *2**   |
                             **       |
                           *          |
                         *            |
                 *   **               |
                       *  *           |
                   *                  |
              *    *                  |
                                      |
                                      |
                                      |
                    *                 |
                                      |
                                      |
 -0.600                               |
     -2.500                                                      2.500
```

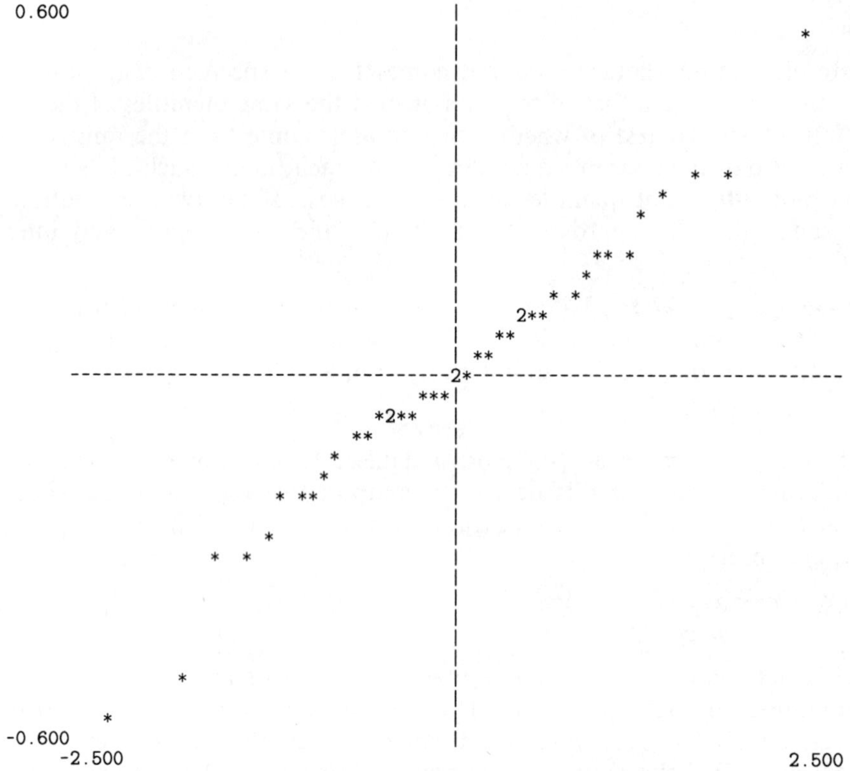

Two-sample Q-Q plots. We can use Q-Q plots to compare the distributions of
two datasets. When the two sample sizes are equal, the sample quantiles are
simply the two sorted datasets and the Q-Q plot is the two sorted datasets
plotted against each other. For example, the BLSS library dataset *kappa* contains in column 1 the heights (in inches) of a group of 30 students at UC
Berkeley and contains in column 3 the heights of a similar group of 30 students at UCLA. To create the Q-Q plot:

```
. load kappa
. sort kappa[1] > h1
. sort kappa[3] > h2
. scat h1 h2
```

When the sample sizes are unequal, the sample quantiles are observed at
different values p of the ecdf. For example, the BLSS library datasets *final84*

and *final85* contain final exam scores for students taking Statistics 2 at UC Berkeley in Summer Sessions 1984 and 1985. The two classes were taught by the same instructor and the two final exams were constructed similarly and graded on an identical scale: 0 through 100. Were the grade distributions the same?

63 students took the final in 1984; 62 took it in 1985. We sort *final84* to form its sample quantiles and construct its ecdf values p as with a one-sample Q-Q plot:

```
. load final84 final85
. sort final84 > q84
. nrows final84 > n
. p = (.5:n-.5)/n
```

The **qdata** command computes the quantiles of a dataset (its first argument) for an array of probabilities (its second argument). See *qdata(BRM)* for more information. We use it to compute the quantiles of *final85* at the values in p:

```
. qdata final85 p > q85
```

The Q-Q plot is given by:

```
. scat q84 q85
```

It appears in Figure 9.4. It too is a fairly straight line except for the top two points and the bottom fifth of the data—the top two students in 1984 scored higher, and the bottom fifth of the students in 1984 scored lower, than the top two and bottom fifth in 1985.

9.5.2   Are the t- and F-Tests Robust Against Non-Normality?

The classical t- and F-tests assume that the underlying data are Normal. In practice this is often not true. A test is said to be *robust* against non-Normality if it behaves similarly for both Normal and (moderately) non-Normal data.[11] One way to examine the robustness of a test is via simulation, using non-Normal data. In this section we present two examples of such simulations, one each for the t- and F-tests.

t-Test for the equality of two means. Our first example is the t-test of the hypothesis that two means are equal. Each simulation consists of two samples of 25 observations from the gamma distribution with $\alpha = 5$ and $\beta = 1$; we use 100 separate simulations. Because many BLSS commands treat columns as separate statistical variables, we organize the data so that each column contains data for a separate simulation. Data for the two samples go into two separate datasets:

```
. rgamma {a=5} {dims=25,100} > g1
. rgamma {a=5} {dims=25,100} > g2
```

---

11. A statistical procedure is said be *robust* if it is insensitive to departures from the assumed model. Of course, *robust* is a relative term (like *big* or *many*): just how insensitive, and to what types and sizes of departures, depends on the context.

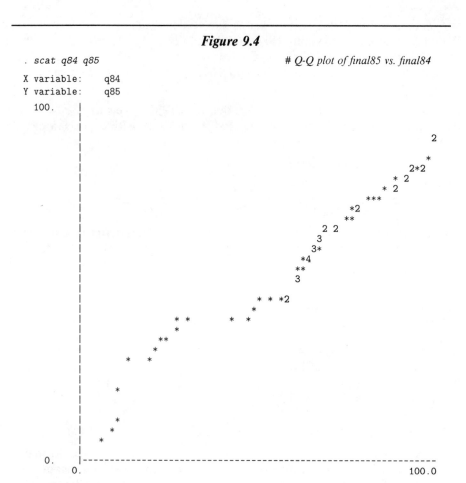

Figure 9.4

```
. scat q84 q85                                    # Q-Q plot of final85 vs. final84
X variable:    q84
Y variable:    q85
   100.    |
           |
           |                                                               2
           |                                                            *
           |                                                       2*2
           |                                                    *  2
           |                                                 *  2
           |                                              ***
           |                                          *2
           |                                         **
           |                                  2  2
           |                               3
           |                               3*
           |                             *4
           |                            **
           |                           3
           |                     *  *  *2
           |                         *
           |            *  *        *  *
           |            *
           |         **
           |          *
           |       *    *
           |
           |      *
           |
           |     *
           |     *
           |    *
     0.    |---------------------------------------------------------
          0.                                                      100.0
```

The **ht** command, presented in Chapter 6, performs the hypothesis test for the difference of the means of corresponding columns in *g1* and *g2*. The output looks like this:

```
. ht g1 g2 {x=1:5; y=1:5} {pool} {terse}
```

Hypothesis Test: two-sample, two-tailed t. Data: g1 and g2.

Col	N	null mean diff	sample mean diff	sample SE (pooled)	df	test stat	P value
1	25,25	0.0000	-0.62012	0.66646	48.0	-0.93	.3570
2	25,25	0.0000	-0.45735	0.64800	48.0	-0.71	.4836
3	25,25	0.0000	1.3002	0.58637	48.0	2.22	.0312
4	25,25	0.0000	1.3798	0.63920	48.0	2.16	.0358
5	25,25	0.0000	-0.23226	0.54919	48.0	-0.42	.6735

We used the {**x=**} and {**y=**} options to limit the output display to the first five simulations (columns) and used the {**terse**} option to suppress the critical values for specific significance levels. What we really want is to examine the P-values for all 100 simulations. We save them, quietly, in the **p@** output dataset:

```
. ht g1 g2 {pool} {quiet} > p@ p
```

When the null hypothesis is true (which in this case it is) and the underlying data is Normal, the test statistic has a t distribution: thus, 5% of the time the P-value of the t-statistics is .05 or less. Is this true for our simulated t-tests? The **count** command (presented in Section 8.7) lets us count the number of observed P-values which meet this condition:

```
. count (p <= .05)

6/100 = 6.00% of the cases meet the condition.
```

More generally, if the observed t-statistics follow the t distribution, the observed P-values have a uniform distribution on [0,1]. The **stemleaf** command, presented in Chapter 3, is one quick way to see the histogram of the observed P-values:

```
. stemleaf p

N = 100, min = 0.01058, 25% = 0.2706, 50% = 0.5229, 75% = 0.7817, max = 0.9977
Leaf digit unit (ldu) = 0.01   (1|2 represents 0.12)

   0|11333466669
   1|2445579
   2|2234457788999
   3|013556677799
   4|00889
   5|0223466779
   6|123344455678
   7|267789
   8|00001344556789
   9|155677799
  10|0
```

Both the quick count above and the observed distribution are not far from what we expect for Normal data. Thus, the t-test appears to be robust against non-Normality. Although this simulation is hardly exhaustive—we considered only one type of non-Normal data—general statistical experience and opinion is that the t-test is indeed reasonably robust against departures from Normality.

F-test for the equality of two variances. The F-test for the equality of population variances for Normal data is based on the fact that, when two Normal distributions have the same variance, the ratio of the sample variances has an F distribution whose degrees of freedom are 1 less than the sample sizes. Either of the observed variances might be larger, so the two-sided test is based on the ratio of the larger variance to the smaller variance. The observed P-value is twice the right-tail probability of this statistic.

Unlike the t-test above, there is no single BLSS command to perform this F-test so we perform it with the following sequence of commands:

```
. colvar g1 > v1
. colvar g2 > v2
. fstat = v1/v2
. fstat = fstat >= 1 ? fstat : 1/fstat
. pf fstat {df=24,24} > pleft          # left tail probability
. p = 2 * (1-pleft)                     # 2 times right tail probability
```

The **colvar** command calculates the variance of each column of a dataset and places the result in a row vector. It is documented in *vecop(BRM)*. The conditional evaluation operator '?:', discussed in Section 8.6, allows us to obtain the ratio of the larger sample variance to the smaller by inverting only those ratios which were less than 1. We calculate the P-values as explained in Section 7.4.4. As before, we can count the number of P-values which are .05 or less:

```
. count (p <= .05)
11/100 = 11.00% of the cases meet the condition.
```

or we can look at the histogram of all the simulated P-values:

```
. stemleaf p

N = 100, min = 7.665e-05, 25% = 0.1951, 50% = 0.3736, 75% = 0.625, max = 0.9792
Leaf digit unit (ldu) = 0.01   (1|2 represents 0.12)

0|000123444555579
1|0234447899
2|0011122234779
3|0001113445569
4|012234566678889
5|15779
6|00023555899
7|024556899
8|0456
9|03378
```

Both the quick count and the stem-and-leaf diagram suggest that there are too many observed P-values near 0 compared to what happens when the data are Normal. In other words, the null hypothesis that the variances are equal is rejected too frequently. As with the previous example, our small simulation is hardly exhaustive but it is consistent with general statistical experience: the F-test is not robust against departures from Normality. It falsely rejects the null hypothesis too often when it is true.

CHAPTER 10

Convenience Features

This chapter describes convenience features provided by BLSS. Although they perform no statistical analyses themselves, these facilities allow you to enter statistical commands—or otherwise get your work done—more quickly.

Sections 10.1 through 10.5 describe features useful to everyone; they require minimal knowledge of BLSS. Sections 10.6 and 10.8 assume that you have read Chapter 4. In addition, you will be able to make much better use of the material in Sections 10.5 through 10.8 if you have learned how to use a text editor. Sections 10.8.3 and, even more so, 10.8.4 are intended for experienced users.

For reasons explained in Section 10.7, the commands presented in this chapter are known as *internal* or *irregular* commands.

10.1 History

BLSS remembers the commands you have given it. This is called the *history* mechanism. The command:

```
. history
```

displays the last few commands on your history list—up to approximately one screenful (23 lines) of commands. If you want to see a different number of commands (either less or more—up to 100) give the number as an argument to **history**. For example, to see your last 10 commands only, type:

```
. history 10
```

Depending on what commands are in your history, the display looks something like this:

```
 6  stemleaf plover
 7  select plover > plov
 8  regress plov {x=1;y=2} > fit res
 9  scat plov[1] plov[2],fit
10  scat plov[1] res
11  select plov (plov[5]) {log} > lived died
12  stemleaf lived[1]
13  stemleaf died[1]
14  ht lived died
15  history 10
```

Each command is preceded by its command number. Commands are numbered, starting at 1, from the beginning of your session.

What is history good for? The **redo** command allows you to repeat any command in your history without retyping it. For example, suppose you want to repeat command 9, in which you generated a scatterplot. Type:

```
. redo 9
```

BLSS then displays command 9 again (without the number or the prompt):

```
scat plov[1] plov[2],fit
```

To confirm that you want to repeat this command, press the RETURN key. BLSS then reruns the command—in this case, the same scatterplot as before appears on your screen. If, instead, you change your mind, press your interrupt key (usually control-C). This interrupts the **redo**, and BLSS gives you a new prompt.

It is also possible to make changes in the command, because **redo** contains a built-in text editor. See the manual entry *history(BRM)* for detailed information.

Note that you do not need to give the **history** command in order to give a **redo** command. The **history** command is necessary only if you want to be reminded what commands are in your history, or what their command numbers are.

If you issue the **redo** command without any number, it **redo**es your most recent command. If you issue the **redo** command with a negative number, it counts backward that far from your current command. For example:

```
. redo -4
```

redoes your fourth most recent command.

10.2 Aliases

The **alias** command allows you to establish abbreviations for commands. For example, suppose you get tired of typing 'show {lpr}' to print files and that, instead, you want to use the command 'print'. Issue the command:

```
. alias print show {lpr}
```

For the remainder of your BLSS session, whenever you give the command 'print' it is just as if you give the command 'show {lpr}'. For example:

```
. print a b c
```

sends the files *a*, *b*, and *c* to the lineprinter. The command:

```
. print {ff} a b c
```

also sends the files to the lineprinter, with the {**ff**} option set, so that each file is printed on a separate page.

To speed up your typing, you can use aliases to make one-letter abbreviations for the commands you use most often. For example, if you use **history** and **redo** a lot, make the aliases:

```
. alias h history
. alias r redo
```

Thereafter, the one-letter command 'h' shows your history, and the one-letter command 'r' invokes **redo**. (Of course, this means that any datasets named *h* or *r* must be shown explicitly with the **show** command—simply typing the name 'h' or 'r' invokes the alias instead of **show**-ing the dataset.)

The general form of the **alias** command is:

alias *name value*

where *name* is a single word and *value* is any number of words.[1] In our first example, the alias name was 'print' and the alias value was 'show {lpr}'; in our second example, the alias name was 'h' and the alias value was 'history'. Once an alias has been established, whenever BLSS sees the alias name used as a command name, it substitutes the alias value in place of the name before executing the command.

The action of substituting some text for some other text is called *text substitution*, or *text expansion* (because the resulting text is usually longer than the starting text—so the text is 'expanded'). This particular text substitution (the alias value in place of the alias name) is called *alias expansion*.

An alias name can be a new name that you make up—as in the examples above—or you can use an alias name to give a new meaning to a standard

---

1. In BLSS, a *word* is any string of characters which is not separated by spaces or tabs, or which is wholly enclosed in a pair of matching ( ), [ ], { }, or "" characters.

BLSS command name. For example, the **scat** command normally makes scatterplots which are 20 lines high and 75 characters wide—a size appropriate for most terminals. The {**height**=} and {**width**=} options override these defaults. Suppose you use a terminal with an extra-large display (in characters) and that you prefer scatterplots larger than the default. You might give the command:

```
. alias scat scat {height=60} {width=130}
```

Thereafter, when you give the command 'scat', the alias expansion mechanism translates it into 'scat {height=60} {width=130}' so that you automatically get larger scatterplots. In this way, you can use aliases to override the standard BLSS command defaults and set your own defaults.

Inhibiting alias expansion. Alias expansion can be inhibited (that is, prevented for an individual command) by preceding the alias name with a backslash '\'. For example, if you have established the 'scat' alias as above, but want to override it on occasion, your conversation with BLSS might look like this:

```
. alias scat scat {height=60} {width=130}
. scat plover {x=1;y=2}        # the alias makes a large scatterplot

  # ... several more commands and scatterplots ...

. \scat plover {x=3;y=4}
  # makes a standard size scatterplot because alias expansion is inhibited

. scat plover {x=3;y=4}        # a large scatterplot this time
```

More about aliases. The command:

```
. alias name
```

shows the value (if any) to which *name* is currently aliased. The command:

```
. alias
```

lists all your currently defined alias names and their values. The command:

```
. unalias name1 name2 ...
```

removes the named aliases—that is, they are no longer in effect. The manual entry *alias(BRM)* contains a summary of the alias commands.

Duration. Normally, the aliases you make are in effect only until the end of your BLSS session. In Section 10.5, we explain how to make aliases permanent.[2]

---

2. The alias features of the BLSS shell are modeled on, and similar to, those of the UNIX C shell. A little-known feature of C shell aliases is that the '\' alias inhibition character, described here for BLSS aliases, works for the C shell too.

10.3   Strings

Strings provide a second type of text expansion, similar to aliases. Whereas aliases allow text substitution to be applied to the first word (only) in a command, strings allow text substitution to be applied to any word in a command. Strings are established using the **set** command. Its general form is the same as that of the **alias** command:

set *name value*

where *name* is a single word and *value* is any number of words. Once a string *name* has been set, any time BLSS sees *name* immediately following a '$' symbol, it substitutes the word(s) in the string *value* for the $*name*. This type of text expansion is called *string expansion*.

Examples. Suppose you have a dataset with a very long descriptive name: *small.plover.chicks*. If you plan to work with this dataset a lot, but retain its long name, you can save on typing by using a short string name. Give the command:

```
. set s small.plover.chicks
```

Thereafter, you can type '$s' in place of the long name:

```
. load $s
. stemleaf $s
. stemleaf log($s)
. t = 10+$s*100
```

And so forth. Note that there is no space between the '$' symbol and the name of the string being expanded.

For a second example, suppose that on occasion (but not frequently) you want to make large scatterplots as in the previous section. Give the command:

```
. set large {height=60} {width=130}
```

Thereafter, when you want to use those options to **scat**, you can type, for example:

```
. scat $large plover
. scat $large $s
```

If, on occasion, you want to force your plots to have certain minimum and maximum X and Y values, set a string which contains the appropriate options. For example:

```
. set bounds {xmin=0;xmax=1;ymin=0;ymax=1}
```

This option string can be used independently of the previous option string or together with it. For example, you might do:

```
. scat data $bounds
```

and then, upon discovering that you like it, make a large version of the same plot:

```
. scat data $bounds $large
```

Notice that using strings saves not only typing, but also the need to remember the precise values of your favorite options (for example, whether your preferred large-format plot width is 120 or 130).

Warning: strings and subscripts.   When a string contains several words, you can access the individual words using string subscripts. For example:

```
. set x a b c d
```

creates a string x which contains four words: 'a', 'b', 'c', and 'd'. Then, $x[1]$ refers to the first word of the string ('a'), $x[2]$ refers to the second word, etc. We mention this feature here not to discuss its applications (which are primarily in macros), but to state a warning: String subscripts conflict with regular BLSS vector and matrix subscripts. For example, the command:

```
. stat $s[2]
```

does not show statistics for column 2 of *small.plover.chicks.* To do that, give instead the command:

```
. stat $(s)[2]
```

In general, use the form:

$$(name)[subscript]$$

whenever you want to use array subscripts together with string expansion of dataset names. Note that the '$' symbol goes *outside* the parentheses, and that there are *no* spaces either before or after the enclosing parentheses.

Strings and aliases compared.   Aliases are simpler to use than strings, because there is no '$' symbol to type. But, unlike aliases, you can use several strings in the same command. Thus, you can piece together different strings in different ways and, with fewer strings, create more different commands. Of course, you can use strings in the same command as an alias. For example, using strings and aliases set in previous examples, we can type:

```
. print $s
```

More about strings.   The command:

```
. set name
```

causes the string *name* to be set, but to have no value. Strings with no value are used as flags under certain circumstances. The command:

```
. set
```

lists all your currently defined strings and their values. This command may list several strings you never set yourself, most of whose names are in uppercase letters. Strings whose names begin with uppercase letters are called *environment strings* and are discussed in the next section. To list only those strings whose names begin with lowercase letters, give the command:

> . *set* -

To find out the value (if any) of the string called *name*, give the command:

> . *echo $name*

Finally, the command:

> . *unset name1 name2*

removes the named strings—they are no longer in effect. The manual entry *set(BRM)* summarizes the string-setting commands.

Duration. As with aliases, the strings you set are in effect only until the end of the BLSS session in which you set them. Section 10.5 explains how to make strings and aliases permanent.[3]

10.4   Environment Strings

Environment string is a UNIX concept which is exploited by BLSS. From the BLSS user's point of view, an environment string is simply a regular string with the special property that its name begins with an uppercase letter. For example, 'B', 'BPATH', and 'Bingo' are environment string names whereas 'b', 'bpath', and 'bANG' are not. All string-handling capabilities described in the previous section—the **set** and **unset** commands, and $ string expansion— apply to environment strings as well.

From the computer's point of view, however, environment strings have a second special property: Once an environment string has been **set**, its value is available not only to you when you *type* BLSS commands, but also *to the command itself, internally.* Thus, environment strings are used to establish global parameters, or option settings, which affect all BLSS commands until the end of the BLSS session, and also UNIX commands which BLSS invokes directly or indirectly.[4]

The manual entry *strings(BRM)* contains a complete list of regular and environment strings which are used for internal purposes by BLSS. You can set or reset some of these to tailor BLSS to your own liking. Other environment strings are used by non-BLSS UNIX facilities. *Warning:* Reset or unset environment strings whose meanings are unknown to you at your own peril.

Example: Resetting your editor. In Section 4.9, we saw that the {**editor**=} option to the **edit** command causes it to use an editor other than the default. To use the same non-default editor without specifying the option every time,

---

3. As with alias expansion, the string expansion features of the BLSS shell are modeled on, and similar to, those of the UNIX C shell. Users of C shell strings should be aware, however, that there are differences between the two. For example, the '=' sign, which is used when setting C shell strings, is *not* normally used when setting BLSS shell strings.

4. In other words, they establish an *environment* within which the command runs, and it is from this notion that environment strings take their name.

you might create an alias for the **edit** command which includes the appropriate option setting. Another way is to set the environment string EDITOR to the editor of your choice. For example:

```
. set EDITOR vi
```

This method is preferable, because not only the **edit** command, but other BLSS and UNIX commands and facilities, recognize the EDITOR environment string.

Example: Resetting your lineprinter. Larger computer systems may have several lineprinters, each identified by its own name. Usually there is a default lineprinter, but it may not be the one you want to use. On our system, for example, printers are named after their room number. The default printer is called *337* (it is in room 337) but we prefer the printer *491* (in room 491), because it is much closer.

On BSD (Berkeley) UNIX systems, the environment string PRINTER specifies the name of the printer to use. To use printer *491* for the remainder of the BLSS session, we give the command:

```
. set PRINTER 491
```

On AT&T System V UNIX, the environment string LPDEST specifies the printer. If our computer ran System V UNIX instead of BSD UNIX, we would instead give the command:

```
. set LPDEST 491
```

Of course, before you can reset your default printer, you must find out the names of the lineprinters on your computer system.

Environment strings and regular strings compared. Environment strings are more powerful than regular strings, in the sense that they are passed on to other commands (such as the UNIX printing commands which BLSS invokes). But there is a slight additional cost associated with this facility. Therefore, use regular strings (whose names begin with lowercase letters) instead of environment strings (whose names begin with uppercase letters) except when you specifically need the latter.

10.5   The '.blss' Startup File

Aliases, strings, and environment strings are in effect only for the BLSS session in which they are defined. To define them permanently, create a text file named *.blss* in your home directory which contains the appropriate **alias** and **set** commands. BLSS executes the commands in this file each time it starts up.[5] For example, a *.blss* file might contain the commands:

---

5. Additional BLSS startup files can be specified via environment strings. See *strings(BRM)* for more information. Note that, because its name begins with a '.' character, the UNIX **ls** command will not list the *.blss* file unless the **−a** option is used.

```
alias    al        alias
al       h         history
al       r         redo
set      EDITOR    vi
set      PRINTER   491          # use your preferred printer name
set      large     {height=60;width=130}
```

Environment strings can be set even before you enter BLSS. UNIX itself provides global startup files which it reads every time you login—you can use these UNIX startup files to set environment strings and they will have the same effect as if set when in BLSS. Because the name of the startup file and the form of the command needed depend on which UNIX shell you use ('%' prompt or '$' prompt), we do not explain how to do this here.

Avoid creating too long a *.blss* (or UNIX) startup file—too long a startup file causes BLSS (or UNIX) to start up noticeably slower. For most people and most computers, one or two dozen **set** or **alias** commands is a reasonable maximum. To create additional special-purpose aliases or strings, it is more efficient to place the appropriate commands in a separate file and **source** that file when it is needed. See the next section.

10.6 Source

Normally, the source of your BLSS commands is the keyboard of your terminal—you simply type in your commands. The **source** command tells BLSS to read and execute commands from a text file, instead. For example, suppose the text file *prin* contains the commands:

```
stat {quiet} x > stan@ xs
svd {quiet} xs > xu xd xv
xprincomp = xu #* diag(xd)
scat xprincomp {x=1;y=2}
scat xprincomp {x=1,2;y=3}
```

Then the BLSS command:

```
.  source prin
```

causes BLSS to execute the commands in *prin* just as if you had typed them yourself. The text file *prin* is called a *command file*.

This example command file computes the principal components of *x* and plots the first three principal components against each other. If you don't know what *principal components* are, don't worry. Just think of this example as an arbitrary sequence of BLSS commands.

If you want to try **source**-ing this command file, create a file *prin* in your work area which contains the commands shown above. Use the **addtext** or **edit** commands explained in Section 4.9. Then load a dataset from the BLSS system data library, such as *steam* or *stackloss*, and rename the dataset as *x* for the purpose of **source**-ing the file. Before you **source** *prin*, make sure that you have no important datasets named *xs, xu, xd, xv,* or *xprincomp*, because they will be overwritten.

The advantage of **source**-ing a command file is that if you want to run the same sequence of commands again, tomorrow or next week, you need not type the entire sequence again. Just give the **source** command again. Note that this works equally well whether you rerun *prin* on the same dataset *x* or whether you use it on some other dataset which you have temporarily named *x* for the purpose.

The bin area and the {bin} ***option.*** Of course, if you want to **source** a file again tomorrow, it must be there tomorrow. So you must **save** it. As explained in Section 4.6, the **save** command normally saves text files in the text area. However, BLSS provides another permanent area, the *bin area*, which is intended for saving text files that are command files. In fact, if the **source** command does not find the named command file in the work area, it automatically looks in the bin area.

To **save** files in the bin area, use the {bin} option. This option is exactly analogous to the {data} and {text} options described in Sections 4.6 and 4.12—it makes the utility command to which it is applied use the bin area instead of the default area. For example, to **addtext** to or **edit** files in the bin area, use the {bin} option. Refer to Section 4.12 for more information.

The **source** command itself does not recognize the {text}, {bin}, or other options. It does not need them. As noted above, if **source** does not find the named command file in the work area, it looks in the bin area automatically.

An alternative to **source**-ing command files is to **run** them as macros. This is described in Section 10.8.

10.6.1   Source and History

You may not know ahead of time what commands you want to put in a file—quite likely, you will type a sequence of BLSS commands and only later decide that they would be useful in a command file. In this case, use **history** with the '–' argument and redirect the output text into a file; then edit the text file to make any changes necessary to **source** it. The '–' argument suppresses the command line numbers in the history display, which you would otherwise need to edit out. Your conversation with BLSS might look like:

```
. history 8 - >! cmdfile     # save the last 8 commands in a text file
. edit cmdfile               # edit it to remove unwanted commands
. source cmdfile             # re-execute the commands as a command file
```

You may want to save the command file in your bin area for later use:

```
. save {bin} cmdfile
```

If, after saving it and exiting and reentering BLSS, you want to re-**source** the command file with variations in what it runs, you can edit it in the bin area:

```
. edit {bin} cmdfile
```

Alternatively, you can **load** *cmdfile* and make any changes to the copy in the work area. The permanent copy (in the bin area) will remain unchanged, and **source** will find the edited copy (in the work area) first.

10.6.2   Source, Aliases and Strings

You can use aliases and strings when **source**-ing a file. Aliases and strings can be created by placing the appropriate commands in a command file and **source**-ing it. An example of this is the *.blss* file in your home directory, which, as explained in Section 10.5, is automatically sourced every time BLSS starts up.

Conversely, aliases and strings which are set before a file is **source**-d affect the commands within it. This fact is useful for constructing a command file which performs more than one simple task. Consider again the file *prin* presented at the beginning of Section 10.6. It operates on a BLSS dataset named *x*. One way to make *prin* operate on different datasets, noted earlier, is to temporarily rename as *x* the dataset of current interest. A better method is to replace the dataset name *x* in the command file with the string $x. In this case, we need only change the first line of *prin*. It becomes:

```
stat {quiet} $x > stan@ xs
```

Each time a new dataset is used, **set** the string *x* to the dataset's name before **source**-ing *prin*. For example, the following sequence of commands runs *prin* on the datasets *steam* and *stackloss* from the BLSS system data area:

```
. load steam stackloss
. set x steam
. source prin          # run the command file on 'steam'
. set x stackloss
. source prin          # run the command file on 'stackloss'
```

The advantage of this method is that you need not rename or copy your BLSS dataset—so you need not remember what *x* is. If any commands in the command file make displays which include the name of the original dataset, this name will appear in the display instead of the mystery name *x*. This is not the case with our example command file *prin*—it always uses the same names for the datasets it creates: *xs*, . . . , *xprincomp*. To include the name of the original dataset in the names of the datasets it creates, modify *prin* as follows:

```
stat {quiet} $x > stan@ $x.s
svd {quiet} $x.s > $x.u $x.d $x.v
$x.prin = $x.u #* diag($x.d)
scat $x.prin {x=1;y=2}
scat $x.prin {x=1,2;y=3}
```

If we **source** this new version of *prin* when the string $x has the value *steam*:

```
. set x steam
. source prin
```

it creates datasets with the names *steam.s, steam.u, ..., steam.prin*, and **scat** labels the scatterplots accordingly. When the command file is re-**source**-d for other datasets (with the string $x set to a different value), it will create datasets with different names. This avoids name conflicts when running it on different datasets.

10.7 The BLSS Shell

When you use BLSS, a program called the *BLSS shell* is responsible for giving you the '.' prompt, reading commands as you enter them, deciding what each command means, executing it, and arranging for any output text redirection. It is the BLSS shell that records your command history, performs alias and string expansion, and reads commands from files via **source**.[6]

Files and datasets exist separately from the BLSS shell. As noted in Chapter 4 (albeit in the footnotes), you can access your BLSS files directly from UNIX. In contrast, aliases, strings, and history are remembered internally by the BLSS shell. That is why aliases and strings must be reinitialized every time BLSS starts up.

Similarly, regular BLSS commands exist separately from the BLSS shell itself. (If necessary, you can access regular BLSS commands directly from UNIX. It's a bit technical, but the manual entry *fromunix(BRM)* explains how.) In contrast, the commands presented in this chapter—**history**, **redo**, **alias**, **set**, **source**, and **run**—are internal to the BLSS shell and are known as *internal* commands. They do not obey the inputs, options, and ouputs conventions of the regular BLSS commands; for this reason, they are also known as *irregular* commands. For example, they do not recognize the {**lpr**} or {**more**} options introduced in Sections 4.3 and 4.8.

On the other hand, because output text redirection is also internal to the BLSS shell, it works with most internal commands. Recall that we used output redirection with **history** in Section 10.6.1.[7]

---

6. The BLSS shell takes its name from the UNIX shells, to which it bears a strong resemblance.

7. An exception: **source** commands, such as:

```
. source xx >> yy
```

do not have their output redirected. But you can redirect text output from commands within a command file.

10.8   Introduction to Macros

Any file of BLSS commands which can be **source**-d can also be **run** as a macro. The difference is that when a command file is sourced, it is read by the current invocation of the BLSS shell. When run as a macro, it is run separately by a new invocation of the BLSS shell.

In order to run a command file, use **run**. For example, in Section 10.6 we created a command file called *prin*. To run it as a macro:

```
. run prin
```

If the named command file exists in your work area, it is run; otherwise BLSS looks in your bin area. (The **run** command is just like the **source** command in this respect.)

If a command file is saved in your bin area (which is generally a good idea), it can be run by simply typing its name—just like a regular BLSS command. For example:

```
. prin
```

Note that command files in the work area can *not* be run this way—recall that when a file in the work area simply named, it is **show**-n. Because macros can be invoked just like regular commands, try to choose command file names which do not conflict with regular BLSS command names.[8]

For many purposes, whether you **source** a command file or **run** it as a macro does not matter. The main differences between sourcing a file and running it as a macro are:

1. Output text redirection (the '>>' and '>!' symbols), which has no effect when used with the **source** command, has the usual effect when running a command file as a macro. See the example in Section 10.8.1.

2. The aliases, strings, and history of your current BLSS shell are not affected by, and do not affect, the macro.

3. Macros ignore *.blss* files.[9]

4. Macros can take arguments—inputs, options, and outputs—just like regular BLSS commands. This is discussed in Sections 10.8.3 and 10.8.4.

Because of the second and third differences, any aliases or strings you want to use in a macro must be explicitly set in it. Of course, environment strings set in your interactive BLSS shell affect your macro. However, environment strings set in a macro do not affect your interactive BLSS shell.

---

8. In case of a name conflict, which command to execute is determined by your BPATH environment string. See *strings(BRM)* for more information about the BPATH.

9. In fact, macros ignore all BLSS startup files, including those described in *strings(BRM)*. Of course, macros can explicitly **source** any startup files desired.

10.8.1   Report Generation

When used together with output text redirection, macros provide a convenient facility for generating reports of your calculations and analyses. Such reports can include: the commands themselves, the output from the commands, and explanatory comments—all automatically interleaved.

To generate a report with a macro, proceed as follows. After you have given a sequence of BLSS commands and you are satisfied with the results, use **history** with the '−' option (as in Section 10.6.1) to save your commands in a command file. For example, if your analysis was spread out over the last 50 commands:

```
. history 50 - >! cmdfile
```

Edit the command file to remove unwanted commands, and then run it as a macro:

```
. run cmdfile >! results
```

The text file *results* contains your report. Remember to **save** your command file in your bin area (use the {**bin**} option) if you want to use it in future BLSS sessions.

The following three features are helpful when creating report files.

The echo mechanism. Setting the string **echo** to have the value '.':

```
. set echo .
```

causes BLSS to display each command before executing it. (Note the space between the 'echo' and the '.'.) This is useful at the top of a command file. You could instead use:

```
. set echo
```

but including the '.' causes the echo mechanism to show a '.' in front of each command it echoes. This makes it easy to visually distinguish commands from output text. The echo mechanism can also echo commands before and after string and alias substitution, and it can echo blank lines before commands to increase readability. These actions are all controlled by the value of the **echo** string. See the manual entry *strings(BRM)* for detailed information.

Comments, first mentioned in Section 4.9.1, can be included in command files. They are introduced by '#' symbols: all text starting at a word that begins with a '#' symbol[10] through the end of the same line is a comment and is ignored. Comments placed by themselves on their own line are shown by the echo mechanism; comments placed to the right of a command are not. For example, from the command file:

```
set echo .
# This comment will be echoed
regress x y       # This comment will not be
```

---

10. With the exception of '#∗', '#^', and '#,', which are matrix operators.

the output is:

```
.  # This comment will be echoed
.  regress x y
```

(output from the regress command)

The TALK environment string. In order to generate a smaller report, you may want to suppress the output display from commands which generate intermediate results. One method is to give the {quiet} option for every such command. Alternatively, the TALK environment string controls the verbosity of all commands. The default TALK value is 2. Setting TALK to 0:

```
.  set TALK 0
```

has the same effect as giving the {quiet} (or the {talk=0}) option for every command which follows it: no output is shown, except for error messages. For commands whose output you *do* wish to see, you can override the TALK setting and get the usual output display by giving the {talk=2} (or the {usual}) option. Or you can **unset** TALK.

An exception: the **show** command always shows its arguments, regardless of the TALK setting—so you can always use it to show results.

Refer to the manual entry *options(BRM)* for more information about TALK.

Warning. Of course, all three features described above work in interactive BLSS shells. This allows you to experiment with them interactively. However, if you **source** a command file that sets the **echo** string or TALK environment string, they will be reset for your current interactive BLSS shell. To unset them, give the appropriate **unset** command:

```
.  unset echo
.  unset TALK
```

(See Section 10.3.)

10.8.2 Strings as Dataset Subscripts

In a command file, you may want to construct expressions that allow a subscript to take on varying values. Recall from Section 8.3.2 that at present BLSS does not (directly) permit datasets or expressions to be used as array subscripts. One way around this restriction is through the **submat** command, described in Section 8.3.4. However, **submat** can be used only for creating a new dataset, not for altering elements of an existing dataset.

A second way around the restriction, which can be used both when creating a new dataset and when altering an existing one, is through strings. Because strings can be expanded anywhere in a command line, they can be expanded in the position of an array subscript. For example, if the string *i* has the text value '17':

```
.  set i 17
```

then, as a result of string expansion, the expression:

    x[$i]

is equivalent to *x*[17].

The **$Ivalue[ ]** string function[11] allows the value of a BLSS dataset or expression (as opposed to a mere string) to be used as an array subscript. It has the form:

$$\textbf{\$Ivalue[}\textit{let-expression}\textbf{]}$$

and returns, as text, the value of *let-expression* (which can be any let-expression as defined in Chapter 8) rounded to the nearest integer.[12] **$Ivalue[ ]** (like any string function) can be used anywhere a string can be used. For example, if the dataset *i* has the value 5:

    . i = 5

then the command:

    . x[$Ivalue[i]] = 16

assigns the value 16 to *x*[5]. The command:

    . x[$Ivalue[2*i-3]] = z

assigns the value *z* to *x*[7].

10.8.3 Positional Macro Arguments

As noted previously, macros are invoked like regular commands. Regular commands can take arguments; so can macros. This section and the next describe methods for accessing the arguments, if any, with which a macro is invoked. These are advanced topics; not all users need this material.

When a macro is invoked, the first argument (if any) is available within the macro as the special string $1, the second argument (if any) as the string $2, and so forth.

For example, suppose we rewrite the command file *prin* of Section 10.6 as follows:

```
stat {quiet} $1 > stan@ $1.s
svd {quiet} $1.c > $1.u $1.d $1.v
$1.prin = $1.u #* diag($1.d)
scat $1.prin {x=1;y=2}
scat $1.prin {x=1,2;y=3}
```

---

11. *String functions* are invoked the same way as regular strings, but they take arguments. They have the general form $*string-function-name*[*argument* ...].

12. Another string function is **$Value[ ]**, which returns the value of a let-expression (not rounded). Additional string functions are discussed in Section 10.8.4.

To run *prin* on the dataset *steam*, type:

```
. prin steam
```

The intermediate datasets are named *steam.c*, *steam.u*, *steam.d*, and *steam.v*; the final result is named *steam.prin*. Note the similarity between this version of *prin* and the version given in Section 10.6.2.

Macro arguments referred to using the strings $1, $2, etc. are called *positional arguments* because they are referred to according to their position in the command line. For instance, in the example above, *steam* was argument number 1. By their nature, positional arguments are of limited utility, because they do not allow you to specify options or tagged input and output datasets—all of which, in regular BLSS commands, are recognized independently of their numerical position.

10.8.4   Options and Tag Arguments

This section briefly discusses more elaborate methods for referring to arguments and options within macros. A full discussion, with extensive examples, is beyond the scope of this book.

Options. If a macro is invoked with an option, the value of the option is available within the macro as the value of the string named **opt_xxx**, where *xxx* is the name of the option itself. If an option is set but has no value, then the corresponding string is set within the macro but has no value. For example, if macro *mac* were invoked as:

```
. mac {width=100} {long}
```

then the value of the {**width=**} option would be available as $**opt_width**, and the string $**opt_long** would be set but have no value (that is, it would be *empty*).

Testing string values. A string can be tested to see if it is set (whether or not it has a value) using the $? string expansion: the form $?*name* expands to the text '1' if the named string is set and expands to the text '0' otherwise. A string can be tested to see if it is set *and* has a value using the $& string expansion: the form $&*name* expands to the text '1' if the named string is set and has a value and expands to the text '0' otherwise.[13]

IF statements. The results of these tests can be used to cause conditional execution of BLSS commands. For example, a macro might contain:

```
IF $?opt_long THEN show x y z
```

---

13. The $? and $& expansions also work for positional arguments. For example, $?4 expands to '1' if there is a 4th positional argument and to '0' otherwise. In addition, $#*argv* expands to the number of arguments with which a macro was called—this, as well as $?1, is used to test whether a macro was called with or without arguments.

The general form of the **IF** statement is:

      IF *text-expression* **THEN** *blss-command* [**ELSE** *blss-command*]

or:

      IF *text-expression* **THEN BEGIN**
         blss-command
         blss-command
         . . .
    END [**ELSE BEGIN**
         blss-command
         blss-command
         . . .
    END]

where *text-expression* is a text expression (*not* a BLSS dataset), *blss-command* is any single BLSS command, and the **ELSE** portion of the statement is, of course, optional. The text expression is most often a result of string expansion (such as $?*name* or $&*name*). It is considered to be true if the resulting text is a non-zero number and false otherwise. The sense of the test can be reversed with the unary negation operator, '!'. For example:

```
IF !$&zstring THEN echo zstring has no value
```

In order to test a condition involving datasets, use the **$True[ ]** string function. It has the form:

      $True[*let-expression***]**

and returns the logical value of *let-expression* (which, just as with **$Ivalue[ ]**, can be any let-expression as defined in Chapter 8) as text ('0' or '1'). For example:

```
IF $True[abs(x) < .005] THEN echo x is too small
```

Tag arguments. The remainder of this section assumes complete familiarity with input and output tags, as explained in Section 11.3.1. Note that the user need not explicitly specify a tag. The general rule for command argument tags also holds for macro argument tags: Untagged input (or output) macro arguments are assigned to unused tags in the order given in the command line until there are no remaining input (or output) arguments.

The name of the input (or output) dataset to a macro corresponding to the tag *tag*@ can be obtained via the string function **$In[***tag***]** (or **$Out[***tag***]**). The resulting text can be used directly in any BLSS command. For example, the following command in a macro sets the string *x* to contain the name of the *x*@ input argument:

```
set x $In[x]
```

This is necessary if the macro refers to the *x*@ input argument more than once because (as a consequence of the general rule) when no *x*@ argument is explicitly tagged as such by the user, each call to **$In[**x**]** returns the next untagged input argument. The same is true for **$Out[ ]** and untagged output arguments. Note that (as another consequence of the general rule) the order in which **$In[ ]** (or **$Out[ ]**) is called for various tags establishes the default order of the tags in the sense of Section 11.3.1.

If **$In[ ]** or **$Out[ ]** is called for a given tag and no command line argument corresponds to that tag, BLSS normally terminates the macro with an explanatory error message. However, **$In[ ]** and **$Out[ ]** take an optional secondary argument which can change this default action. Here are the possible secondary arguments and the actions they cause when there is no appropriate command line argument to the macro:

-**ERR**   (The default.) Terminate the macro with an explanatory error message.

-**RET**   Return an empty string (a string with no value). No error is generated.

-**TAG**   (**$Out[ ]** only.) Return the tag name itself.

name   (**$Out[ ]** only.) Return the string *name.tag*, where *tag* is the *tag* argument (primary argument) of the string function. *Name* can be any legitimate name (including the result of a string expansion). The most useful name is that of the primary input dataset to the macro.

Note that the minus sign is needed with -**ERR**, -**RET**, and -**TAG**. It serves to distinguish the special actions from any possible *name*. The secondary argument must be separated from the first by at least one space or tab character. For example, to set the string *y* to contain the name of the *y@* output argument if it is given, and to the empty string otherwise:

```
set y $Out[y -RET]
```

If the primary input dataset to a macro is called $*x*, then to set the string *y* to contain the name of the *y@* output argument if it is given, and to the name $*x.y* otherwise:

```
set y $Out[y $x]
```

The exit and abort commands. The **exit** command can be used anywhere within a macro to make it exit before its end. It can print an explanatory message if one is supplied:

exit [*explanatory-message*]

The **abort** command causes a BLSS macro to exit with the message 'Abort' followed by an explanatory message if one is supplied. This is useful for exiting on error conditions.

For examples of BLSS macros, you can look at several BLSS commands. Simple examples are **binomial**, **poisson**, **rbinorm**, and **rmultinorm**; more complex examples are **cluster**, **editdata**, and **transfer**; interactive examples are the demonstration modules **demo.let1**, **demo.stat2**, etc. discussed in *demo(BRM)*. You can copy any of these and adapt them to your own purposes. They are in the BLSS system bin area: On most systems, this is the directory *~blss/bin*; from within BLSS, it can always be addressed as $*BLSS_SYS/bin*. For example, the command file for **poisson** is *~blss/bin/poisson* or $*BLSS_SYS/bin/poisson*.

At the time of this writing, more detailed information on the BLSS macro facility is in preparation.

CHAPTER 11

General Forms and Conventions

Part One of this book discussed many specific BLSS commands and showed how to use them with many detailed examples. However, there are just too many BLSS commands, with too many options, to cover them all in this expository fashion.

Part Two of this book is the *BLSS Reference Manual.* It contains complete, detailed write-ups (or *entries*) for every BLSS command. To save space, and for convenience in reference, BLSS manual entries follow a common form and observe certain conventions.

This chapter explains the general forms and conventions used in the manual entries and throughout BLSS. Once you have learned them, it will be easy for you to read the manual entries and learn how to use new BLSS commands.

11.1 Typographical Conventions Revisited

We repeat here the typographical conventions which are used throughout this book and all BLSS documentation.

The notation *xxx(BRM)* refers to the *BLSS Reference Manual* entry whose name is *xxx*; the notation *xxx(BDL)* refers to the *BLSS Dataset Library* entry whose name is *xxx*.

Bold type indicates literal text: words which must be typed exactly as they appear in the text. Thus command names, option names, and tag names (introduced in Section 11.3) are all shown in bold.

Italic type indicates prototype text: text for which you may substitute your own names or words. Thus dataset names, text file names, and other UNIX file names appear in italics. We also use italics to denote places where you may substitute specific text of your own (for example, specific command names or option names).

Examples use `slanted monospace` font for what you type and `monospace` font for what the computer types. Within examples, comments are shown in *italics* to the right of '#' signs (comment characters). You need not type comments. If you do, BLSS ignores them.

11.2   BLSS Manual Entries

Part Two of this book, the *BLSS Reference Manual*, consists of a large number of entries arranged in alphabetic order—rather like a small encyclopedia. Most entries discuss specific BLSS commands, or closely related groups of commands. The remaining entries discuss specific topics—such as *strings(BRM)*, which lists all strings, including environment strings, that have special meanings to BLSS.

All manual entries are available on-line within BLSS via the **help** and **man** commands. Section 4.11 discusses these two commands.

Manual entries for BLSS commands share a common format and common section headings, as follows.

NAME. The command name and a one-line description of what it does. (The word 'NAME' itself appears only in the on-line manual entries, not the printed entries.)

SYNOPSIS. A one-line summary of how to use the command.

DESCRIPTION. What the command does and how to use it.

INPUTS. Lists and describes the input dataset(s) in their default order.

OPTIONS. Lists each of the command's options, what type of values they take (if any), and their effect. Options which apply to all BLSS commands (such as {**lpr**} and {**more**}, introduced in Chapter 4) are listed in the manual entry *options(BRM)* and are not repeated in the OPTIONS section for individual commands.

OUTPUTS. Lists and describes the output dataset(s) in their default order.

EXAMPLES. Typical uses of the command.

LIMITATIONS. Limitations of the command, if any.

REFERENCES. References to related statistical literature.

SEE ALSO. References to related BLSS commands, manual entries, and software documentation.

The SYNOPSIS, INPUTS, OPTIONS, and OUTPUTS sections use additional conventions which arise from the general form of a BLSS command. These conventions are explained in the next section.

11.3   The General Form of Regular BLSS Commands

Regular BLSS commands take the following general form:[1]

   command-name [*inputs*] {*options*} [> *outputs*] [>> *output-textfile*]

In this general form, anything in square brackets [] or curly braces {} is optional. The square brackets are not actually typed when specifying the inputs, outputs, or output text file. The curly braces *are* typed when specifying options.

The SYNOPSIS section of a manual entry shows the general form for that particular command and summarizes its inputs, options, and outputs. For example, the SYNOPSIS for *sort(BRM)* is:

        sort x [i] {**r;s=;x=;shind**} [> *sort o*]

This tells us that the **sort** command has two possible input datasets which, for the sake of discussion in the manual entry, are called x and i. The first dataset, x, is required but the second, i, is optional. The command has four options—{**r**}, {**s=**}, {**x=**}, and {**shind**}—and two optional output datasets.

If a command has more inputs, options, or outputs than can be listed in a one-line synopsis, only those most commonly used are shown in the SYNOPSIS. For example, the SYNOPSIS for *regress(BRM)* is:

   regress x [y w] {**x=;y=;noint;rstat;anova;**...} [> *fit res b t p rstat anova* ...]

The ellipses '...' indicate that additional options and outputs were omitted to save space. A complete list is in the body of the manual entry, of course, and can also be shown on-line by using the {**usage**} option as explained in Section 11.4.

Now let us consider the individual pieces of the general form.

Command-name is the name of the BLSS command.

Inputs are the BLSS datasets that supply the input data for the command to act upon. Most commands only use one input dataset. However, a few (such as **sort** and **regress**) allow more than one. When giving more than one input dataset to such a command, you must take care specifying which is which. For information on how to do so, see Section 11.3.1 below on *Input and Output Tags*.

---

1. **Let** commands, described in Chapter 8, and *internal* (or *irregular*) commands, described in Chapter 10, do not use this general form.

Utility commands are an exception. Most utility commands take any number of input arguments—whether BLSS datasets or text files—and process them all identically. As we have seen, for example, **show** simply shows all its arguments. This is indicated by the ellipses '...' in the inputs position of its SYNOPSIS:

> **show** *files* ... {**cols**=;**ff**;**width**=;**shape**; **format**=;**f**=;**f**;**g**=;**g**;**i**=;**i** ... }

The **addtext**, **edit**, **help**, **list**, **load**, **man**, **remove**, and **save** commands work similarly—each processes all its arguments identically. For more information and a complete list of utility commands, see Section 4.12.

Options modify the behavior of commands. They are always enclosed in curly braces. For example, we have seen the {**g**} option to the **show** command, and the {**lpr**}, {**more**}, and {**quiet**} options, which apply to all commands.

Some options take values that must be specified using an '=' sign, such as the {**x**=**1**} and {**y**=**2**} options to the **scat** command. Some options even take lists of values, such as the {**x**=**1,2,3**} option we saw with the multiple regression example. Most option values introduced in Part One of this book have been *integers* (that is, whole numbers). Certain options take on other sorts of values, such as *floating point numbers*, or *floats* for short (that is, numbers with decimal points and fractional parts)—examples include the option {**level**=**.95**} used in Chapter 6 and the option {**m**=**3.5**} used in Chapter 7.

Here is a table which shows the different types of option values, the way in which they are denoted in manual entries, and an example of each.

Option Value Type	Denoted	Example
Integer	*INT*	3
List of integers	*INTLIST*	1,3,5
Float	*FLOAT*	.05
List of floats	*FLOATLIST*	-1,.4,-2.3,0.92
Character	*CHAR*	a
List of characters	*CHARLIST*	a,b,c,+,-,#
String of characters	*STRING*	wxyz

Dataset names may be used to supply numerical option values; however, expressions may not be used.[2] For example, to obtain a quantile of the t distribution with degrees of freedom given by element [2,2] of the dataset *anova*:

```
.  n = anova[2,2]
.  qt .95 {df=n} > t
```

When the option value is a list of numbers, dataset names cannot be used as part of the list value. Instead, the dataset must contain the entire option list. For example, the {**df**=} option to the **qf** command needs two values. To obtain a quantile of the F distribution with numerator degrees of freedom in

---

2. The restrictions described here may be relaxed in a future version of BLSS.

element [1,2] of *anova* and denominator degrees of freedom in element [2,2]:

```
. mn = anova[1,2],anova[2,2]
. qf .90 {df=mn} > f
```

Placement of options. When several options are specified, they may be separated by semicolons within the curly braces or they may be enclosed in separate sets of curly braces. Options may be given in any order, and they may be placed anywhere with respect to the input datasets. For example, the following commands are all equivalent:

```
. scat plover {x=1;y=2}
. scat {y=2;x=1} plover
. scat plover {x=1} {y=2}
. scat {x=1} {y=2} plover
. scat {y=2} plover {x=1}
```

We have seen examples of such variations throughout Part One of this book.

Outputs are the BLSS datasets in which the numerical results of commands or computations are to be placed. They are almost always optional. Outputs must be placed to the right of a single '>' symbol (which in this context is called the *output dataset separator*); the '>' itself must be placed to the right of all inputs and options.

If a command has only one possible output dataset, then you need only specify its name—as with the **read** command:

```
. read > mydata
```

If a command has several possible output datasets, refer to Section 11.3.1 on *Input and Output Tags*, below.

Output datasets overwrite any preexisting file (whether dataset or text file) of the same name.

Output-textfile. As explained in Sections 4.5 and 4.10.1, any regular command may be followed by an output text redirection symbol, followed in turn by an output text file name. The following redirection symbols cause the output display which is normally sent to the terminal to be sent instead to the text file whose name follows the symbol:

Symbol	Action
>! *textfile*	Write output text (excluding error messages) to *textfile*.
>> *textfile*	Append output text (excluding error messages) to *textfile*.
>& *textfile*	Write all output text (including error messages) to *textfile*.
>>& *textfile*	Append all output text (including error messages) to *textfile*.

To *write* to a file means to overwrite any preexisting contents it might have. To *append* to a file means to add new text to the end if it already exists. Both actions cause the file to be created if it does not already exist.

Output text redirection can also be used with most irregular commands. It cannot be used with **let** commands.

Be careful to distinguish between *output datasets*, which are BLSS datasets, and *output text files*, which contain text that would otherwise have been displayed on your terminal. Sections 1.4 and 4.10 discuss this distinction. Note that some commands produce output datasets but not output text; some produce output text but not output datasets; and some produce both.

11.3.1   Input and Output Tags

Certain commands allow for multiple input or (more commonly) multiple output datasets. To correctly indicate which dataset corresponds to which input or output,[3] you must do one of the following:

Either: Specify the inputs or outputs in the default order—that is, the order in which they are presented in the manual entry.

Or: Use the correct input tags or output tags, as given in the manual entry.

Use of input and output tags in a BLSS command always takes the form:

tag@ dataset-name

where *tag* is a specified tag for the input or output in question and *dataset-name* is the name of the dataset you wish to use as the corresponding input or output. On most terminal keyboards, the @ symbol is located on the '2' key.

The INPUTS and OUTPUTS sections of a command's manual entry list all possible input and output datasets for the command, according to their tags, in the default order. The SYNOPSIS section uses the tag name (without the @ symbol) to refer to the input and output datasets.

The rules for input tags and output tags are identical. We present examples of each.

Examples of input tags.   Look at the manual entry *sort(BRM)*. Under the INPUTS heading we see that **sort** has two possible inputs: the first has tag **x@** and contains the dataset to be sorted; the second has tag **i@** and contains an input sorting vector.[4] These tag names correspond to the prototype dataset names *x* and *i* given in the SYNOPSIS. The square brackets which enclose the *i* in the SYNOPSIS tell us that the **i@** input is optional.

Thus, to sort a dataset called *alpha* in ascending order, invoke the **sort** command just as in Section 3.4:

```
. sort alpha
```

---

3. Throughout Section 11.3.1, the term *input* means *input dataset* and the term *output* means *output dataset.*

4. The input sorting vector can be used to give a specific order in which to sort the dataset—the manual entry describes it in detail, of course. If no input sorting vector is given, the dataset is sorted in ascending order—or, if the {**r**} option (for 'reverse') is given, in descending order.

To sort *alpha* according to an input sorting vector called *beta*, each of the following commands is correct:

```
. sort alpha beta
. sort x@ alpha i@ beta
. sort i@ beta x@ alpha
```

However, the command:

```
. sort beta alpha
```

incorrectly treats *beta* as the **x@** input dataset and *alpha* as the **i@** input dataset.[5]

Examples of output tags. The OUTPUTS section of *sort(BRM)* lists two output datasets. The first has tag **sort@** and contains the sorted data; the second has tag **o@** and contains the output sorting vector. These tag names agree with the prototype dataset names *sort* and *o* in the SYNOPSIS. As with most BLSS commands, the **sort** command's output datasets are optional—the square brackets which enclose them in the SYNOPSIS tell us so.

If only one untagged output is specified, then only the first output dataset is created. For example, the command:

```
. sort alpha > gamma
```

saves the **sort@** output into the dataset called *gamma*. If two untagged output datasets are specified, then the first one specified gets the first output and the second one specified gets the second output. For example, the command:

```
. sort alpha > gamma delta
```

puts the **sort@** output into the dataset called *gamma*, and the **o@** output into the dataset called *delta*.

The next two commands both produce the same effect as the last command—but by using tags:

```
. sort alpha > sort@ gamma o@ delta
. sort alpha > o@ delta sort@ gamma
```

If you want the output sorting vector only, you must say so using the tag **o@**:

```
. sort alpha > o@ delta
```

Output tags are useful when you want to save only a few outputs from a command which has many. For example, the **stat** command has seven possible output datasets—these correspond to various summary statistics. The

---

5. If you want to try out the examples in this section, you can **load** the BLSS dataset *finalexam*, **rename** it as *alpha*, and create an input sorting vector *beta* which contains the numbers 62 through 1:

```
. load finalexam
. rename finalexam alpha
. beta = 62:1
```

covariance matrix is the fourth output and has tag **cov@**. To get it, you need not remember its position (four), but only its tag name. For example:

```
. stat plover > cov@ plov.cov
```

A general rule. When you specify output datasets without tags, you may omit any number of final output datasets if you don't want them. Untagged output datasets are assigned to unused tags in the order given in the command line until there are no more output datasets. For example, these four commands all produce the same result as the example shown above—the sorted dataset (**sort@**) goes into *gamma* and the output sorting vector (**o@**) goes into *delta*:

```
. sort alpha > sort@ gamma delta
. sort alpha > delta sort@ gamma
. sort alpha > gamma o@ delta
. sort alpha > o@ delta gamma
```

The same rule is true of input datasets; however, most commands require that at least one input dataset be given.

11.3.2 Words and Comments

We repeat here the definition of the term *word* in the context of BLSS. To the BLSS shell, a *word* is any string of characters which is not separated by white space (spaces or tabs), or which is wholly enclosed in a pair of matching parentheses ( ), square brackets [ ], curly braces { }, or double-quotes "".

If the first character in any word is a '#', the entire remainder of that line is a *comment* and is ignored by BLSS. We have seen comments using this syntax in examples throughout this book. The '#*', '#^', and '#,' symbols are exceptions: they denote matrix operations and do not introduce comments.

Section 10.8.1 discusses the use of comments in command files.

11.4 The {usage} Option

It can be helpful to see, on-line, a list of all possible inputs, options, and outputs for a command. The {**usage**} option, which is recognized by every regular BLSS command,[6] shows the complete list—and causes the command to do nothing else. For example, with the **sort** command the {**usage**} option shows:

```
. sort {usage}

sort:
Input tags:     x@  i@
Options:        r   s=   shind   x=
Output tags:    sort@  o@
```

Use *alpha* and *beta* for the examples.

6. The {**usage**} option is not recognized by **let** or the irregular commands. Not all macros recognize {**usage**}; those which do display a message similar to a command synopsis.

The input and output tags are listed in their default order—the same order as in the manual entry.

Notice that the {**usage**} option does not list options recognized by all BLSS commands such as {**lpr**}, {**more**}, or {**usage**} itself. As noted in Section 11.2, neither does the OPTIONS manual entry section. All such options are listed in *options(BRM)*.

By now, you have probably encountered error messages which result from giving a BLSS command with a bad option. For example, the **sort** command has no {**y=**} option. If you use it, you get an error message:

```
. sort alpha {y=1}

ERROR:  Bad usage of "sort".
Invalid option:        y=

Valid input tags:      x@  i@
Valid options:         r   s=   shind   x=
Valid output tags:     sort@  o@
```

If you use an invalid input or output tag, you get a similar error message. It tells you which input tag, option, or output tag is bad, and it shows a complete list of the good ones in a display similar to the {**usage**} option.

11.5 Conclusion to Part One

This chapter concludes Part One of this book. The following entries in Part Two contain additional general information.

Intro(BRM) is a concise technical introduction and overview of BLSS. It summarizes much of the general information in this and other chapters, but at a higher level. *Blssf(BRM)* contains information about the BLSS *frontend*—which is responsible for starting up the BLSS system each time it is invoked—and explains how BLSS utilizes UNIX directories as BLSS 'areas'.

An alphabetical list of BLSS manual entries is in *index(BRM)*. A list of manual entries arranged by topic is in *topics(BRM)*. These two manual entries, along with *intro(BRM)*, are located at the beginning of Part Two.

Part Three of this book is the *BLSS Data Library*. It describes datasets in the BLSS system data library, which are used in examples throughout this book. Its introduction, *data(BDL)*, contains a list of the datasets in the library. Individual datasets are described in manual entries with names of the form *data.xxx(BDL)*, where *xxx* is the name of the dataset. Such manual entries can be seen on-line by typing 'help data.*xxx*'.

PART TWO

BLISS Reference Manual

Release 4.0

Introduction
to the Manual

Part Two of this book is the *BLSS Reference Manual*, or *BRM*. It consists of about one hundred short entries. Most describe specific commands or groups of closely related commands; the remainder discuss specific topics. Individual entries in this manual are referred to by the form *entryname(BRM)*.

All but the first five entries are arranged in alphabetical order. The first five entries are:

blss   General description and overview of BLSS.

intro   Concise introduction to BLSS—explains conventions used throughout BLSS documentation. This is the most technical of all BLSS manual entries.

index   Main index of BLSS manual entries and commands—lists, in order, the contents of the *BLSS Reference Manual*.

index2   Supplementary BLSS command index—lists those commands which do not have their own manual entries and, instead, are documented as subtopics in other manual entries.

topics   List of BLSS manual entries, arranged by topic.

New users of BLSS should begin with the *BLSS User's Guide*, or *BUG*—Part One of this book.

Note. This reference manual applies to BLSS Release 4.0. Your site may have a newer or older release of BLSS. In case of discrepancies, consult the on-line documentation via the **help** command. The **news** command summarizes changes between different releases of BLSS.

---

blss — Berkeley Interactive Statistical System

SYNOPSIS

To enter BLSS from UNIX, type 'blss'. To exit BLSS, type 'exit'.

DESCRIPTION

BLSS is a highly interactive statistics system which runs on UNIX. It provides: the common tools for data analysis (including descriptive statistics, plots, multiple regression, cross-tabulation, confidence intervals, hypothesis tests, probability functions, EDA routines, etc.); a number of advanced statistical procedures (such as spectral analysis and estimation of optimal transformations); and good facilities for algebraic and matrix operations.

BLSS is easy to learn and use, yet is fast and efficient—hence it is ideal for instructional purposes. It has many convenience features, a macro facility which allows existing commands to be combined together into new commands, and subroutine support for user-written compiled programs—hence it is also suitable for certain advanced applications.

DOCUMENTATION

Hardcopy

BLSS—The Berkeley Interactive Statistical System (1988). Part One: Introduction and User's Guide. Part Two: BLSS Reference Manual (descriptions of individual commands and topics, arranged alphabetically). Part Three: BLSS Data Library. Individual entries in Parts Two and Three are referred to by the forms *entry(BRM)* and *entry(BDL)*, respectively.

Installing and Running BLSS (1988). Administrator's guide to BLSS. How to install and run BLSS at a new site.

On-Line

Individual manual entries may be seen on-line within BLSS via its **help** command. For example, type 'help help' to see the manual entry *help(BRM)* on the **help** command itself.

Type 'help' in BLSS for a quick introduction to the on-line help system. Type 'help topics' for a list of manual entries organized by topic. Type 'help index' for an alphabetical list. Type 'news' for news of recent changes to BLSS.

Movies

Several self-explanatory modules which demonstrate both BLSS itself and elementary statistical methods are available. For more information, type 'help demo' in BLSS.

CONTROL CHARACTERS

In BLSS the '@' and '#' characters have important meanings which conflict with the line-kill and character erase functions they are often assigned. If your regular UNIX line-kill character is '@' then when you enter BLSS it is reset to control-U; if your regular erase character is '#' it is reset to control-H.

When you leave BLSS, your original line-kill and erase characters are restored. BLSS informs you of these changes, but it can be confusing at first.

SEE ALSO

index(BRM) and *topics(BRM)*, for the lists of manual entries referred to above.
intro(BRM) for a concise introduction to the use of BLSS.
blssf(BRM) for a description of the BLSS frontend program.

CONSULTING

Type 'help consult' to find out about consulting on BLSS at your site (if there is any). A certain amount of consulting is available directly from the BLSS Project itself — send electronic mail to *blss@bach.berkeley.edu* (INTERNET) or *blss@ucbbbach.bitnet* (BITNET) or *ucbvax!bach!blss* (UUCP).

Comments and bug reports on BLSS are welcome. To the extent feasible, attempts are made to correct known problems with the next distribution of BLSS.

---

intro — Concise technical introduction to BLSS

SYNOPSIS

> *commandname* [*inputs*] {*options*} [> *outputs*]   # *regular BLSS commands*
> [let] *dataset* = *let-expression*   # *let-commands*
> *internal-command*   # *internal (irregular) commands*

DESCRIPTION

This manual entry is a concise introduction to BLSS. It provides a technical overview and contains many definitions. See the *BLSS User's Guide* (*BUG*) for extensive explanations and examples. See the manual entry *blss*(*BRM*) for a general description of BLSS.

BLSS Datasets

BLSS stores data in three-way arrays of single precision floating point numbers known as *BLSS datasets*. The three dimensions are respectively known as *sheets*, *rows*, and *columns*. For example, a 3-by-4-by-5 array is said to have 3 sheets, 4 rows, and 5 columns.

Scalars, *vectors*, and *matrices* are special cases of BLSS datasets which suffice in most applications: the sheet index (and perhaps other indices) are not referenced. As a further convention, many statistics commands assume that columns of matrices represent statistical variables and that rows of matrices represent *cases* (observations).

BLSS dataset names may consist of letters, digits, and the period '.' and underscore '_' characters; may be up to 14 characters long; and should begin with a letter. (These rules are not strictly enforced in all cases, but it is a good idea to stick to them.) Upper- and lowercase letters are distinct.

BLSS datasets are stored as UNIX disk files, and are sometimes referred to as 'files' if the context warrants. The data are stored in binary form for efficient handling.

Text Files

BLSS can also manipulate text files: these may contain, for example, documentation, macros, or text output from commands. BLSS text files are no different from UNIX text files. (Note that BLSS datasets are *not* text files.)

The BLSS Shell

BLSS datasets and text files are manipulated via an interactive monitor known as the *BLSS shell* (so called because of its strong resemblance to UNIX shells). The BLSS shell recognizes three types of commands: *regular* commands, which follow a common syntax; *let-commands* and *let-expressions*, which include algebraic operations, array subscripts, etc.; and *internal* (or *irregular*) commands, each of which has its own special syntax.

Regular Commands

The general syntax of the regular commands is:

> *commandname* [*inputs*] {*options*} [> *outputs*]

In syntax statements such as this, and in SYNOPSIS sections at the beginning of manual entries, anything in square brackets [] or curly braces { } is optional. The square brackets are not actually typed, but the curly brackets *are* typed. Unless otherwise noted, *white space* (spaces or tab characters) is used to separate the different parts of a command.

Commandname is the name of the command. Commands are listed in *index(BRM)* (arranged alphabetically) and *topics(BRM)* (arranged by topic).

Inputs are the names of the datasets that supply the input data for the command to act upon. *Let-expressions* (see below) may be used in place of input dataset names.

Options are always enclosed in curly braces { }. Each option specified may be enclosed in its own set of curly braces, or a list of options separated by semi-colons ';' may be enclosed in one set of braces. The order of options with respect to each other and with respect to input datasets does not matter, but they must be given to the left of the '>' sign (if any).

Some options are simply indicators; their presence causes a specific action. Other options take user-specified values using the form {*opt=value*}. There are seven possible option value types:

Option Value Type	Denoted	Example
Character	CHAR	a
Character list	CHARLIST	a,b,c,+,−,#
Integer	INT	3
Integer list	INTLIST	1,3,5,3
Float	FLOAT	.05
Float list	FLOATLIST	−1,.4,1.3
String	STRING	wxyz

Datasets may be used to supply numerical option values (both single values and lists of values); however, let-expressions may not be used as option values. (This restriction may be relaxed in a future version of BLSS.)

An exception: ascending or descending sequences of numerical list-valued options may be indicated using single-colon notation. For example, the option value '5:3,1' is equivalent to '5,4,3,1'; the option value '1.5:3.5' is equivalent to '1.5,2.5,3.5'. Dataset names cannot be used as part of an option value list and cannot be combined with colon notation in option values. Instead, the dataset must contain the entire option value list.

White space in option values is normally ignored. Double-quote characters "" are necessary to protect any space characters in a *STRING* option value: for example, {opt="This is four words"}.

Outputs specify the names of the BLSS datasets in which the numerical results of the command are to be stored. They always appear to the right of a single '>' character, which in this context is known as the *output dataset separator*.

Note that *output datasets*, which contain numerical results, are distinct from the *output text* (or *output display*) which is normally shown on the terminal.

Tags

Regular commands can take several optional input and output datasets. To facilitate this, datasets in the command line may be preceded by *tags* to indicate which dataset corresponds to which input or output specification. Tags take the form:

tag@ datasetname

For example, the output specification:

command inputs > ... anova@ iris.a ...

says that the output corresponding to the tag **anova@** should be put in the BLSS dataset *iris.a.* Untagged datasets are assigned to unused tags in the order in which they appear in the command line; the unused tags are assigned in the order in which they are listed in the command's manual entry.

An Example

Here is a simple BLSS session. Prompts are shown, but not the output text displayed by BLSS.

```
% blss
. load iris
. regress {x=1:3; y=4} iris > fit resid rstat@ r
. scat fit resid
. exit
%
```

The UNIX command **blss** invokes BLSS. Its prompt is the '.' character. The BLSS command **exit** exits from BLSS.

The first BLSS command in the example loaded the dataset *iris* from the BLSS system data area into the work area. The **regress** command regressed the fourth column of *iris* on the first through third columns. The fitted values and residuals were saved in two column vectors (output datasets) called *fit* and *resid*; the output corresponding to the **rstat@** tag was saved in the output dataset *r*. Finally, the **scat** command plotted the regression residuals against the fitted values.

Let Commands and Expressions

Arithmetic, matrix operations, array subscripting, and a number of mathematical functions (e.g., **abs**, **sqrt**) are performed by the **let** command. **Let** may be accessed either *explicitly* in order to assign the value of an expression to a dataset, or *implicitly* in order to pass the value of an expression directly to a regular BLSS command. Here is an example of each:

```
. let root = sqrt((rutabaga+turnip)/2)   # explicit let-command
. scat 5*x[1] log(y[2])                   # two implicit let-expressions
```

The first command creates a new BLSS dataset, *root*. The second command makes a scatterplot of 5 times the first column of *x* against the natural log of the second column of *y*.

The terms *let-expression* and *let-function* refer, respectively, to any expression or function recognized by **let**. **Let** is discussed in *let(BRM)* and in *BUG* Chapter 8.

Internal (Irregular) Commands

Several *internal* commands (so called because they are executed internally by the BLSS shell) make life easier. Each has its own special syntax which differs from the regular command syntax; hence, they are also known as *irregular* commands. The most useful of these are **alias**, **echo**, **history**, **redo**, **run**, **set**, **source**, and **unix**.

Output Text Redirection

The text output from regular BLSS commands (and most irregular commands, but not **let** commands) which is normally displayed at the terminal may be redirected into a text file in a manner analogous to that of the UNIX shells. Four output text *redirection symbols*, '>!', '>>', '>&', and '>>&', have meanings identical to those of the C shell. See *outputs(BRM)* or *BUG* Sections 4.5 and 11.3 for more information.

Words and Comments

In the BLSS shell, a *word* is any string of characters which is not separated by white space, or which is wholly enclosed in a pair of matching (), [], { }, or "" characters. If the first character in any word is a '#', the entire remainder of that line is a *comment* and is ignored by the BLSS shell, as in the examples above. '#*', '#^', and '#,' are exceptions: they are matrix operators and do not introduce comments.

Areas

BLSS creates its own protected environment in which to function. Each time it is invoked, the user is placed in a (normally unique) temporary active *work area* which contains any BLSS datasets and text files created in the course of the session. Upon a normal exit from BLSS, the work area and its contents are automatically removed (this does not happen for abnormal exits—e.g., system crashes). Each user is also given several permanent areas including a *data area* and a *text area* into which valuable files may be saved. Files are loaded from a data or text area into the work area using the **load** command; files in the work area may be saved in the user's data or text areas using the **save** command. See *blssf(BRM)* for more information about the BLSS areas and environment. See *data(BDL)* for a list of datasets in the BLSS system data area.

UNIX Pathnames

UNIX pathnames which begin with '/', './', '../', or '~' may generally be used wherever file names are expected (for example, as command names, text file names, dataset names, or *STRING* option values) with the exception that UNIX pathnames may not be used in let-expressions. The '~' character invokes home directory expansion as in the C Shell: ~ is an abbreviation for the invoking user's home directory; and *~xxx* is an abbreviation for the home directory of user *xxx*.

Grammar and Macros

The BLSS shell provides a grammar which allows for conditional and iterated execution of BLSS commands and a macro facility which permits existing commands to be combined together into new regular commands. Brief descriptions of these are in *BUG* Section 10.8; longer descriptions are in preparation.

Manual Entries

Each BLSS command has a manual entry (closely related commands may share the same manual entry). Certain specific topics also have manual entries. Manual entries are available on-line within BLSS via the **help** command. On-line entries contain the date last modified and the BLSS release number. Entries for commands contain the following sections, not all of which always appear.

NAME. The name(s) of the command(s) and a one-line description. (The word 'NAME' itself appears only in the on-line entries, not the printed entries.)

SYNOPSIS. The syntax of the command. Items enclosed in square brackets [ ] are optional; the square brackets are not typed. The curly braces { } which enclose options *are* necessary and must be typed.

DESCRIPTION. What the command does and how to use it.

INPUTS. Lists and describes the input datasets in their default order.

OPTIONS. The option names, what type of values they take if any, and their effect. The preceding section on options lists the possible value types. See *options(BRM)* for a list of options common to all regular commands; these are not repeated in the individual manual entries. See *area(BRM)* for a list of options common to the *utility* commands (**load**, **save**, **help**, etc.).

OUTPUTS. Lists and describes the output datasets in their default order.

EXAMPLES. Typical uses of the command, or nonobvious but useful invocations.

LIMITATIONS. Limitations of the command, if any.

REFERENCES. Pointers to pertinent statistical literature.

SEE ALSO. Pointers to related BLSS documentation.

Typographical Conventions

The following typographical conventions are used throughout BLSS documentation. **Bold** type indicates literal text: text which must be typed just as it appears. *Italic* type indicates prototype text: text for which may be substituted one's own words or phrases. Examples use `slanted monospace` font for what the user types and `monospace` font for what the computer types. Within examples, comments are shown to the right of '#' signs (comment characters), in *italics*.

BLSS documentation is referred to in the following manner:

BUG	Refers to the *BLSS User's Guide* (Part One of the book).
entry(BRM)	Refers to *entry* in the *BLSS Reference Manual* (Part Two).
entry(BDL)	Refers to *entry* in the *BLSS Dataset Library* (Part Three).
entry(BLR)	Refers to *entry* in *BLSS Library Routines*.
entry(BML)	Refers to *entry* in the *BLSS Mathematical Library*.

SEE ALSO

BUG Chapters 1 and 11, which present the material in this manual entry in greater detail.

index(BRM), for an alphabetical list of entries in the *BLSS Reference Manual*.

topics(BRM), for a list of entries in the *BLSS Reference Manual* arranged by topic.

---

index — Main index of BLSS manual entries and commands

DESCRIPTION

This is a complete list of entries in the *BLSS Reference Manual*, in the order in which they appear. Most entries document specific commands or groups of closely related commands. The remainder discuss specific topics. These entries may be seen on-line by typing **help** *entryname*.

See *index2(BRM)* for a list of commands that do not have their own manual entries and, instead, are documented as subtopics in other manual entries.

Name	*Brief Description*
blss	Short general description of BLSS.
intro	Concise technical introduction to BLSS.
index	Main index of BLSS manual entries and commands.
index2	Supplementary BLSS command index.
topics	List of BLSS manual entries arranged by topic.
$\tilde{}$ ace	Optimal transformation$\tilde{}$s for multiple regression.
addtext	Add to text files or create new text files.
alias, unalias	Define (or remove) alias names for commands.
anovapr	Print an anova table.
area	Area options understood by BLSS utility commands.
arrop	Simple array operations.
autcor, autcov	Auto- and cross-correlation and covariance of time series.
autreg	Autoregressive model fitting for time series.
binomial	Binomial distribution probabilities and cdf.
biweight	Robust estimates of center and spread.
blssf	BLSS frontend invocation from UNIX.
box	Draw tickets from a box model and display sample statistics.
boxplot	Tukey box-and-whisker plots.
cdf	Continuous probability distribution functions and inverses.
chisq	Chi-square (x^2) tests of independence and goodness-of-fit.
chol	Cholesky decomposition of a matrix.
ci	Confidence intervals for means and differences of means.
cluster	Create and print a cluster analysis tree.
code	Symbolic display of data.
coin	Coin-tossing simulation.
compare	Comparative boxplots.
confid	Confidence interval illustration via simulation.
consult	How to contact BLSS consultants (on-line only).
count	Count rows (cases) which meet a specified condition.
demo	Introductory demonstration modules.
demod	Complex demodulation of a time series.
demoiv	Probability histogram of sum of 2^n draws from a box model.
dft	Discrete Fourier transform.
diff	Difference a time series.
echo, prompt	Echo arguments.

Name	Brief Description
edit	Edit text files.
editdata	Edit BLSS datasets.
eigen	Eigenvalue-eigenvector decomposition of a matrix.
filter	Filter a time series.
freq	Frequency distributions.
fromunix	How to execute BLSS commands directly from UNIX.
help, man	Access on-line documentation.
history, redo	Show command history; repeat or edit a command.
ht	Hypothesis tests (z-tests, t-tests) for means and differences of means.
imprsk	Compute impulse response functions from transfer functions.
indic	Construct indicator variables.
lag	Lag a time series.
let	Mathematical functions and matrix operations.
list	List contents of work, data, text, ... areas.
load	Retrieve saved datasets and text files.
mdrace	Optimal transformations for dimensionality reduction.
medmad	Median and median absolute deviation.
medpolish	Median polish of two-way tables.
news	Current news about BLSS (on-line only).
oneway	Least squares and robust one-way anova.
options	Options understood by all regular BLSS commands.
output	How to redirect text output to a text file.
pgrm1	Periodogram of a univariate time series.
pgrmk	Periodograms and cross-periodograms of k-variate time series.
poisson	Poisson distribution probabilities and cdf.
polar, rect	Transform between polar and rectangular coordinates.
print	How to send files to a printer.
qdata	Quantiles of a dataset.
qr, rank	QR decomposition and rank of a matrix.
random	Random numbers from continuous probability distributions.
rbinorm	Generate bivariate Normal random numbers.
read	Enter data into a BLSS dataset.
redim	Redimension an array.
regress	Least squares multiple regression and weighted regression.
remove	Remove datasets and text files.
rename	Rename a dataset or text file.
rmultinorm	Generate multivariate Normal random numbers.
robust	Robust weighted least squares regression.
sample	Random sampling with and without replacement.
save	Save datasets or text files for future use.
scat	Scatterplots.
scatter	Interactive video scatterplots.
seed	Reset the seed for the random number generator.
select	Select rows (cases) which meet a specified condition.

Name	Brief Description
set, unset	Define (or remove) strings and environment strings.
show	Display or print datasets and text files.
smooth	Tukey robust smoothers (running medians, etc.).
solve, inv, det	Solve systems of linear equations; inverse and determinant of a matrix.
sort	Sort rows of a dataset.
source, run	Execute a command file or macro.
spec1	Spectral analysis of univariate time series.
speck	Spectral and cross-spectral analysis of k-variate time series.
stat	Summary statistics: mean, SD, correlation, covariance, quartiles, etc.
stemleaf	Tukey stem-and-leaf diagrams.
strings	Strings and environment strings with special meanings.
submat	Choose a submatrix from a matrix.
svd	Singular value decomposition of a matrix.
topics	List of manual entries arranged by topic.
transfer	Compute the transfer function of a filter.
trnfrk	Estimate pairwise transfer functions from k-variate spectra.
tsop	Simple time series operations: remove means, trends, etc.
twoway	Least squares and robust two-way anova.
unix	How to execute UNIX commands from within BLSS.
vecop	Functions (sum, max, ...) applied to vectors within datasets.
xtab	Cross-tabulation.
xvalid	Cross-validated estimate of regression R^2 and RSS.

SEE ALSO

help(*BRM*), *index2*(*BRM*), *topics*(*BRM*).

---

index2 — Supplementary BLSS command index

DESCRIPTION

This is a complete alphabetical list of BLSS commands and let-functions that do not have their own manual entry and are, instead, documented as subtopics in other manual entries.

Manual entries for these commands can be seen on-line by typing **help** *name*, where *name* is either the command name itself or the name of the manual entry in which it is documented.

Name	Entry	Brief Description
abs	let	Absolute value.
acos	let	Arc cosine.
amax	vecop	Absolute maximum of array elements.
asin	let	Arc sine.
asum	vecop	Absolute sum of array elements.

Name	*Entry*	*Brief Description*
atan	let	Arc tangent.
atanh	let	Arc hyperbolic tangent.
autcov	autcor	Autocovariance of a time series.
BesJ0, ...	let	Bessel functions.
ceil	let	Ceiling function.
colamax	vecop	Absolute maximum of column elements.
colasum	vecop	Absolute sum of column elements.
colindex	arrop	Create a column index array.
colmad	vecop	MAD of column elements.
colmax	vecop	Maximum of column elements.
colmean	vecop	Mean of column elements.
colmed	vecop	Median of column elements.
colmin	vecop	Minimum of column elements.
colnorm	vecop	Norm of column elements.
colprod	vecop	Product of column elements.
colsd	vecop	SD of column elements.
colss	vecop	Sum of squares of column elements.
colsum	vecop	Sum of column elements.
colvar	vecop	Variance of column elements.
const	arrop	Create a constant array.
cos	let	Cosine.
cosh	let	Hyperbolic cosine.
cumprod	arrop	Cumulative product of array elements.
cumsum	arrop	Cumulative sum of array elements.
dbeta	cdf	Beta density.
dcauchy	cdf	Cauchy density.
dchisq	cdf	χ^2 (chi-square) density.
det	solve	Determinant of a square matrix.
dexp	cdf	Exponential density.
df	cdf	Fisher's F density.
dgamma	cdf	Gamma density.
dgau	cdf	Gaussian (Normal) density.
diag	arrop	Construct diagonal matrix or extract main diagonal.
dims	arrop	All three dimensions of an array.
dlogis	cdf	Logistic density.
dt	cdf	Student's t density.
duni	cdf	Uniform density.
exp	let	Exponential function.
fac	let	Factorial function.
floor	let	Floor function.
gam	let	Gamma function, $\Gamma(x)$.
idn	arrop	Create an identity matrix.
int	let	Integer part function.
inv	solve	Inverse of a matrix.
iota	arrop	Create an element index array.
lfac	let	Log factorial function.
lgam	let	Log gamma function.
ln, log	let	Natural logarithm.

Name	*Entry*	*Brief Description*
log10	let	Base 10 logarithm.
ltri	arrop	Create an upper triangular matrix.
mad	vecop	MAD (median absolute deviation) of array elements.
man	help	Show manual entries on-line.
max	vecop	Maximum of array elements.
mean	vecop	Mean of array elements.
med	vecop	Median of array elements.
min	vecop	Minimum of array elements.
ncols	arrop	Number of columns in array.
neg	let	Negative part function.
norm	vecop	Norm of array elements.
nrows	arrop	Number of rows in array.
nsheets	arrop	Number of sheets in array.
pbeta	cdf	Beta cdf.
pcauchy	cdf	Cauchy cdf.
pchisq	cdf	χ^2 (chi-square) cdf.
pexp	cdf	Exponential cdf.
pf	cdf	Fisher's F cdf.
pgamma	cdf	Gamma cdf.
pgau	cdf	Gaussian (Normal) cdf.
plogis	cdf	Logistic cdf.
pnt	cdf	Noncentral t cdf.
pos	let	Positive part function.
prod	vecop	Product of array elements.
prompt	echo	Display a message without a newline.
pt	cdf	Student's t cdf.
puni	cdf	Uniform cdf.
qbeta	cdf	Inverse beta cdf.
qcauchy	cdf	Inverse Cauchy cdf.
qchisq	cdf	Inverse χ^2 (chi-square) cdf.
qexp	cdf	Inverse exponential cdf.
qf	cdf	Inverse Fisher's F cdf.
qgamma	cdf	Inverse gamma cdf.
qgau	cdf	Inverse Gaussian (Normal) cdf.
qlogis	cdf	Inverse logistic cdf.
qt	cdf	Inverse Student's t cdf.
quni	cdf	Inverse uniform cdf.
rank	qr	Rank of a matrix.
rbeta	random	Beta random numbers.
rcauchy	random	Cauchy random numbers.
rchisq	random	χ^2 (chi-square) random numbers.
rect	polar	Transform from polar to rectangular coordinates.
redo	history	Redo a command.
rexp	random	Exponential random numbers.
rf	random	Fisher's F random numbers.
rgamma	random	Gamma random numbers.
rgau	random	Gaussian (Normal) random numbers.
rlogis	random	Logistic random numbers.

Name	Entry	Brief Description
rowamax	vecop	Absolute maximum of row elements.
rowasum	vecop	Absolute sum of row elements.
rowindex	arrop	Create a row index array.
rowmad	vecop	MAD of row elements.
rowmax	vecop	Maximum of row elements.
rowmean	vecop	Mean of row elements.
rowmed	vecop	Median of row elements.
rowmin	vecop	Minimum of row elements.
rownorm	vecop	Norm of row elements.
rowprod	vecop	Product of row elements.
rowsd	vecop	SD of row elements.
rowss	vecop	Sum of squares of row elements.
rowsum	vecop	Sum of row elements.
rowvar	vecop	Variance of row elements.
rt	random	Student's t random numbers.
run	source	Run a command file as a macro.
runi	random	Uniform random numbers.
sd	vecop	SD (standard deviation) of array elements.
sgn	let	Signum function.
sheetamax	vecop	Absolute maximum of sheet vector elements.
sheetasum	vecop	Absolute sum of sheet vector elements.
sheetindex	arrop	Create a sheet index array.
sheetmad	vecop	MAD of sheet vector elements.
sheetmax	vecop	Maximum of sheet vector elements.
sheetmean	vecop	Mean of sheet vector elements.
sheetmed	vecop	Median of sheet vector elements.
sheetmin	vecop	Minimum of sheet vector elements.
sheetnorm	vecop	Norm of sheet vector elements.
sheetprod	vecop	Product of sheet vector elements.
sheetsd	vecop	SD of sheet vector elements.
sheetss	vecop	Sum of squares of sheet vector elements.
sheetsum	vecop	Sum of sheet vector elements.
sheetvar	vecop	Variance of sheet vector elements.
sin	let	Sine.
sinh	let	Hyperbolic sine.
sqrt	let	Square root.
ss	vecop	Sum of squares of array elements.
sum	vecop	Sum of array elements.
tan	let	Tangent.
tanh	let	Hyperbolic tangent.
trace	let	Trace of a matrix.
unalias	alias	Remove an alias.
unset	set	Remove a string or environment string.
utri	arrop	Create a lower triangular matrix.
var	vecop	Variance of array elements.

SEE ALSO
index(BRM).

topics — List of BLSS manual entries arranged by topic

DESCRIPTION

This is a list of BLSS manual entries, organized by general topic. Type **help** *entryname* to see the manual entries on-line.

Note that some commands do not have their own manual entries and are instead documented as subtopics in other manual entries. See *index2*(*BRM*) for a list of such commands.

General Topics

area	Area options understood by BLSS utility commands.
blss	Short general description of BLSS.
consult	How to contact BLSS consultants (on-line only).
data	Description of BLSS system data library.
demo	Introductory demonstration modules.
index	Main index of BLSS manual entries and commands.
index2	Supplementary BLSS command index.
intro	Concise technical introduction to BLSS.
news	Current news about BLSS (on-line only).
options	Options understood by all regular BLSS commands.
output	How to redirect text output to a text file.
print	How to send files to a printer.
topics	This list of manual entries arranged by topic.

Utility Commands

addtext	Add to text files or create new text files.
edit	Edit text files.
editdata	Edit BLSS datasets.
help, man	Access on-line documentation.
list	List contents of work, data, text, . . . areas.
load	Retrieve previously saved datasets and text files.
read	Enter data into a BLSS dataset.
remove	Remove datasets and text files.
rename	Rename a dataset or text file.
save	Save datasets or text files for future use.
show	Display or print datasets and text files.

General Data Analysis

chisq	Chi-square (χ^2) tests of independence and goodness-of-fit.
ci	Confidence intervals for means and differences of means.
cluster	Create and print a cluster analysis tree.
code	Symbolic display of data.
freq	Frequency distributions.
ht	Hypothesis tests (z-tests and t-tests) for means and differences of means.
qdata	Quantiles of a dataset.

| stat | Summary statistics: mean, SD, correlation, covariance, quartiles, etc. |
| xtab | Cross-tabulation. |

Pictures and Plots

boxplot	Tukey box-and-whisker plots.
compare	Comparative boxplots.
scat	Scatterplots.
scatter	Interactive video scatterplots.
stemleaf	Tukey stem-and-leaf diagrams.

Matrix Manipulation and Mathematical Operations

arrop	Simple array operations.
chol	Cholesky decomposition of a matrix.
count	Count rows (cases) which meet a specified condition.
eigen	Eigenvalue-eigenvector decomposition of a matrix.
indic	Construct indicator variables.
let	Mathematical functions (sin, log, ...) and matrix operations (subscripts, transpose, ...).
polar, rect	Transform between polar and rectangular coordinates.
qr, rank	QR decomposition and rank of a matrix.
redim	Redimension an array.
select	Select rows (cases) which meet a specified condition.
solve, inv, det	Solve systems of linear equations; inverse and determinant of a matrix.
sort	Sort rows of a dataset.
submat	Choose a submatrix from a matrix.
svd	Singular value decomposition of a matrix.
vecop	Functions (sum, max, ...) applied to vectors within datasets.

Probability and Random Numbers

binomial	Binomial distribution probabilities and cdf.
cdf	Continuous probability distribution functions and inverses.
poisson	Poisson distribution probabilities and cdf.
random	Random numbers from continuous probability distributions.
rbinorm	Generate bivariate Normal random numbers.
rmultinorm	Generate multivariate Normal random numbers.
sample	Random sampling with and without replacement.
seed	Reset the seed for the random number generator.

Least Squares and Robust Fitting

ace	Optimal transformations for multiple regression.
anovapr	Print an anova table.
biweight	Robust estimates of center and spread.
mdrace	Optimal transformations for dimensionality reduction.
medmad	Median and median absolute deviation.
medpolish	Median polish of two-way tables.
oneway	Least squares and robust one-way anova.

regress	Least squares multiple regression and weighted regression.
robust	Robust weighted least squares regression.
twoway	Least squares and robust two-way anova.
xvalid	Cross-validated estimate of regression R^2 and RSS.

Time Series

autcor, autcov	Auto- and cross-correlation and covariance of time series.
autreg	Autoregressive model fitting for time series.
demod	Complex demodulation of a time series.
dft	Discrete Fourier transform.
diff	Difference a time series.
filter	Filter a time series.
imprsk	Compute impulse response functions from transfer functions.
lag	Lag a time series.
pgrm1	Periodogram of univariate time series.
pgrmk	Periodograms and cross-periodograms of k-variate time series.
smooth	Tukey robust smoothers (running medians, etc.).
spec1	Spectral analysis of univariate time series.
speck	Spectral and cross-spectral analysis of k-variate time series.
transfer	Compute the transfer function of a filter.
trnfrk	Estimate pairwise transfer functions from k-variate spectra.
tsop	Simple time series operations: remove means, trends, etc.

Instructional Commands

box	Draw tickets from a box model and display sample statistics.
coin	Coin-tossing simulation.
confid	Confidence interval illustration via simulation.
demoiv	Probability histogram for sum of 2^n draws from a box model.

Convenience Features (Internal Commands)

alias, unalias	Define (or remove) alias names for commands.
echo, prompt	Echo arguments.
history, redo	Show command history; repeat or edit a command.
set, unset	Define (or remove) strings and environment strings.
source, run	Execute a command file or macro.
strings	Strings and environment strings with special meanings.

Command Interface to UNIX

blssf	BLSS frontend invocation from UNIX.
fromunix	How to execute BLSS commands directly from UNIX.
unix	How to execute UNIX commands from within BLSS.

$\sim$

---

ace — Optimal transformations for multiple regression

SYNOPSIS

ace x [y w px] {**x**=;**y**=;**type**=;**wnd**=; ... } [> tx ty py sd ...]

DESCRIPTION

Ace uses the *alternating conditional expectation* (ACE) technique of Breiman and Friedman (1982, 1985) to estimate the optimal univariate transformations which linearize the relationship between a single dependent variable Y and several independent variables X. If a vector of weights w is specified, **ace** uses weighted regression to find the transformations.

The estimated standard deviations (SDs) of the transformed X variables are displayed. Because the SD of the transformed Y variable is 1, the SD of a transformed X variable is a measure of how strongly it enters into the model for the given Y transformation.

Ace can predict the response, based on the optimal transformations, from a set of values px of the independent variables.

INPUTS

x@ n-by-k array (or n-by-$(k+1)$ if no **y@** dataset is specified) whose columns are the observations on the independent variables X. If no **y@** input is specified, by default it is the last column of the **x@** input dataset.

y@ n-vector of observations on the dependent variable Y.

w@   Optional n-vector of weights for the observations.

px@   Optional r-by-k array of values of the independent variables for which a prediction is desired.

Missing values are not allowed in the **x@**, **y@**, or **w@** inputs. Rows in **px@** which contain missing values are ignored.

OPTIONS

x=*INTLIST*

Columns of the **x@** array to use as the independent variables.

y=*INT*

Column of the **y@** array (or **x@** array, if **y@** is not specified) to use as the dependent variable.

In the following options, if the {**x**=} option is in effect then the i-th predictor variable is the column of x specified by the i-th element of the {**x**=} array.

type=*INTLIST*

$k+1$ integers with values as below. The first k elements are the value types for the k predictor variables. The $k+1$st element is the value type for the response variable.

Type Meaning
  0   Variable is not used.
  1   Variable transformation type is unrestricted (default).
  2   Variable assumes circular (periodic) values (see {**circ**=} below).
  3   Variable transformation must be monotonic.
  4   Variable transformation must be linear.
  5   Variable assumes categorical values.

For prediction, the transformation of the response variable must be monotone (3 or 4). If the {**x**=} option is in effect, no value should be 0.

The following options may be used in place of the {**type**=} option. A variable may not be specified more than once in any option. The *i*-th predictor variable is specified by its index *i*; the response variable is specified by 0. For prediction, the response variable must be listed in {**mon**=} or {**lin**=}.

circ=*INTLIST*
  Specifies variables which assume circular (periodic) values in [0, 1] with period 1.

mon=*INTLIST*
  Specifies variables which must be transformed monotonically.

lin=*INTLIST*
  Specifies variables which must be transformed linearly.

cat=*INTLIST*
  Specifies variables which assume categorical values.

The following options affect the *super smoother* algorithm used within the ACE routine.

alpha=*FLOAT*
  The *alpha* parameter of *super smoother*. If *alpha* < 0 or *alpha* > 10, it has no effect. The default is *alpha* = 0.

wnd=*FLOAT*
  Size of the smoother window. Must be in the range [0, 1]; reasonable values are in the range [.3, .5]. If *wnd* = 0 then *super smoother* uses a self-adjusting variable window size. The default is *wnd* = .4 if $n \le 40$ and 0 otherwise.

OUTPUTS
  tx@   The transformed X variables.

  ty@   The transformed Y variable.

  py@   The predicted responses corresponding to the values in *px*. The response is back-transformed; hence a prediction is possible only if the predicted response transformation is monotone (*type* = 3 or 4).

  sd@   The estimated SDs of the transformed X variables.

rsq@ R^2 for each iteration.

z@ The transformed residuals.

The **tx@** and **ty@** output datasets are always created. The **py@** output dataset is created whenever **px@** is specified. The **sd@** and **rsq@** output datasets are created whenever {**quiet**} is in effect. If an output dataset is created but no name is specified by the user, the dataset is named *x.tag*, where x is the name of the **x@** input dataset and *tag* is the output tag name.

EXAMPLE

In the following example, data are created according to the model

$$y_i = \exp[\sin(x_i) + e_i/2]$$

where the x_i are i.i.d. uniform on $[0, 2\pi]$ and the e_i are i.i.d. Normal (0,1). **Ace** is employed to discover the transformations needed to linearize the relationship between x and y. The plots of *tx* against x and *ty* against y display the transformations which must be applied to the x and y variables, respectively, to linearize the relation between the two. The plot of *ty* against *tx* displays the extent to which this relationship is linear.

```
.  runi {dims=200,1} > u    # (200,1) array of uniform (0,1) random number
.  rgau u > e               # (200,1) array of Normal (0,1) random numbers
.  x = 2*PI*u
.  y = exp(sin(x) + e/2)
.  ace x y > tx ty
.  scat x tx
.  scat y ty
.  scat tx ty
```

REFERENCES

Leo Breiman and Jerome H. Friedman (1982). *Estimating optimal transformations for multiple regression and correlation.* Technical Report No. 9, Department of Statistics, University of California, Berkeley.

Leo Breiman and Jerome H. Friedman (1985). *Estimating optimal transformations for multiple regression and correlation.* **JASA**, **80**, 580-598.

Jerome H. Friedman and Werner Stuetzle (1982). *Smoothing of scatterplots.* Technical Report ORION006, Department of Statistics, Stanford University. (Reference for *super smoother*.)

SEE ALSO

mdrace(BRM).

---

addtext — Add to text files or create new text files

SYNOPSIS
> **addtext** *files* ... {**clobber;terse;** ... }

DESCRIPTION
> **Addtext** adds text to the end of the named ASCII text files. If no files are specified, you are asked for a file name. If the named file does not exist, it is created.

> For each file named, text is read from the keyboard. To indicate the end of keyboard input into a file, press the RETURN key followed by a control-D (hold down the CTRL key while typing 'd').

> **Addtext** acts upon text files in your work area unless some other area is specified as an option. UNIX pathnames may also be used as file names.

> Note that text files cannot be used as BLSS datasets. To create a dataset from a text file, use the **read** command.

OPTIONS
> **clobber**
>> Do not ask for permission before overwriting a dataset which already exists.

> **bin, class, data, doc, help, home, sys, text**
> **dirs**=*dirname,* **tree**=*treename,* **area**=*areaname*
>> These options may be used to specify an alternate area (or directory) in which to create files. For a full description, see *area(BRM)*. The default values are:
>>
>> | | |
>> |---|---|
>> | **dirs** | your work area |
>> | **tree** | your *blss* tree (normally *~/blss*) |
>> | **area** | the text area |

> **terse** Fewer diagnostic messages.

EXAMPLES
> To create a text file in your work area and convert it to a BLSS dataset using the **read** command:

```
. addtext mytext
Creating text file "mytext" in your work area.
Type your text; finish with RETURN and CTRL-D.
1 2 3 4 5
6 7 8 9 10
Control-D
. read mytext > mydata
```

> To create a file in your help area which describes the BLSS dataset *mydata*:

```
. addtext {help} mydata
Creating text file "mydata" in your help area.
Type your text; finish with RETURN and CTRL-D.
```

```
Mydata is a BLSS dataset with 2 rows and 5 columns.
It contains the integers 1 through 10.
Control-D
```

The command 'help mydata' displays the help file you have just created.

SEE ALSO
BUG Section 4.9.2.
area(*BRM*), *edit*(*BRM*), *help*(*BRM*), *read*(*BRM*).

---

alias, unalias — Define (or remove) alias names for commands

SYNOPSIS
alias
alias *name*
alias *name value*
unalias *name1 name2 ...*

DESCRIPTION
Alias and **unalias** are internal commands to the BLSS shell which manipulate *aliases*—that is, new names for commands. The command **alias** *name value* aliases *name* to *value*. *Name* must be a single word; *value* may be a list of words. When *name* is typed as the first word in a BLSS command, *value* is substituted in its place. This is known as *alias expansion*. Aliasing is most often used to provide abbreviations for commonly used commands or command-and-option combinations.

Aliasing as described above is known as *simple aliasing*. The BLSS shell also provides **-aliasing*, which allows a word to be alias expanded anywhere in a command line; and *%-aliasing*, which allows alias expansions to take arguments. These are described in the *The BLSS Shell* (in preparation).

The command **alias** with no arguments lists all currently defined aliases.

The command **alias** *name* shows the value to which *name* is aliased.

The command **unalias** *name1 name2 ...* eliminates the named aliases.

An alias name may be escaped (that is, alias expansion of the name inhibited) by preceding it with a backslash.

The *alias* feature of the BLSS shell is very similar to that of the UNIX C shell.

EXAMPLES
The following alias, named *print*, sends its arguments to the default line-printer:

```
. alias print show {lpr}
```

To change the default behavior of a BLSS command, use its name as an alias whose value is the command with the chosen options. For example, to cause the **stat** command to compute the maximum likelihood estimates of the variance and SD (variance divisor = N), do the following:

```
. alias stat stat {dn=0}
```

Then, as desired, the following command automatically invokes the {**dn=0**} option:

```
. stat tractor
```

To obtain the default estimates (variance divisor = N–1) without **unalias**-ing (that is, to escape the alias):

```
. \stat tractor
```

SEE ALSO
  set(*BRM*).
  BUG Sections 10.2 (more examples and discussion), 10.5 (explains how to use
    the *.blss* file to define aliases permanently), 10.6.2 (interaction with the
    source command).

---

anovapr — Print an anova table

SYNOPSIS
  anovapr *anova*

DESCRIPTION
  Anovapr displays as an anova table the **anova@** output from the **regress**, **robust**, **oneway**, and **twoway** commands. The display is the same as that produced by the commands themselves.

  The column headings are:

Source	Source of variation
df	Degrees of freedom
SS	Sum of squares
MS	Mean sum of squares ($= SS / df$)
F	F statistic and its degrees of freedom
P-Value	P-value for the F statistic

  All possible anova table row labels and sources of variation are listed below. Note that any given anova table has some but not all of these. The type of row is coded as an integer in the first column of the *anova* array.

Label	*Source of Variation*	*Code*
Grand Total	Total SS about the origin	9000
Mean	SS for the mean	9001
Total	Total SS about the mean	9002
Residual	Residual SS	9003
Fit	SS for the fit (that is, the model)	9004
Columns	Column SS	9005
Rows	Row SS	9006

INPUTS
> **anova@**
>> An **anova@** output dataset containing the anova table.

EXAMPLE
```
. load stackloss
. regress {quiet} stackloss {x=1,2;y=4} > anova@ a
# ... other commands ...
. anovapr a

Anova Table: a
  Source      df        SS         MS          F                  P-Value
  Fit          2      1880.4     940.22     89.642 (df=2,18)     0.0000
  Residual    18      188.80     10.489
  Total       20      2069.2
  Mean         1      6448.8
  Grand Total 21      8518.0
```

---

area — Area options understood by BLSS utility commands

SYNOPSIS
> *utility-command* {**bin;class;data;home;sys;text;** ... }

DESCRIPTION
This manual entry describes area options used by the BLSS utility commands **addtext, doc, edit, help, list, load, man, read, remove, rename, save, show.** Here is a brief summary of the areas in your *blss* tree which these commands normally access, options for accessing them, and their use:

Area	*Option*	*Use*
bin area	{**bin**}	Permanent storage of command files.
data area	{**data**}	Permanent storage of datasets.
help area	{**help**}	Permanent storage of help files.
home directory	{**home**}	Where you first login to UNIX.
text area	{**text**}	Permanent storage of text files (other than command or help files).
work area	(none)	Temporary active work area.

The manual entries for the individual commands explain more precisely how the area options affect each command. The remainder of this manual entry provides more detailed information.

Each of the utility commands listed above has a built-in searchlist of areas (which are UNIX directories) where it looks when accessing files outside of the active work area (the current directory). A command searches through this list until the named file is found. An exception is: Files named by UNIX pathnames that begin with '/', './', '../', or '~' are processed directly, instead of through this searchlist mechanism. See *intro*(BRM) for more about UNIX pathnames and ~ expansion.

OPTIONS

The following options may be used with BLSS utility commands to specify alternate searchlists of areas. Each of the option values *dirlist*, *treelist*, and *arealist* is a list of directory names separated by colons or commas. UNIX pathnames (including those beginning with ~) can be used as values in a *dirlist* or *treelist*.

dirs=*dirlist*
> If the {**dirs**=} option is given, the specified *dirlist* becomes the search-list. This option overrides all others.

home Equivalent to {**dirs**=$HOME}; that is, only the user's home directory is searched.

tree=*treelist*
area=*arealist*
> If no {**dirs**=} option is specified, the searchlist is constructed from a built-in list of trees and areas. Each name in the *treelist* is paired with each name in the *arealist*. Each pair becomes an element (in the form *tree/area*) of the searchlist. The {**tree**=} and {**area**=} options specify, respectively, alternate tree and area lists.

sys Equivalent to {**tree**=$BLSS_SYS}; that is, the BLSS system tree is searched.

class Equivalent to {**tree**=$BCLASS}; that is, the class *blss* tree is searched. The BCLASS environment string is normally set by an instructor in the context of a class. This option is ignored (with a warning) if BCLASS is not set.

bin Equivalent to {**area**=bin}; that is, the bin area is searched.

data Equivalent to {**area**=data}.

doc Equivalent to {**area**=doc}.

help Equivalent to {**area**=help}.

text Equivalent to {**area**=text}.

dircode
> Display the complete list of areas (directories) to be searched.

Environment Strings

Diagnostic messages printed by the utility commands normally use BLSS col-loquial names such as 'your data area'. If the environment string PATH-NAMES is set, UNIX pathnames (such as '../data') are used instead. See *set*(BRM) for information on manipulating your environment from within BLSS.

EXAMPLES

The **load** command normally searches the data and text areas in the user's

own *blss* tree and then the BLSS system tree. That is, **load** without any
options is normally equivalent to:

```
. load {tree=..,$BLSS_SYS;area=data,text}
```

To load a file from your home directory, use:

```
. load {home}
```

To search the bin and text areas for files to load, use any of the following:

```
. load {bin;text}
. load {area=bin:text}
. load {area=bin,text}
```

To load from the data area in only the BLSS system tree, use either of the following:

```
. load {sys}
. load {tree=$BLSS_SYS}
```

SEE ALSO
BUG Section 4.12.
options(*BRM*), *set*(*BRM*), *dircode*(*BLR*).

---

arrop — Simple array operations

SYNOPSIS

const [*c*] [*d*] {**dims**=} > *y*	# *create a constant array*
colindex [*d*] {**dims**=} > *y*	# *create column index array*
rowindex [*d*] {**dims**=} > *y*	# *create row index array*
sheetindex [*d*] {**dims**=} > *y*	# *create sheet index array*
iota [*d*] {**dims**=} > *y*	# *create element index array*
cumsum *x* > *y*	# *cumulative sum*
cumprod *x* > *y*	# *cumulative product*
idn [*d*] {**dims**=} > *y*	# *create identity matrix*
ltri [*d*] {**dims**=} > *y*	# *create upper triangular matrix*
utri [*d*] {**dims**=} > *y*	# *create lower triangular matrix*
ncols *d* > *y*	# *number of columns in an array*
nrows *d* > *y*	# *number of rows in an array*
nsheets *d* > *y*	# *number of sheets in an array*
dims *d* > *y*	# *all three dimensions of an array*
diag *x* > *y*	# *construct diagonal matrix*
	or extract main diagonal

DESCRIPTION

These commands perform simple array operations.

Const creates a constant array: all elements have the value *c*. If *c* is not given,
the value 0 is used.

Colindex, **rowindex**, and **sheetindex** create arrays whose elements contain their column (or row, or sheet) index: column 1 (or row 1, or sheet 1) contains 1's; column 2 (or row 2, or sheet 2) contains 2's; etc.

Iota creates an array whose elements have the values 1, 2, 3, ... The element positions are numbered in *row-major order* (that is, going along the rows).

Cumsum generates the cumulative sum of the elements of *x*; **cumprod** generates the cumulative product. The output *y* has the same shape as the input *x*. The elements are accumulated (added or multiplied) in row-major order.

Idn creates an identity matrix; **ltri** creates a lower triangular matrix of 1's; **utri** creates an upper triangular matrix of 1's. If the new matrix is not square (see below), it is filled out with 0's.

Ncols gives the number of columns in the dataset *d*; **nrows** gives the number of rows; **nsheets** gives the number of sheets. **Dims** gives a row vector of length 3 containing the number of sheets, rows, and columns (in that order).

If *x* is a matrix, **diag** extracts its main diagonal as a vector; if *x* is a vector, **diag** constructs the diagonal matrix *y* with main diagonal *x*.

SPECIFYING DIMENSIONS

The {**dims**=} option and the (optional) **d@** input are common to most of these commands.

The {**dims**=} option specifies the dimensions of the dataset to create. Generally, if less than three numbers are given in {**dims**=}, 1's are assumed to the left. (Thus, if two numbers are given, the command creates a matrix; if one number is given, the command creates a row vector of that length.) However, for the **idn**, **utri**, and **ltri** commands, if only one number is given in {**dims**=}, the command creates a square matrix of that size.

If {**dims**=} is not given, then these commands use the dimensions of the optional **d@** input dataset. The contents of **d@** are ignored; only its dimensions are used.

If neither {**dims**=} nor **d@** is given, the dimensions (1,1,1) are used.

EXAMPLES

To construct a 5-by-3 matrix *a* that consists entirely of 0's, and a second matrix *b* of the same dimensions that consists entirely of 1's:

```
. const 0 {dims=5,3} > a
. const 1 a > b
```

To construct an identity matrix *I* of rank 7 (dimensions 7-by-7):

```
. idn {dims=7} > I
```

The **rowindex**, **colindex**, and **sheetindex** commands make it easy to construct matrices whose elements are functions of their indices without explicit use of loops. For example, to construct a matrix X of dimensions $(3,10)$ such that $X[i,j] = i^2 + j$:

```
. rowindex {dims=3,10} > i
. colindex i > j
. X = i^2 + j
```

To construct a $(10,10)$ band matrix B whose center three diagonals are 1:

```
. rowindex {dims=10,10} > i
. colindex i > j
. B = (abs(i-j) <= 1)              # logical operator: results are 0's and 1's
```

The identity, lower triangular, and upper triangular matrices can also be constructed this way, but they are important enough to merit special commands of their own.

Any construction that uses **rowindex**, **colindex**, or **sheetindex** can also be made using index vectors and dimension-expansion (see *BUG* Section 8.3.5). For example, to create B as above:

```
. i = 1:10                         # row vector of column indices
. j = i'                           # column vector of row indices
. B = (abs(i-j) <= 1)              # dimension-expansion applied to i and j
```

Or simply:

```
. B = (abs((1:10)-(1:10)') <= 1)
```

SEE ALSO
   BUG Section 8.3.1.
   redim(*BRM*), *vecop*(*BRM*).

---

autcor, autcov — Auto- and cross-correlation and covariance of time series

---

SYNOPSIS
   autcor x [y] {o=} [> *cor cov pcor*]
   autcov x [y] {o=} [> *cov cor pcor*]

DESCRIPTION
   Autcor and **autcov** compute autocorrelations, autocovariances, and partial autocorrelations. The default output of **autcor** is autocorrelations. The default output of **autcov** is autocovariances.

   If exactly two input series are provided, **autcor** (or **autcov**) computes cross-correlations or cross-covariances of the two series; see below.

INPUTS
   x@   Vector or array whose columns contain the x series.

   y@   Vector or array whose columns contain the y series.

   The input datasets x and y must have the same length. Missing values are not allowed.

OPTIONS

o=*INT*

The maximum order of the estimated correlations or covariance. The default is the minimum of 21 and the integer part of .75*(series length).

OUTPUTS

cor@ A matrix containing the estimates of the auto- and cross-correlations (if appropriate) for orders 0 through *o*.

cov@ A matrix containing the estimates of the auto- and cross-covariances (if appropriate) for orders 0 through *o*.

pcor@

A matrix containing the estimates of the partial autocorrelations for orders 0 through *o*.

The **cor@** and **cov@** output datasets normally consist of one column for each input series. If x has exactly two columns or if both x and y are vectors, the cross-correlation and cross-covariance functions are also produced, as listed below for the **cov@** output dataset:

Column	Contains
1	$\mathrm{cov}(x_t, x_{t+l})$
2	$\mathrm{cov}(y_t, y_{t+l})$
3	$\mathrm{cov}(x_t, y_{t+l})$
4	$\mathrm{cov}(y_t, x_{t+l})$

where t is the time index (that is, row index of the data) and l is the lag index (which is one less than the row index of the result—for example, row 1 of **cov@** contains the lag 0 autocovariance, etc.).

When the command is invoked as **autcov**, the positions of the **cor@** and **cov@** tags are reversed, as implied in the description section above.

If no output datasets are specified to **autcor**, it puts the autocorrelations in *x*.**cor**, where *x* is the name of the **x@** input dataset. If no output datasets are specified to **autcov**, it puts the autocovariances in *x*.**cov**.

REFERENCE

Keith A. Haycock and David R. Brillinger (1985). *LIBDRB: A subroutine library for elementary time series analysis.* Technical Report No. 48, Department of Statistics, University of California, Berkeley.

---

autreg — Autoregressive model fitting for time series

SYNOPSIS

    autreg x {o=;**wide**} [> *a aic v*]

DESCRIPTION

    Autreg fits autoregressive time series models to the time series x. The default display shows the residual variance, the Akaike information criterion, and the estimated coefficients for each model.

    Algorithm. Recursive solution of the Yule-Walker equations.

INPUT

    x@   A vector containing the time series. Multiple series must be placed in separate sheets. Missing values are not allowed.

OPTIONS

    o=*INT*

         The maximal order autoregression to be fit. Default is 5. **Autreg** recursively fits models of order 0 through that specified.

    wide   The display is about 130 columns wide, suitable for sending to line-printers. For higher-order models, this allows more coefficients to be displayed on each line.

OUTPUTS

    a@   An array whose columns are the estimated coefficients for models of order 0 through o. The array is filled out with zeros.

    aic@   The Akaike information criterion for the fitted models of order 0 through o.

    v@   The residual variance for the fitted models of order 0 through o.

    All output datasets are created. If no name for a dataset is specified by the user, the dataset is named *x.tag*, where x is the input series name and *tag* is the output tag name.

REFERENCE

    Keith A. Haycock and David R. Brillinger (1985). *LIBDRB: A subroutine library for elementary time series analysis.* Technical Report No. 48, Department of Statistics, University of California, Berkeley.

---

binomial — Binomial distribution probabilities and cdf

SYNOPSIS
> **binomial** *n p* [> *pdf cdf*]

DESCRIPTION
> **Binomial** computes the probabilities that a binomial random variable with parameters *n* and *p* assumes the values 0, 1, . . . , *n*. It can also compute the cumulative distribution function (cdf).

INPUTS
> **n@**   The *n* parameter of the binomial distribution.
>
> **p@**   The *p* parameter of the binomial distribution.

OUTPUTS
> **pdf@**   Row vector that contains the probabilities.
>
> **cdf@**   Row vector that contains the cumulative distribution function.

NOTE
> This command is implemented as a macro. It illustrates macro features described in *BUG* Section 10.8.4, and it uses methods described in Section 9.2.1 to compute the probabilities and the cdf.

---

biweight — Robust estimates of center and spread

SYNOPSIS
> **biweight** *x* {**x=;c=;eps=**} [> *biw w*]

DESCRIPTION
> **Biweight** computes robust estimates of center (location) and spread (scale) for the columns of *x*. The location estimate is an adaptive weighted mean (an M-estimate); the scale estimate is the square root of the corresponding weighted sum of squared deviations.

> At each iteration the observations are reweighted: if an observation is far from the current estimate of the center, it gets a lower weight. The bisquared weight function is used:

$$\text{weight}(z) \;=\; \begin{cases} \left[1 - \left[\dfrac{z}{k}\right]^2\right]^2 & \text{if } |z| \le k \\[2ex] 0 & \text{if } |z| > k \end{cases}$$

> where *z* is the data value minus the current estimate of center. The value of *k* is *c*∗*s* where *s* is the current estimate of spread and *c* is a user-supplied robustness parameter.

> Values of *c* between about 4 and 10 are reasonable. The larger *c* is, the closer

the biweight estimates are to the sample mean and SD. Note that data points beyond $c*s$ from the center are ignored (given 0 weight).

The bisquared weight function is zero outside the interval $(-k, k)$; it is symmetric about 0 and attains a maximum value of 1 at 0. The corresponding influence function, influence$(z) = z *$ weight(z), is zero at 0 and outside $(-k, k)$; it decreases on $(-k, -k/\sqrt{5})$, increases on $(-k/\sqrt{5}, k/\sqrt{5})$, and decreases on $(k/\sqrt{5}, k)$.

INPUTS

x@   The input data. Missing values are ignored.

OPTIONS

x=*INTLIST*
> Columns of x for which to compute **biweight**. Default is all columns.

c=*FLOAT*
> Robustness parameter c used in the bisquared function. Default is $c = 6$.

eps=*FLOAT*
> Iterations stop when the absolute change in the center estimate divided by the spread estimate is less than *eps*. Default is .01.

OUTPUTS

biw@   One row for each column of x, as follows:

Column	Contains
1	N, the number of non-missing values.
2	Center.
3	Spread.
4	Sum of weights.

> If the {**quiet**} option is given, the **biw@** output dataset is always created. If no output name is specified, the dataset is named x.**biw**, where x is the name of the input dataset.

w@   Final weights for each observation. These are not normalized, because the sum of the weights is also of interest.

EXAMPLE

To obtain the **biweight** estimates for each column of the data in *chickwts*:

```
. biweight chickwts
```

REFERENCE

Donald R. McNeil (1977). *Interactive Data Analysis*. Wiley, New York. Chapter 7, pp. 152-160, discusses this general approach. The specific algorithm McNeil presents is an older one which uses the mean absolute deviation as the estimate of spread and is less numerically stable.

SEE ALSO

medmad(BRM), *stat(BRM)*.

---

blssf — BLSS frontend invocation from UNIX

SYNOPSIS (from UNIX)

blss [**–a –b***blss_sys* **–B –d***dir* **–m***macro* **–n –s** ...] [*arguments*]

DESCRIPTION

Blssf is the BLSS frontend program; it is executed when users type the UNIX
command **blss**. It creates a protected environment within which BLSS functions. Specifically:

1. It changes to the user's home directory (as defined by the environment
string HOME) and creates several directories if they do not already exist: *blss*,
blss/bin, *blss/data*, *blss/doc*, *blss/help*, and *blss/text*. These are the 'bin area',
'data area', and so forth used for permanent storage by the BLSS **save** and
load commands.

2. It also creates a temporary directory, *blss/tmpN*, where *N* is the least digit
not already in use by an existing directory. (If single digits are exhausted,
names of the form *blss/tmpNN* are used.) This directory is known as the
'work area' and becomes the user's working directory for the BLSS session.
Thus, each invocation of BLSS has a unique work area; but each user has just
one data area, text area, etc.

3. It establishes an environment string BLSS_SYS (unless it already exists)
which contains the pathname of the BLSS tree. (In order to make it easy to
move the BLSS tree, *blssf* is the only BLSS program with the BLSS tree pathname compiled in. All other BLSS programs look at the BLSS_SYS environment string instead.)

4. If the erase character is a printing character (e.g., #), it is changed to
control-H. If the kill character is a printing character (e.g., @), it is changed
to control-U.

Blssf then forks an interactive BLSS shell, *blssh –i*. When *blssh* terminates,
blssf resets the erase and kill characters (if changed) and removes the temporary directory. If the other directories are empty, they too are removed.

CRASH RECOVERY

If *blssf* terminates abnormally (for example, the system crashes), the directory
cleanup described above does not happen; a ghost *blss/tmpN* directory
remains behind. When invoked, *blssf* checks for the existence of such ghost
directories and asks the user about those which are sufficiently old:

1. If *blssf* can determine when the system came up, it asks about *blss/tmpN*
directories which are older than the uptime or older than one day. (*Blssf* may
be recompiled to use either of two methods to find the uptime. Neither
method is foolproof on all systems. See the source code for recommenda-
tions.)

2. If *blssf* cannot determine when the system came up, it asks about
blss/tmpN directories older than 15 minutes.

OPTIONS

The following options are recognized by the BLSS frontend. Observe that they are specified using UNIX option syntax, not BLSS option syntax.

-a   Ask about old *blss/tmpN* directories regardless of their age. Useful if BLSS terminates abnormally due to a broken phone connection, broken remote connection, etc.

-b*blss_sys*

Instead of the compiled-in default, use the directory *blss_sys* as the BLSS_SYS environment string (and hence the base of the BLSS tree). This option overrides the environment string BLSS_SYS if it is already set.

-B   Echo the value of BLSS_SYS to be used.

-d*dir* Change to the directory *dir* (instead of the HOME directory) before creating the *blss* directories; return to original directory when done. This option allows you to maintain and use separate BLSS trees in separate directories—each of which can be devoted to a separate project.

-i*blssh*

Use the specified BLSS shell instead of the default ~*blss/bin/blssh*.

-m*macro*

Instead of an interactive BLSS session, run the BLSS macro file *macro*. The file *macro* may be in the directory from which *blssf* was invoked or in any directory on the BLSS_SYS (for example, the user's directory *blss/bin*).

-n   Never ask about old *blss/tmpN* directories. This option is recommended for accounts which are shared among many naive users—but then someone must periodically clean the account of any old *blss/tmpN* directories.

-s   Stay still: do not change, create, or remove directories. The current directory is used as the BLSS work area.

-Z   Zap: silently and unconditionally remove any old *blss/tmpN* directories. This option is not recommended for general use.

ARGUMENTS

Any arguments to *blssf* specified after the options are passed on to *blssh*. Thus they may serve as arguments to a macro. Because *blssf* is invoked from the UNIX shell, arguments to a macro must be specified using UNIX syntax, not BLSS syntax. See *fromunix*(BRM) for more information and examples.

FILES

~/blss/*

/etc/mtab   (for the default uptime scheme)

/dev/mem

/dev/kmem	(for the alternate uptime scheme)
/unix	(/vmunix on 4BSD systems)
~blss/help/blssf.menu	(menu when asking about old *blss/tmpN* directories)
~blss/bin/blssh	(BLSS shell)

SEE ALSO
BUG Section 10.8.

---

box — Draw tickets from a box model and display sample statistics

---

SYNOPSIS
> box [*box*] {ndraw=;nsamp=;evse;sumavg; ... } [> *box samps stat sumavg*]

DESCRIPTION
> **Box** draws tickets at random, with replacement, from a *box model*. (In standard vocabulary, it samples randomly from a finite discrete distribution.) It displays the average and SD of the box (that is, the mean and SD of the population—the variance divisor is N), draws a number of samples, and then computes summary statistics for each sample using the **stat** command.

> This command illustrates concepts from Chapters 16 through 18 of *Statistics*, by Freedman, Pisani, and Purves. *BUG* Section 7.1.2 also gives a brief description of box models.

INPUTS
> **box@** An optional dataset which contains the box model. If not specified, **box** asks the user to type in a box at the keyboard. **Box** reads it using the **read** command and saves it for future use (see below).

> If the input dataset has exactly two columns and at least two rows, **box** treats the first column as a list of tickets and the second column as a list of frequencies or probabilities for the tickets. In this case, it is an error for the box to contain missing values, negative frequencies, or for all the frequencies to be zero.

> Otherwise, **box** treats its input dataset as a list of tickets and assigns equal probability to each. Missing values are ignored.

OPTIONS
> **ndraw**=*INT*
>> The number of tickets to draw per sample. The default is 1. If set to 0, **box** exits after showing the average and SD of the box.

> **nsamp**=*INT*
>> The number of samples to draw. The default is 1. If set to 0, **box** exits after showing the average, SD, and SEs (if requested).

> If neither {**ndraw**=} nor {**nsamp**=} is given, **box** asks the user for both.

dn=*INT*

> Change the variance divisor used by **stat** when computing summary statistics for the samples. See *stat*(*BRM*) for specifics. The default is {**dn**=**-1**}: divide by *N*-1.

evse   Display the expected value (EV) for the sum of the draws and the standard error (SE) for the sum and average of the draws.

show   Show the sample values as they are drawn.

stat   Display summary statistics for each sample using the **stat** command. This is the default.

nostat

> Do not display summary statistics for the samples. This option overrides {**stat**}.

sumavg

> Display the sum and average of the draws (for a 0-1 box, the number and proportion of 1's) for each sample. This option turns off the (default) {**stat**} display unless the {**stat**} option is explicitly given.

OUTPUTS

box@ The dataset in which to save the box if entered at the keyboard. If not specified, a name of the form *boxNN* is used.

samps@

> The samples. If not specified, a name of the form *sampsNN* is used.

stat@ The summary statistics for the samples. It contains one row for each sample; each row has the same layout as the **stat** command's **stat@** output dataset.

sumavg@

> The information displayed by the {**sumavg**} option. It contains one row for each sample; each row contains, in order: the sample number, the number of draws in the sample, the sum of the draws, and the average of the draws.

EXAMPLE

> The following commands simulate 10 rolls of a fair die and show the results:

```
. load die                 # contains the values: 1 2 3 4 5 6
. box die {ndraw=10;show}
```

REFERENCES

> David Freedman, Robert Pisani, and Roger Purves (1978). *Statistics*. Norton, New York.

SEE ALSO

> *BUG* Section 7.1.2.

---

boxplot — Tukey box-and-whisker plots

SYNOPSIS

   boxplot *x* {**fi**=;**fo**=;**x**=;**width**=}

DESCRIPTION

   Boxplot displays John Tukey's *box-and-whisker* plot (for short, *boxplot*) of a univariate dataset. It is interpreted as follows:

   The ends of the box correspond to the lower and upper quartiles. The * inside the box indicates the median. The X's at ends of the lines which extend beyond the box are the data points furthest from the box, but no more than 1 interquartile distance away. (The locations 1 interquartile distance above and below the quartiles are known as the 'inner fences'.) Points from 1 to 1.5 interquartile distances away are marked by o's. Points more than 1.5 interquartile distances away (beyond the 'outer fences') are marked by *'s. Multiple data points at these locations are indicated by numbers below the o's and *'s. The minimum and maximum data values are printed above the plot.

INPUTS

   x@   The input data. One boxplot is made for each column of *x*. Missing values in *x* are ignored.

OPTIONS

   fi=*FLOAT*

      The factor used to obtain the inner fences from the interquartile range. Default value is 1. Note, however, that many authors use 1.5.

   fo=*FLOAT*

      The factor used to obtain the outer fences from the interquartile range. Default value is 1.5. Note, however, that many authors use 3.0.

   x=*INTLIST*

      Specifies columns to be plotted. Default is all columns.

   width=*INT*

      The width of the plot, in characters. Default is 79. The WIDTH environment string can also be used; the {**width**=} option overrides it.

EXAMPLES

   Precip contains the average annual precipitation, in inches, for 69 weather stations in the U.S. To make its boxplot:

```
. load precip
. boxplot precip {width=65}
```

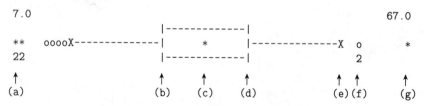

The symbols in the boxplot have the following meanings:

Key	*Meaning*
(a)	The minimum value, 7.0.
(b)	The first quartile, 29.1.
(c)	The median, 36.2.
(d)	The third quartile, 42.8.
(e)	The furthest data point less than 1 interquartile distance above the box, 54.7.
(f)	The '2' signifies that there are two data points at this location.
(g)	The maximum value, 67.0.

Rivers contains the lengths of 141 major rivers in North America. Inverse square roots make the distribution of the lengths nearly symmetric.

```
. load rivers
. boxplot rivers
. let ir = rivers ^ -.5
. boxplot ir
```

REFERENCES

Donald R. McNeil (1977). *Interactive Data Analysis.* Wiley, New York. Chapter 1.

John W. Tukey (1977). *Exploratory Data Analysis.* Addison-Wesley, Reading, Mass. Chapter 2.

SEE ALSO

compare(BRM), for comparative boxplots.

stat(BRM), *stemleaf(BRM)*.

---

cdf — Continuous probability distribution functions, inverses, and densities

---

SYNOPSIS

pbeta q {a=;b=;eps=} > p	# *beta cdf*
pcauchy q {a=;b=} > p	# *Cauchy cdf*
pchisq q {df=;eps=} > p	# χ^2 *(chi-square) cdf*
pexp q {l=;b=} > p	# *exponential cdf*
pf q {df=;eps=} > p	# *Fisher's F cdf*
pgamma q {a=;l=;b=;eps=} > p	# *gamma cdf*
pgau q {m=;s=;v=} > p	# *Gaussian (Normal) cdf*
plogis q {a=;b=} > p	# *logistic cdf*
pnt q {df=;ncp=;eps=} > p	# *noncentral t cdf*
pt q {df=;eps=} > p	# *Student's t cdf*
puni q {a=;b=} > p	# *uniform cdf*
qbeta p {a=;b=;eps=} > q	# *inverse beta cdf*
qcauchy p {a=;b=} > q	# *inverse Cauchy cdf*
qchisq p {df=;eps=} > q	# *inverse χ^2 (chi-square) cdf*

qexp *p* {l=;b=} > *q*	# *inverse exponential cdf*
qf *p* {df=;eps=} > *q*	# *inverse Fisher's F cdf*
qgamma *p* {a=;l=;b=;eps=} > *q*	# *inverse gamma cdf*
qgau *p* {m=;s=;v=} > *q*	# *inverse Gaussian (Normal) cdf*
qlogis *p* {a=;b=} > *q*	# *inverse logistic cdf*
qt *p* {df=;eps=} > *q*	# *inverse Student's t cdf*
quni *p* {a=;b=} > *q*	# *inverse uniform cdf*
dbeta *x* {a=;b=;eps=} > *f*	# *beta density*
dcauchy *x* {a=;b=} > *f*	# *Cauchy density*
dchisq *x* {df=;eps=} > *f*	# χ^2 *(chi-square) density*
dexp *x* {l=;b=} > *f*	# *exponential density*
df *x* {df=;eps=} > *f*	# *Fisher's F density*
dgamma *x* {a=;l=;b=;eps=} > *f*	# *gamma density*
dgau *x* {m=;s=;v=} > *f*	# *Gaussian (Normal) density*
dlogis *x* {a=;b=} > *f*	# *logistic density*
dt *x* {df=;eps=} > *f*	# *Student's t density*
duni *x* {a=;b=} > *f*	# *uniform density*

DESCRIPTION

Cdf calculates continuous distribution functions (cdf's), their inverses, and their densities. Each type of cdf, inverse cdf, and density has its own command name. Cdf commands have names of the form **p***xxx*; inverse cdf commands have names of the form **q***xxx*; density commands have names of the form **d***xxx*.

If an input value is out of range, the corresponding output value is an NA.

INPUTS

q@   For **p***xxx* commands, the values (quantiles) at which to evaluate the cdf.

p@   For **q***xxx* commands, the probabilities at which to evaluate the inverse cdf.

x@   For **d***xxx* commands, the values at which to evaluate the density.

OPTIONS

eps=*FLOAT*

The required accuracy of the result. For the **d***xxx* and **p***xxx* commands, the absolute accuracy. For the **q***xxx* commands, the relative accuracy—for example, *eps* = .0005 gives three significant digits.

The default is *eps* = 1.0e−8. This option is not available for the **cauchy**, **exp**, **gau**, **logis**, or **uni** distributions. Refer to *cdf(BML)* for more information about accuracy.

a=*FLOAT*, **b**=*FLOAT*, **d**=*FLOAT*, **df**=*FLOATLIST*, **l**=*FLOAT*,
m=*FLOAT*, **s**=*FLOAT*, **v**=*FLOAT*

Parameter values of the cdf's and densities. For their meanings, refer to *BUG* Table 7.1 or to the on-line table in 'help cdftable'. For the F distribution, the {**df**=} option takes two values to specify the two degrees of freedom parameters. All other parameter options take a single value.

OUTPUTS

p@   For the **p**xxx commands, the probabilities which correspond to the values (quantiles) *q*.

q@   For the **q**xxx commands, the quantiles which correspond to the probabilities *p*.

f@   For the **d**xxx commands, the values of the density which correspond to the values *x*.

If no output dataset is specified, the results are displayed.

EXAMPLES

To find the standard Gaussian (Normal) probabilities of the values in *z* and place the results in *p*:

    `. pgau z > p`

To find the P-value for the result *F* of an F-test with $(8, 35)$ degrees of freedom:

    `. pf F {df=8,35} > p`
    `. pval = 1 - p`

To find the quantiles which correspond to the array *p* of probabilities for the logistic distribution with parameters $a = 0$ and $b = 2$ and store them in *q*:

    `. qlogis p {b=2} > q`

To display the gamma density with parameters $a = 5$ and $l = 5$ at the values $x = 0, .1, .2, \ldots, 3.5$:

    `. x = 0:3.5:.1`
    `. dgamma x {a=5;l=5}`

SEE ALSO

BUG Sections 7.3 and 7.4.
random(BRM), for random number generators for these cdf's.
cdf(BML).

---

chisq — Chi-square (χ^2) tests of independence and goodness-of-fit

---

SYNOPSIS
chisq *obs* [*exp*] {**nfit**=;**long**} [> *chisq row col contr exp*]

DESCRIPTION
Tests of Independence
Chisq performs tests of independence for two-way tables and displays the χ^2 test statistic, its degrees of freedom, and its P-value.

Tests of Goodness-of-Fit
If the second input, *exp*, is present, **chisq** performs a goodness-of-fit test and displays the χ^2 test statistic, its degrees of freedom, and its P-value.

INPUTS
obs@ *nr*-by-*nc* matrix which contains the observed cell frequencies.

exp@ *nr*-by-*nc* matrix which contains the expected cell frequencies for a goodness-of-fit test.

Missing values are not allowed in either input dataset.

OPTIONS
nfit=*INT*
> The number of parameters estimated when computing the expected values for a goodness-of-fit test. The degrees of freedom for the test are then $nr*nc - 1 - nfit$. By default, *nfit* is 0.

long In addition to the default output, show contributions to the χ^2 test statistic from individual cells. If the test is a test of independence, also show the computed row and column proportions.

OUTPUTS
chisq@
> The χ^2 statistic, the degrees of freedom, and the P-value.

row@ The computed row proportions for test of independence.

col@ The computed column proportions for test of independence.

contr@
> Contributions to the χ^2 test statistic from individual cells.

exp@ The table of expected entries for tests of independence.

EXAMPLES
Test of Independence
Mendenhall (1983) gives the following data on the occurrence of flu and the number of flu vaccinations for a random sample of 1000 people.

		Vaccinations		
		Zero	*One*	*Two*
Flu?	*Yes*	24	9	13
	No	289	100	565

Are number of vaccinations and contracting the flu independent? Suppose that these data have been entered into the 2-by-3 dataset *flu*. The command:

    . chisq flu > chisq rows cols

produces the χ^2 statistic and P-value for testing the hypothesis.

Test of Goodness-of-Fit

Hoel (1971) gives the following example. An experimenter breeding a certain species of flower obtained 120 magenta flowers with a green stigma, 48 magenta flowers with a red stigma, 36 red flowers with a green stigma, and 13 red flowers with a red stigma. Genetic theory predicts that flowers of these types should occur in the ratios 9 : 3 : 3 : 1. Are these results compatible with the theory?

The input matrix of the observations is:

$$colors \quad\quad 120 \quad\quad 48 \quad\quad 36 \quad\quad 13$$

Since there are 217 observations and the predicted probabilities of the various color combinations are 9/16, 3/16, 3/16, and 1/16, the expected cell frequencies are:

$$expected \quad 122.06 \quad 40.69 \quad 40.69 \quad 13.56$$

The command:

    . chisq colors expected {nfit=0}

gives the χ^2 statistic and its P-value for testing the hypothesis.

Expected frequencies are computed from expected ratios as follows. The ratios 9 : 3 : 3 : 1 are converted to proportions by dividing each by the sum of the ratios: 9/16 : 3/16 : 3/16 : 1/16. Because the total number of observations is 217, the expected number of observations is 217 * (expected proportion of observations). Thus, the BLSS command to calculate the expected cell frequencies is:

    . expected = 217 * (9, 3, 3, 1) / 16

Arithmetic on lists of numbers as used in this example is discussed in *BUG* Section 2.2.

REFERENCES

Paul G. Hoel (1971). *Introduction to Mathematical Statistics,* 4th edition. Wiley, New York. Page 230.

William Mendenhall (1983). *Introduction to Probability and Statistics,* 6th edition. Duxbury, Boston. Page 502.

SEE ALSO

xtab(BRM), to construct tables of observed cell frequencies from bivariate data.

---

chol — Cholesky decomposition of a matrix

SYNOPSIS
   chol *a* {**lower;pivot;pivot**= } [> *r e eae*]

DESCRIPTION
   Chol computes the Cholesky decomposition of the real symmetric positive definite or positive semidefinite matrix **A** (contained in the dataset *a*) and places the results in the following datasets:

   r   The upper triangular Cholesky square root **R** of the input matrix **A**. When pivoting is not in effect, **R′R** = **A**. Otherwise, **R′R** = **E′AE**. **R** has nonnegative diagonal elements. Thus, if **A** is positive definite, the decomposition is unique.

   e   The permutation matrix **E** when pivoting is in effect; otherwise, the identity matrix.

   eae The matrix **E′AE** (that is, the permuted **A** matrix).

   If the matrix **A** is not symmetric, **chol** attempts to symmetrize it. If both the [*i,j*]-th and [*j,i*]-th elements of **A** are missing, its Cholesky decomposition cannot be obtained. If the dataset *a* contains more than one sheet and its decomposition cannot be obtained, then the corresponding sheets of the datasets *r*, *e*, and *eae* are filled with missing values.

   Algorithm. This command uses the LINPACK subroutine *schdc*.

INPUTS
   a@   Dataset which contains the matrix **A** to be decomposed. If it contains multiple sheets, **chol** decomposes each sheet separately.

OPTIONS
   lower Normally, the output dataset *r* contains the upper triangular Cholesky square root **R**. This option causes it to contain the lower triangular Cholesky square root **R′**.

   pivot=*INTLIST*
   pivot Enable pivoting: the rows and columns of **A** are permuted in order to obtain the most numerically stable decomposition. In this case, the decomposition satisfies **E′AE** = **R′R**, where **E** is the permutation matrix and **E′AE** is **A** with its rows and columns permuted.

   If *INTLIST* is specified, it controls pivoting as follows. Rows and columns of **A** corresponding to positive elements of *INTLIST* are moved to the leading part of **A** before the decomposition begins and are frozen in place. Rows and columns of **A** corresponding to negative elements are moved to the trailing part of **A** before the decomposition begins and frozen in place. Rows and columns of **A** corresponding to zero elements are allowed to move freely during the decomposition. If *INTLIST* is not specified, all rows and columns are allowed to move freely. See the *LINPACK Users' Guide*, pp. 8.3-8.4, for details.

OUTPUTS
   r@   The Cholesky square root **R** (or **R′** if the {**lower**} option is given).

e@   The permutation matrix **E** for pivoting.

eae@   The permuted **A** matrix, **E′AE**.

If no output datasets are specified, the Cholesky square root is put in *a*.**r**, where *a* is the name of the input matrix. If pivoting is in effect but no outputs are specified, the Cholesky square root is put in *a*.**r** and the **E′AE** matrix in *a*.**eae**.

EXAMPLES

If *z* is a random row vector of i.i.d. Normal(0, 1) elements, then $x = z\mathbf{T}$ is multivariate Normal with mean 0 and covariance matrix $\mathbf{T}'\mathbf{T} = \mathbf{A}$. Given **A**, **T** may be obtained using Cholesky decomposition. The following sequence generates 100 multivariate Normal pseudorandom vectors of length 5 with mean 0 and covariance matrix **A**:

```
.  chol a > t
.  rgau {dims=100,5} > z
.  x = z #* t
```

If instead *z* is a random column vector, the lower triangular Cholesky square root is more convenient to use. It can be obtained as the transpose of the upper triangular square root, or it can be obtained directly using the {**lower**} option:

```
.  chol a {lower} > lt
.  rgau {dims=5,1} > z
.  x = lt #* z
```

The result is a single random Normal column vector with mean 0 and covariance **A**.

To obtain the Cholesky decomposition of the covariance of *stackloss*:

```
.  load stackloss
.  stat {quiet} stackloss > cov@ stack.cov
.  chol stack.cov > r@ stack.r
```

If a dataset contains missing values, the covariance matrix obtained via *stat*(*BRM*) using the {**pairx**} option (pairwise exclusion of missing data) may not be positive definite. In this case, it may be desirable to enable pivoting. In the following example, the {**pivot=**} specification causes the third row and column of *t.cov* to be moved to the trailing part of *t.cov* and frozen there, and the first, second, and fourth rows and columns to be moved freely. This setting of {**pivot=**} was chosen because the third column of *turbines* contains missing values and thus the estimates in the covariance matrix which use the third column of *turbines* may not be as reliable as the others.

```
.  stat {pairx} turbines > cov@ t.cov
.  chol t.cov {pivot=0,0,-1,0} > r e eae
```

REFERENCE

J. J. Dongarra et al. (1979). *LINPACK Users' Guide*. SIAM, Philadelphia. Chapter 8.

SEE ALSO

eigen(*BRM*), *qr*(*BRM*), *linpack*(*BML*).

---

ci — Confidence intervals for means and differences of means

SYNOPSIS
　　ci *x* [*y*] {b;level=;nx=;ny=;p;pool;sdx=;sdy=;x=;y=;z; ... } [> *mean* ...]

DESCRIPTION
Ci computes confidence intervals for a population mean, probability of success or proportion of successes, and the difference between two population means, probabilities of success, or proportions of successes. It and the **ht** command are closely related; many of their inputs, options, and outputs are identical.

One-Sample Confidence Intervals
If *x* is specified but *y* is not, **ci** calculates confidence intervals for the mean of the population from which the sample *x* is drawn. If {**nx=**} and {**sdx=**} are specified, *x* contains sample means instead of data. If *x* contains several columns, **ci** calculates separate confidence intervals for each column.

Two-Sample Confidence Intervals
If both *x* and *y* are specified, **ci** calculates confidence intervals for the difference of the means of the populations from which *x* and *y* are drawn. If *x* and *y* contain several columns, **ci** calculates confidence intervals for the difference between the means of the corresponding columns.

Normal or t Distribution?
If the population SDs are known and specified via the {**sdx=**} and {**sdy=**} options, the confidence intervals are based on the Normal distribution. Otherwise, **ci** estimates the population SDs and computes confidence intervals based on the t distribution by default; or on the Normal distribution if the {**z**} option is specified.

Confidence Intervals for Proportions
A random variable which is 1 with probability p and 0 with probability $1-p$ is called *Bernoulli* with parameter p. The sum of n Bernoulli random variables with parameter p is called *binomial* with parameters n and p. If the {**b**} option is specified, *x* is a binomial observation. If the {**p**} option is specified, *x* is the observed proportion of 1's in a Bernoulli sample of size n. In either case, n must be specified using the {**nx=**} option. If {**b**} or {**p**} is specified and *y* is given, it contains a second sum or proportion; the Bernoulli sample size for *y* must be specified by {**ny=**}.

The Output Display
The output display shows: the number of observations; the sample mean (or mean difference); the sample or known SE; the degrees of freedom (df) in the SE, if appropriate; the confidence level(s) $1-\alpha$; the $1-\alpha/2$ quantile of the distribution the confidence interval is based on; and the endpoints of the confidence interval(s).

INPUTS

x@   Sample(s) or sample mean(s) from a population. If {**nx**=} is specified and x is a column vector, it is treated as a row vector. Otherwise, if x is a row vector or a sheet vector, it is treated as a column vector.

y@   Sample(s) or sample mean(s) from a population. If {**ny**=} is specified and y is a column vector, it is treated as a row vector. Otherwise, if y is a row vector or a sheet vector, it is treated as a column vector.

c@   If the {**ht**} option is given, this input is used as the **ht c@** input.

OPTIONS

level=*FLOATLIST*
alpha=*FLOATLIST*
a=*FLOATLIST*

        Ci computes confidence intervals for each of the specified confidence levels, $1-\alpha$. Default values are {**level**= .5, .95, .99} or, equivalently, {**alpha**= .5, .05, .01} or {**a**= .5, .05, .01}.

        Note: For confidence intervals, the confidence level is $1-\alpha$; whereas for hypothesis tests, the significance level is α.

b, p   The input x (and y, if given) is the sum (under the {**b**} option) or the average (under the {**p**} option) of n Bernoulli observations. Under the {**b**} option, if x or y is between 0 and n but is not an integer, it is rounded to the nearest integer. The confidence intervals are based on the Normal approximation, as follows. For one sample the interval is:

$$\hat{p} \pm \Phi^{-1}(\alpha/2)\sqrt{\hat{p}(1-\hat{p}) / n}$$

where $\hat{p} = x/n$ under the {**b**} option or x under the {**p**} option, and $\Phi^{-1}(x)$ denotes the inverse standard Normal cdf. For two samples:

$$(\hat{p}_1 - \hat{p}_2) \pm \Phi^{-1}(\alpha/2)\sqrt{\hat{p}_1(1-\hat{p}_1) / n_x + \hat{p}_2(1-\hat{p}_2) / n_y}$$

where $\hat{p}_1 = x/n_x$ and $\hat{p}_2 = y/n_y$ under the {**b**} option or $\hat{p}_1 = x$ and $\hat{p}_2 = y$ under the {**p**} option.

exact   For one-sample Bernoulli data (under the {**b**} or {**p**} option), calculate exact confidence intervals for p based on the binomial distribution instead of approximate confidence intervals based on the Normal approximation, as follows:

$$\underline{p} = I^{-1}_{(\alpha/2)}(x+1, n-x), \quad \bar{p} = I^{-1}_{(1-\alpha/2)}(x+1, n-x)$$

where n is the number of observations, x is the observed number of 1's, and $I_p^{-1}(a,b)$ denotes the inverse of the incomplete beta function $p = I_x(a,b)$ regarded as a function of x for fixed parameters a and b (see *BUG* Table 7.1).

nx = *INTLIST*
   The number of observations if x contains means; the number of Bernoulli observations if {**p**} is specified; or the n parameter of the binomial distribution if {**b**} is specified. In all cases, one value of *nx* may be specified for each element of x. If only one value of *nx* is given, it is used for each element of x.

ny = *INTLIST*
   If y contains means, or {**b**} or {**p**} is specified, *ny* has a meaning for y analogous to that of *nx* for x.

sdx = *FLOATLIST*
sdy = *FLOATLIST*
   SDs of the populations from which the columns of x and y are drawn: one for each column of x or y. Normally, {**sdx** = } and {**sdy** = } are taken to be the true population SDs, and the confidence interval is based on the Normal distribution. However, if the {**nx** = } and {**ny** = } options are also given and the {**z**} option is *not* given, then {**sdx** = } and {**sdy** = } are taken to be estimates based on the given number of observations and the unbiased variance estimator (divide by $n-1$), and the confidence interval is based on the t distribution.

   One SD may be given for each column of x or y. If only one value is specified, every column of the corresponding dataset is assumed to have that SD.

pool   If both x and y are specified and their population SDs are not, use a pooled estimate of the population SDs (which assumes that they are equal). If {**pool**} is not specified, the degrees of freedom for the t distribution used to compute the confidence interval are computed using Welch's approximation:

$$df \quad = \quad \frac{\left[\hat{SE}_x^2 \; + \; \hat{SE}_y^2\right]^2}{\dfrac{\hat{SE}_x^4}{n_x-1} \; + \; \dfrac{\hat{SE}_y^4}{n_y-1}}$$

z   Calculate the confidence intervals based on the Normal distribution even when the population SDs are unknown. This is appropriate only when the number of observations is large.

ht = , **ht**, **cc**
   {**ht**} and {**ht** = } cause **ci** to display the corresponding hypothesis test information for the confidence intervals. These three options have the same meanings as described in *ht*(*BRM*).

OUTPUTS
   mean@, **se@**, **df@**
      These outputs have the same meanings as the corresponding outputs described in *ht*(*BRM*).

ci@   The endpoints of the calculated confidence intervals. This is an array with dimensions (number of columns of x) by ($2 *$ number of values of *level*). The endpoints of the confidence interval with confidence level *level*[j] for column i of the data are contained in array elements $ci[i, 2j-1]$ and $ci[i, 2j]$.

t@, p@, cv@, h@
Output datasets for the corresponding hypothesis test. See *ht*(*BRM*) for details.

EXAMPLES

To construct 50%, 95%, and 99% confidence intervals for the mean salary and age of full professors at UC Berkeley in spring 1984 based on a random sample of size 27 (columns 1 and 2 of the library dataset *ucprofs*):

```
. ci ucprofs
Confidence Interval for the mean.  Data: ucprofs.

Col    N       sample    sample   df    level    t      confidence interval
               mean      SE                             lower end  upper end
 1     27     4.267e+04  1342.6   26.0  .5000   0.68    -918.41      918.41
                                        .9500   2.06    -2759.8      2759.8
                                        .9900   2.78    -3730.7      3730.7
 2     27      51.963    1.5382   26.0  .5000   0.68    -1.0522      1.0522
                                        .9500   2.06    -3.1618      3.1618
                                        .9900   2.78    -4.2742      4.2742
```

To construct 90% and 95% confidence intervals for the difference between the mean petal lengths, petal widths, sepal lengths, and sepal widths of the flower species iris versicolor (in the library dataset *iris.ve*) and iris virginica (in *iris.vi*):

```
. ci iris.ve iris.vi {level=.9, .95}
```

HYPOTHESIS TESTS

Ci can also perform hypothesis tests for the mean of a population or difference between the means of two populations. The **c@** input and the {**ht**}, {**ht=**}, and {**cc**} options are recognized and have the same meanings as in *ht*(*BRM*).

SEE ALSO

BUG Sections 6.1 and 6.2.
cdf(*BRM*), for commands which generate P-values and confidence coefficients.
confid(*BRM*), for confidence interval simulations.
ht(*BRM*), for hypothesis tests.

REFERENCES

B. L. Welch (1938). *The significance of the difference between two means when the population variances are unequal.* **Biometrika, 29,** 350-62. Presents Welch's approximation.

---

cluster — Create and print a cluster analysis tree

SYNOPSIS
> **cluster** *sim* {**dis;right;method**=;**full;height**=} [> *tree*]

DESCRIPTION
> **Cluster** produces a numerical and graphical representation of a cluster tree from a similarity matrix (such as a correlation matrix) or dissimilarity matrix, *sim*.

INPUTS
> **sim@** Symmetric similarity or dissimilarity matrix from which to form the tree.

OPTIONS

> **Algorithm Options**
> > **dis**   The input is a dissimilarity matrix. By default, the input is a similarity matrix.
> >
> > **right** Do tie breaking right-associatively. Default is left-associatively.
> >
> > **method**=*STRING*
> > > Specifies the clustering method to use. Use one of the following:
> > > | | |
> > > |---|---|
> > > | **average** | Arithmetic averages (default). |
> > > | **single** | Nearest neighbor clustering. |
> > > | **complete** | Total linkage clustering. |
> > > | **weighted** | Weighted averages based on cluster size. |

> **Printing Options**
> > **full**   Print a full scale along the picture margin. Default is a partial scale.
> >
> > **height**=*INT*
> > > Approximate height (in lines) of the picture, including the variable labels at the bottom. (Actual height may differ by one or two lines.) Default is 20.
> >
> > **long**   Print out the *tree* matrix.

OUTPUTS
> **tree@** Dataset whose rows are nodes in the tree and whose columns have the following meaning:
>
> | Column | Contains |
> |---|---|
> | 1 | Index of the left child. |
> | 2 | Index of the right child. |
> | 3 | Index of the parent. |
> | 4 | Similarity level of the cluster. |
> | 5 | Size of the cluster. |

EXAMPLES

In this example, **cluster** is used to perform a cluster analysis on *ozone* using the correlation matrix.

```
. stat ozone > cor@ c
. cluster c

Cluster tree:
```

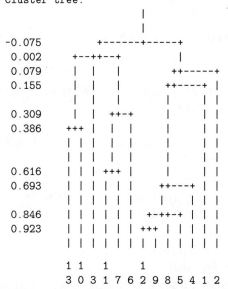

```
                              |
                              |
  -0.075         +------+-----+
   0.002     +--++--+          |
   0.079     |  |   |       ++-----+
   0.155     |  |   |       ++----+ |
            |  |   |       |     | |
   0.309    |  | ++-+      |     | |
   0.386   +++ | |  |      |     | |
            | | | |  |      |     | |
            | | | |  |      |     | |
   0.616    | | | +++  |     |     | |
   0.693    | | | | |  |   ++---+ | |
            | | | | |  |   |   | | |
   0.846    | | | | |  | +-++-+ | | |
   0.923    | | | | |  | +++ | | | | |
            | | | | | | | | | | | | |
           1 1  1     1
           3 0 3 1 7 6 2 9 8 5 4 1 2
```

This tree shows, for example, that variables 5, 8, 9, and 12 are quite closely related (as far as correlations are concerned). Compare this with the coded correlation matrix for *ozone* shown in *code(BRM)*.

---

code — Symbolic display of data

SYNOPSIS

code *data* { **cen;scale;std;eps**=;**high**=;**low**=;**vhigh**=;**vlow**=;**cor;hs**=;**vs**=;**x**= }

DESCRIPTION

Code gives a simple visual display of *data*. Positive values and negative values are respectively displayed as '+' and '–'; large positive values and large negative values as '#' and '='. Values close to zero are displayed as dots; missing values as blanks.

Code is useful when trying to discover patterns in two-dimensional residuals, when presenting large correlation matrices, etc.

OPTIONS

cen   Center the data (subtract off the mean) before coding it.

scale Scale the data (divide by the SD; variance divisor $= N - 1$) before coding it.

std   Standardize the data (first center it; then scale it) before coding it. Equivalent to {**cen;scale**}.

eps=*FLOAT*
   Low and *high* are respectively set equal to *–eps* and *eps*. Default is *eps*=0.5.

high=*FLOAT*
   Values greater than *high* (but not greater than *vhigh*) are displayed as '+'. Default is *eps*. Overrides {**eps**=}.

low=*FLOAT*
   Values less than *low* (but not less than *vlow*) are displayed as '–'. Default is *–eps*. Overrides {**eps**=}.

vhigh=*FLOAT*
   Values higher than *vhigh* are displayed as '#'. Default is 5*eps*. Overrides {**eps**=}. Ignored if less than or equal to 0.

vlow=*FLOAT*
   Values lower than *vlow* are displayed as '='. Default is *–5*eps*. Overrides {**eps**=}. Ignored if greater than or equal to 0.

cor   Equivalent to {**vhigh**=.8;**high**=.4;**low**=–.4;**vlow**=–.8}: suitable for displaying coded correlation matrices. This option overrides all options listed above.

hs=*INT*
   Horizontal spacing: number of spaces between the displayed characters. Default is 2.

vs=*INT*
   Vertical spacing: number of blank lines between each row of the display. Default is 0.

x=*INTLIST*
   Columns of *data* to display. Default is all columns.

EXAMPLES

Finding Patterns in Residuals
   Consumption contains personal consumption expenditures for the U.S. in five categories (rows) for the five years 1940, 1945, 1950, 1955, and 1960 (columns). The coded display of residuals from **twoway** shows a nonlinear trend.

```
. load consumption
. twoway consumption > resids
. code {scale} resids
```

```
=  -  .  +  +
-  -  .  .  +
+  .  .  .  .
+  +  .  -  -
+  +  .  -  -
```

This kind of structure can be removed by taking logs. See the references for an explanation.

```
. twoway log(consumption) > lresids
. code {scale} lresids

+  +  .  -  -
+  -  .  .  -
.  -  .  .  +
+  +  -  -  .
=  .  +  +  +
```

Coded Correlation Matrices

Ozone contains Los Angeles ozone pollution data. To display its coded correlation matrix:

```
. load ozone
. stat ozone {quiet} > cor@ c
. code {cor} c

#  .  .  .  .  .  .  .  .  .  .  .  .
.  #  .  .  .  .  .  .  .  .  .  .  .
.  .  #  .  .  .  .  .  .  .  .  .  .
.  .  .  #  +  .  +  +  +  -  .  +  -
.  .  .  +  #  .  #  #  -  .  #  .
.  .  .  .  .  #  .  .  .  .  .  .  .
.  .  .  +  .  .  #  .  .  .  +  .  -
.  .  .  +  #  .  #  #  -  .  #  -
.  .  .  +  #  .  #  #  -  .  #  -
.  .  .  -  -  .  -  -  #  .  -  .
.  .  .  .  .  .  +  .  .  .  #  .  .
.  .  .  +  #  .  #  #  -  .  #  -
.  .  .  -  .  .  -  -  -  .  .  -  #
```

This display shows strong correlations between variables 5, 8, 9, and 12; mild positive correlations between variable 4 and certain other variables; and mild negative correlations between variables 10 and 13 and certain others.

REFERENCES

Donald R. McNeil (1977). *Interactive Data Analysis.* Wiley, New York. Chapter 5, Section 4.

John W. Tukey (1977). *Exploratory Data Analysis.* Addison-Wesley, Reading, Mass. Chapter 11.

coin — Coin-tossing simulation

SYNOPSIS
   coin { **p**=;**ntoss**=;**nsamp**=;**show**;**noshow**;**se**;**mystery**= } [> *toss group cum*]

DESCRIPTION
   Coin tosses a (simulated) coin, counts the number of heads, and summarizes the results. This command illustrates elementary concepts of probability, as in Chapters 16, 17 and 21 of *Statistics*, by Freedman, Pisani, and Purves.

   By default, the user is queried for information such as the probability of heads, the number of tosses, etc. In this case the command is said to be *interactive*. If either the {**ntoss**} or {**nsamp**} option is specified, the command is *noninteractive* and the user is not queried.

OPTIONS
   p=*FLOAT*
      The probability of landing heads. If {**p**=} is not specified and the command is interactive, the user is asked for the probability of heads. Otherwise, the probability is set to .5.

   ntoss=*INT*
      Number of times to toss the coin between reporting results.

   nsamp=*INT*
      Number of samples (repetitions) of *ntoss* tosses each. Default is 1 if the command is noninteractive.

   show
   noshow
      Respectively show, or do not show, the results of the individual tosses. The default when interactive is to ask the user; the default when noninteractive is not to show the tosses.

   terse Show the cumulative results but not the results for the last group of tosses.

   se   Display the estimated and true standard errors.

   mystery=*STRING*
      Toss a mystery coin (i.e., unknown *p*). *STRING* is an integer immediately followed by one of the characters 'x', 'y', or 'z'. The true (unknown) value of *p* is a function of the integer, the character, and the invoking user; but otherwise is constant. The default integer is 0; the default character is 'x'—i.e., {**mystery**=**0x**}.

OUTPUTS
   toss@ Dataset which contains the results of the individual tosses, encoded as 1 (for heads) and 0 (for tails). If the command is interactive, this output dataset is a column vector. If noninteractive, it is a matrix with *ntoss* columns and *nsamp* rows.

group@
> *Nrep*-by-11 dataset which contains the results for each group of tosses.

cum@ *Nrep*-by-11 dataset which contains the cumulative results.

The **group@** and **cum@** output datasets contain the following:

Column	Contains
1	Number of tosses.
2	Observed number of heads.
3	Expected number of heads.
4	Chance error.
5	Estimated standard error for the observed number.
6	True standard error for the observed number.
7	Observed proportion of heads.
8	Probability of heads.
9	Relative chance error.
10	Estimated standard error for the observed proportion.
11	True standard error for the observed proportion.

EXAMPLES
Try each of these commands:

```
. coin
. coin {se}
. coin {ntoss=8}
. coin {ntoss=1000;nsamp=10;p=.2}
. coin > tosses
. coin >> results
. coin {mystery=3x}
. coin {mystery=3z}
```

REFERENCES
David Freedman, Robert Pisani, and Roger Purves (1978). *Statistics*. Norton, New York.

SEE ALSO
BUG Section 7.1.1 shows several additional examples, including a sample interactive session.

---

compare — Comparative boxplots

SYNOPSIS
    compare *x* {fi=;fo=;x=;height=}

DESCRIPTION
    Compare makes parallel boxplots of the columns of the dataset *x*, all on the
    same scale. This display is useful for comparing the distributions of several
    sets of data. See *boxplot(BRM)* for more information on boxplots.

INPUTS
    x@   The input data. Missing values in *x* are ignored.

OPTIONS
    fi=*FLOAT*
            The factor used to obtain the inner fences from the interquartile range.
            Default is 1.

    fo=*FLOAT*
            The factor used to obtain the outer fences from the interquartile range.
            Default is 1.5.

    x=*INTLIST*
            Columns to be plotted. Default is all columns.

    height=*INT*
            Height of the plot in lines. Default is 21. 60 gives a full-page plot for a
            lineprinter.

EXAMPLES
    Warpbreaks contains the number of breaks per unit length of yarn observed in
    a weaving experiment. Each of its six columns contains observations on a
    different type of yarn. To compare the number of breaks in the six types:

```
. load warpbreaks
. compare warpbreaks

 70.00         x
               |
               |
               |
               |
 55.00        -+-
              |*|
              | |
              | |
              | |        o   x   x
 40.00        | |            |  -+-
              | |    x   x   |  | |
              | |    |   |   |  | |
              | | -+- | -+- | |
              | | | | -+- |*| |*|   x
 25.00       -+- | | |*| | | | | | | |
              |*| | | | | | | -+- -+-
              -+- -+- -+- | | | |
               |   |   |  x  -+-
               x   |   x      x
 10.00         x
```

Illit is a two-way table of median illiteracy rates for nine different regions of the U.S. (rows) and six different years (columns): 1900, 1920, 1930, 1950, 1960, and 1970. To compare the different years:

```
. load illit
. compare illit
```

To compare the different regions, take the transpose:

```
. compare illit'
```

REFERENCES

Donald R. McNeil (1977). *Interactive Data Analysis.* Wiley, New York. Chapter 2.

John W. Tukey (1977). *Exploratory Data Analysis.* Addison-Wesley, Reading, Mass. Chapter 4.

SEE ALSO

oneway (*BRM*), *twoway* (*BRM*).

---

confid — Confidence interval illustration via simulation

SYNOPSIS

confid [*box*] {**bounds;dist**=;**m**=;**s**=;**level**=;**ndraw**=;**nsamp**=;**wide;** ... } [> *ci* ...]

DESCRIPTION

Confid illustrates the definition of a confidence interval via simulation. It draws a series of samples from a distribution and displays the sample mean, sample standard error (SE), and a diagram of the corresponding confidence interval for each sample. After drawing the samples, **confid** reports how many sample confidence intervals covered the mean.

Confid can sample from the Normal (Gaussian), logistic, or double exponential distributions, a user-supplied box model (in the *box* input dataset), or a mystery box. If none of these are specified, then by default **confid** asks the user to enter a box model from the keyboard in the same manner as the **box** command.

INPUTS

box@ The dataset which contains the box model (if used). **Confid** constructs its box of tickets in the same fashion as **box**. See *box(BRM)* for a description of the input box.

OPTIONS

level=*FLOAT*

alpha=*FLOAT*

Compute the confidence intervals at the specified confidence level, $1-\alpha$. The default value is {**level**= .95} or, equivalently, {**alpha**= .05}.

ndraw=*INT*

The number of tickets to draw per sample. If not specified, **confid** asks the user.

nsamp=*INT*

The number of samples to draw. If not specified, **confid** asks the user.

bounds

Display the upper and lower bounds of the sample confidence interval instead of the sample mean and estimated SE.

wide Display the estimated mean, SE, upper and lower bounds of the confidence interval for each sample, and a diagram of the corresponding confidence interval in a format which is approximately 130 columns wide (suitable for a lineprinter).

dist=*CHAR*

Sampling distribution. *CHAR* must be one of the following:

b	Use a box model. Default.
d	Double exponential distribution (density = $\frac{1}{2}e^{-|x|}$).
l	Logistic distribution.
m	Use a mystery box.
n	Normal (Gaussian) distribution.

m=*FLOAT*

Specify the population mean when sampling from the double exponential, logistic, or normal distributions. If not specified, **confid** uses a random number between –10 and 10.

s=*FLOAT*

Specify the population SD when sampling from the double exponential, logistic, or normal distributions. If not specified, **confid** uses a random number between .1 and 10.

OUTPUTS

ci@ Contains one row for each sample, as follows:

Column	Contains
1	Sample mean.
2	Estimated SE.
3	Lower bound of the confidence interval.
4	Upper bound of the confidence interval.

box@ If a box model is used, the dataset in which to save it.

EXAMPLE

With no inputs or options **confid** runs interactively, as in this session:

```
. confid
Either:  enter the numbers on the tickets in the box;
   Or:  on separate lines, enter each ticket's number and frequency.
Finish with RETURN and CTRL-D.
0 2 3 4 6
Control-D
Box saved in:  box03
Number of draws per sample? 100
Number of samples? 10
```

```
10 samples of 100 draws each are made from box03.
Population mean is 3.  Population SD is 2.

Sample Mean   Estimated SE            95% Confidence Interval
   2.9700        0.1951       |----------*----------|
   3.2800        0.2021              |----------*----------|
   2.9900        0.1899       |----------*----------|
   3.1400        0.1923            |----------*----------|
   2.8000        0.2005     |-----------*----------|
   3.4400        0.1961                   |----------*---------->
   2.9100        0.1832       |---------*----------|
   3.3000        0.2028            |-----------*-----------|
   3.0000        0.1770        |---------*---------|
   3.0900        0.1907        |-----------*----------|

True mean + and - 4 SEs:    ......................x........................

9 out of 10 confidence intervals (or 90.00%) covered the population mean.
```

In the diagrams, the sample means are denoted by '*' characters and the end-points of the confidence intervals by '|' characters. If a confidence interval extends beyond the left or right edge of the picture area, that fact is denoted by a '<' or '>' character, respectively.

REFERENCES

David Freedman, Robert Pisani, and Roger Purves (1978). *Statistics.* Norton, New York. Chapters 21, 23, and 24 discuss confidence intervals. In particular, see Figure 1 on page 349.

SEE ALSO

ci(BRM), for computing confidence intervals from real data.

---

count — Count rows (cases) which meet a specified condition

SYNOPSIS

count *cond* {**log;nm;x=**} [> *n*]

DESCRIPTION

Count reports the number of rows of *cond* out of the total number which meet a specified criterion. By default, *cond* is treated as a logical dataset: the number and proportion of rows which are non-zero and non-missing is reported. If *cond* is a row vector, **count** treats it as a column vector.

INPUTS

cond@

> The input array to be examined. If *cond* has more than one sheet, the count is given for each sheet separately. Each sheet of *cond* must be a vector unless the {**nm**} option is set.

OPTIONS

log   Logical: count the number of rows which are non-zero and non-missing. This is the default.

nm   Count the number of rows of *cond* which contain no missing values.

x=*INTLIST*
> Columns of *cond* to check for missing values. Default is all columns. This option is available only when used with the {**nm**} option.

OUTPUTS

n@   *Ns*-by-1-by-2 dataset which contains the counts and proportion of elements which meet the criterion: one sheet for each sheet of *cond*.

EXAMPLES

A logical dataset can be created by an implicit or explicit **let** command—see *let*(*BRM*).

To count the number of rows of *turbines* which contain no missing data:

```
. load turbines
. count turbines {nm}
```

To generate 100 Normal(0,1) random numbers and count how many fall in various intervals:

```
. rgau {dims=100} > x
. count (x > 0)                  # positive values
. count (x >= -1 && x <= 1)      # values in the interval [-1, 1]
. count (x < -2 || x > 2)        # values outside the interval [-2, 2]
```

SEE ALSO

BUG Section 8.6 (more information on logical operators which may be used to form the *cond* vector); Section 8.7 (more complex examples of **count** using **let**).

select(*BRM*) for conditional selection of cases.

---

demo — Introductory demonstration modules

SYNOPSIS

demo.*name*

DESCRIPTION

A number of demonstration modules are available to acquaint the beginning user with basic features of BLSS. They provide both explanations and examples of actual BLSS commands. To run any module, enter BLSS and then type its name.

Module	*Demonstrates*
demo.let1	Elementary variable transformations using the **let** command.
demo.let2	Variable transformations to attain symmetry.
demo.medp	Median polish for a two-way table.
demo.plot1	Visual displays of univariate datasets.
demo.plot2	Visual displays of multivariate datasets.

demo.read	Create a BLSS dataset by entering data at a terminal.
demo.reg1	Multiple regression by ordinary least squares.
demo.reg2	Robust multiple regression by iteratively weighted least squares.
demo.reg3	Comparison of fits made by the **regress** and **robust** commands when the data contain outliers.
demo.smo	Smooth a sequence of observations.
demo.stat1	Numerical summaries of univariate datasets.
demo.stat2	Numerical summaries of multivariate datasets.
demo.util	Utility commands.

These modules are implemented as BLSS macros. Instructors can copy these modules (from the directory *~blss/bin*) and adapt them to their own purposes.

---

demod — Complex demodulation of a time series

SYNOPSIS
demod x [y] {**freq**=;**l**=;**d**=} [> *amp phase aux*]

DESCRIPTION
Demod performs complex demodulation of a real or complex valued time series whose real part is in x and imaginary part is in y. The i-th column of y corresponds to the i-th column of x. If y is not given, the series is assumed to be real valued.

INPUTS
x@   An n-by-k array which contains the real part of the time series. Each of the k columns is treated as a separate time series of length n.

y@   An n-by-k array which contains the imaginary part of the time series.

The input datasets x and y must have the same dimensions. Missing values are not allowed.

OPTIONS
freq=*FLOAT*
   The frequency at which to demodulate. The default is zero.

l=*INT* An integer which defines the length of the smoothing filter. The filter has $2*l+1$ coefficients. The default value of l is 10% of the length n of the time series.

d=*INT*
   Decimate the output to d points; that is, estimate the output at d equally spaced time points between 1 and n inclusive. The value specified may not exceed n. If $d = 0$, no decimation takes place. The default value is the minimum of 101 and n.

OUTPUTS
All output datasets described below are created. If no output dataset name is specified, the name *x.tag* is used, where *x* is the name of the first input dataset and *tag* is the tag name of the output dataset.

amp@
An *n*-by-*k* array whose columns contain the estimated $\log_{10}$ instantaneous amplitude for each series at equispaced time points which cover the entire time span of the input series.

phase@
An *n*-by-*k* array whose columns contain the estimated instantaneous phase for each series at equispaced time points which cover the entire time span of the input series.

aux@ A column vector of 5 numbers:

Element	Contains
1	The series length.
2	The frequency of demodulation.
3	The length of the smoothing filter.
4	The bandwidth of the smoothing filter.
5	A number which, when added to the $\log_{10}$ instantaneous amplitude, gives an estimate of one half times the $\log_{10}$ instantaneous power.

REFERENCES
Peter Bloomfield (1976). *Fourier Analysis of Time Series: An Introduction.* Wiley, New York. Chapter 6.

Keith A. Haycock and David R. Brillinger (1985). *LIBDRB: A subroutine library for elementary time series analysis.* Technical Report No. 48, Department of Statistics, University of California, Berkeley.

---

demoiv — Probability histogram of sum of 2^n draws from a box model

---

SYNOPSIS
demoiv [*box*] {**n**=} [> *box dist*]

DESCRIPTION
Demoiv calculates the distribution of the sum of draws from a box of tickets and displays its histogram. At each cycle, **demoiv** convolves the current distribution for the sum of 2^j draws from the original box with itself to get the distribution for the sum of 2^{j+1} draws. It displays a scaled histogram of the significant parts of the new distribution and then asks whether it should continue. If not, type 'n', 'N', 'q', or 'Q'.

This command provides an elementary illustration of the central limit theorem for discrete distributions. It supplements Chapter 18 of *Statistics*, by Freedman, Pisani, and Purves.

INPUTS

box@ The dataset which contains the input box. If not specified, **demoiv** asks the user to type in a box at the keyboard.

The input box is treated just as in the **box** command. See *box(BRM)* for complete information.

OPTIONS

n=*INT*

By default, **demoiv** asks the user whether to continue after each cycle. If this option is set, **demoiv** displays the histograms for $1, 2^1, 2^2, \ldots, 2^n$ draws and then stops. As a special case, setting {**n**=**0**} displays the histogram for 1 draw (only) from the box.

OUTPUTS

box@ The name of the dataset in which to save the box, if entered at the keyboard. If no dataset name is specified, a name of the form *boxNN* is used.

dist@ The name of the dataset in which to save the distribution of the sum after the last cycle.

REFERENCES

David Freedman, Robert Pisani, and Roger Purves (1978). *Statistics.* Norton, New York.

SEE ALSO

BUG Section 7.1.3 and Figure 7.2, for an interactive example and the histograms it produces.

---

dft — Discrete Fourier transform

SYNOPSIS

dft *x* [*y*] {**inv**} [> *dft*]

DESCRIPTION

Dft computes the discrete Fourier transform of a series:

$$z(\omega_j) = \sum_{t=0}^{n-1}(x_t + iy_t)e^{-i\,\omega_j t}$$

evaluated at the *n* Fourier frequencies $\omega_j = 2\pi j/n$ for $j = 0, 1, \ldots, n-1$. (Note that indices in the formula run from 0 to $n-1$, whereas the corresponding BLSS array indices run from 1 to *n*.) Fourier transforms of several series may be computed by placing each series in a separate sheet.

INPUTS

x@   An array which contains the real part of the data.

y@   An array which contains the imaginary part of the data. May be omitted if the series has no imaginary component.

X and *y* must be vectors of the same length *n*. *N* should be reasonably composite: if n fails the factorization criterion, **dft** reports the next largest and smallest acceptable values of n.

Missing values are not allowed in the input datasets.

OPTIONS

inv   Compute the inverse transform. Divides by n (and uses i, not $-i$, in the complex exponent). Hence, the result of applying **dft** to a series and then **dft** {**inv**} (or vice versa) is the original series.

OUTPUTS

dft@   The discrete Fourier transform. The first column contains the real part of the transform and the second the imaginary part. If no output dataset is specified, the transformed dataset is put in x.**dft**, where x is the name of the **x@** input dataset.

REFERENCES

Peter Bloomfield (1976). *Fourier Analysis of Time Series: An Introduction.* Wiley, New York. Chapter 3.

David R. Brillinger (1975). *Time Series: Data Analysis and Theory.* Holt, New York. (Expanded ed., 1981, Holden-Day, San Francisco.) Chapter 3.

Keith A. Haycock and David R. Brillinger (1985). *LIBDRB: A subroutine library for elementary time series analysis.* Technical Report No. 48, Department of Statistics, University of California, Berkeley.

---

diff — Difference a time series

SYNOPSIS

diff x {**diff**=;**nona**} [> *diff*]

DESCRIPTION

Diff differences each column of x. By default, **diff** performs first differencing: $x[i,]$ is replaced by $x[i,] - x[i-1,]$.

Missing values are not allowed in x.

INPUTS

x@   Array whose columns are to be differenced.

OPTIONS

diff=*INTLIST*

Specify non-default differencing. The series x is differenced repeatedly, once for each value in the list. The values give the spacing used: If a value is l, $x[i,]$ is replaced by $x[i,] - x[i-l,]$, where x is the input series adjusted by any previous differencing. The value l may be negative. The default value is 1.

nona   Missing values at the start or end of the differenced series are omitted.

OUTPUTS

diff@ The differenced series.

If no output dataset is specified, the differenced series is put in *x*.**diff**, where *x* is the name of the input dataset.

SEE ALSO

lag(*BRM*), *tsop*(*BRM*).

---

echo, prompt — Echo arguments

SYNOPSIS

echo *args*
prompt *args*

DESCRIPTION

Echo and **prompt** are BLSS shell internal commands. Both commands display their arguments on the terminal, separated by spaces. **Echo** adds a newline. **Prompt** adds a space but no newline.

EXAMPLES

To add a comment to a text file:

```
. echo Now we regress fum on foo >> textfile
. regress foo fum >> textfile
```

To see the values of the string variables *money* and *honey*:

```
. echo $money $honey
```

Warning: the symbols >, $, and # have special meanings to the BLSS shell. To **echo** these characters, you must remove their special meanings, either by preceding them by a backslash \, or by wholly embedding them in double-quote characters "". The following examples work:

```
. echo "a >> b"
. echo regress x y \> e
```

The $< String

The **prompt** command is often used together with the $< string, whose value is one line of input read from the terminal. Commands such as the following are useful in interactive BLSS macros:

```
prompt What dataset shall we use?
set name $<
echo OK, we'll use $name.
```

SEE ALSO

BUG Sections 4.9.1 and 10.8.

---

edit — Edit text files

SYNOPSIS
    edit *files* ...   {**editor**=;**terse**; ... }

DESCRIPTION
    Edit invokes the editor on the specified text files. If no file name is specified, it asks what you want to edit. If a named file does not exist, **edit** creates it. Unless another area is specified, the file edited is in your work area. UNIX pathnames may be used as file names.

    Edit cannot be used to edit datasets. Use **editdata** to do so.

OPTIONS
    editor=*STRING*
        Specifies the editor to use. The EDITOR environment string can also be used; the {**editor**=} option overrides it. The default editor is normally *vi*; local sites may establish different defaults. You can select the editor of your choice by setting the EDITOR environment string—see *set(BRM)*.

    bin, class, data, doc, help, home, sys, text
    dirs=*dirname,* **tree**=*treename,* **area**=*areaname*
        These options specify an alternate area (or directory) in which to edit. For a full description, see *area(BRM)*. The default values are:

dirs	your work area
tree	your *blss* tree (normally *~/blss*)
area	the text area

    terse Fewer diagnostic messages.

EXAMPLES
    Edit the text file named *fuerte* in your work area:

      `.  edit fuerte`

    Edit the text file named *haas* in your text area:

      `.  edit {text} haas`

    Edit the text file named *mymacro* in your bin area:

      `.  edit {bin} mymacro`

    (This is one way to write your own macros. The command 'mymacro' then runs the macro. See *BUG* Section 10.8 for more information.)

SEE ALSO
    BUG Section 4.9.3, for more examples and discussion.
    addtext(BRM), *area(BRM)*.
    Appendix B: Basic Vi Command Summary.

---

editdata — Edit a BLSS dataset

SYNOPSIS

   editdata *a* {**editor**=; *show-format-options*} [> *b*]

DESCRIPTION

   Editdata allows you to edit the BLSS dataset *a*. If the output dataset *b* is specified, the new dataset is called *b*; *a* is unchanged. Otherwise, the new dataset is called *a*; **editdata** creates a backup copy of the old *a* called *a*.**bak**.

   Because BLSS datasets are stored in a binary format, **editdata** first invokes the **show** command to convert *a* into a text file. It then invokes the editor on the text file. Finally, it uses the **read** command with the {**autodims**} option to convert the text file back into a BLSS dataset. Thus, whatever changes you make to the text representation of the dataset must be acceptable as input to the **read** command.

OPTIONS

   editor=*STRING*

   > Specifies the editor to use. The EDITOR environment string can also be used; the {**editor**=} option overrides it. The default text editor is normally *vi*; local sites may establish different defaults. You can select the editor of your choice by setting the EDITOR environment string—see *set(BRM)*.

   Show Format Options

   The default **show** format used for creating the text representation of the dataset is {**g**=7}. Any **show** format option may be used to specify a different format. However, be aware that a format with too few decimal places will cause precision to be lost.

LIMITATIONS

   BLSS datasets with too many columns will not fit into an 80 column (or even 132 column) text format; screen wraparound occurs. For datasets with many columns, you should choose the narrowest **show** format with sufficient precision; **editdata** has no ability to do this.

SEE ALSO

   BUG Section 4.10.3.
   edit(BRM), *show(BRM)*.
   Appendix B: Basic Vi Command Summary.

---

eigen — Eigenvalue-eigenvector decomposition of a matrix

SYNOPSIS
>   **eigen** *x* [> *l v*]

DESCRIPTION
>   **Eigen** computes the eigenvalue-eigenvector decomposition of the real sym-
>   metric *n*-by-*n* matrix **X** (contained in the dataset *x*) and places the results in
>   the following datasets:

>   *l*   Row vector λ of the *n* eigenvalues, in descending order.
>   *v*   *n*-by-*n* matrix **V** whose columns contain the *n* eigenvectors.

>   These matrices satisfy **X** = **VΛV$'$** or, equivalently, **XV** = **VΛ**, where **Λ** =
>   *diag*(λ), that is, the *n*-by-*n* diagonal matrix with entries λ. In terms of the
>   individual eigenvalues λ_i and eigenvectors $\mathbf{v}_i$ (columns of **V**), $\mathbf{Xv}_i = \lambda_i \mathbf{v}_i$.

>   If the matrix **X** is not symmetric, **eigen** attempts to symmetrize it. If both the
>   [*i,j*]-th and [*j,i*]-th elements of **X** are missing, the eigenvalues and eigenvec-
>   tors cannot be obtained. If the dataset *x* contains more than one sheet and its
>   decomposition cannot be obtained, then the corresponding sheets of the
>   datasets *l* and *v* are filled with missing values.

>   *Algorithm.* This command uses the EISPACK subroutines *tred2* and *tql2*.

INPUTS
>   **x@**   Dataset which contains the matrix **X** to be decomposed. If it contains
>   multiple sheets, **eigen** decomposes each sheet separately.

OUTPUTS
>   **l@**   The row vector λ of eigenvalues.
>   **v@**   The matrix **V** of eigenvectors.

>   If no output datasets are specified, the eigenvalues are put in *x*.**l** and the
>   eigenvectors in *x*.**v**, where *x* is the name of the input matrix.

EXAMPLE
>   To obtain the eigenvalues and eigenvectors of a random symmetric matrix:

>   ```
>   . rcauchy {dims=5,5} > x
>   . x = (x+x')/2
>   . eigen x > l v
>   ```

>   To verify the equations above:

>   ```
>   . show (x - v#*diag(l)#*v')
>   ```

REFERENCE
>   B. T. Smith et al. (1972, 1976). *Matrix Eigensystem Routines* — *EISPACK
>   Guide.* Springer-Verlag, New York.

SEE ALSO
>   *chol(BRM), svd(BRM), eispack(BML).*

---

filter — Filter a time series

SYNOPSIS
  filter *x filter* {**nona**} [> *f*]

DESCRIPTION
  Filter filters the time series *x* using the filter coefficients in *filter*:

$$f_t = \sum_{i=1}^{nf} filter_i * x_{t+nf-i}$$

  where *f* is the filtered series and *nf* is the number of filter coefficients.

  Each column of *x* is filtered using the corresponding column of *filter* until the last column of *x* or *filter* is used, whichever comes first. The final column of whichever array has fewer columns is used repeatedly with the remaining columns of the other array. As special cases, if *x* is a matrix and *filter* is a single column (or vice versa), each column of *x* is filtered by *filter* (or vice versa).

  Normally, filtering produces missing values at the end of the series; the {**nona**} option causes the missing values to be omitted.

INPUTS
  x@   The time series to be filtered. If *x* is a row vector or sheet vector, it is treated as a column vector.

  filter@
          The filter coefficients. If *filter* is a row vector or sheet vector, it is treated as a column vector.

  Missing values are not allowed in the input datasets.

OPTIONS
  nona Missing values at the end of the output series are omitted.

OUTPUT
  f@   The filtered series.

  If no output dataset is specified, the filtered series is put in *x*.**f**, where *x* is the name of the input series.

REFERENCE
  Keith A. Haycock and David R. Brillinger (1985). *LIBDRB: A subroutine library for elementary time series analysis.* Technical Report No. 48, Department of Statistics, University of California, Berkeley.

---

freq — Frequency distributions

SYNOPSIS

    freq x {**x**=;**width**=;**cum**;**all**;**quart**;**end**=;**n**=;**p**=; ... } [> *count freq cum*]

DESCRIPTION

    Freq displays frequency distributions for the columns of x. For each column, it shows: the values the observations take on, the number (or count) of observations which take on each value, and the fraction (or frequency) of observations which take on each value.

INPUTS

    x@   The input dataset. If x is a row vector or a sheet vector, it is treated as a column vector.

OPTIONS

    x=*INTLIST*

        Columns of x for which to compute frequency distributions. Default is all columns.

    nona   Do not tabulate NAs (missing values).

    width=*INT*

        Maximum width of the output display. If the table exceeds this width, it is printed in blocks. The WIDTH environment string can also be used; the {**width**=} option overrides it. Default width is 79, or 132 if the {**lpr**} option is set.

Cell Contents

    The following options control the contents of the cells in the table. They are listed in the order in which they appear in the cells.

    nocount

        Omit the cell counts from the table. By default, the cell counts are tabulated.

    nofreq

        Omit the cell frequencies (proportions). By default, the cell frequencies are tabulated.

    cum   Tabulate cumulative counts.

    all   Tabulate and display all possible cell contents.

Determining the Categories

    The following options specify the categories. They are listed in order of precedence, from high to low. (That is, if conflicting options are specified, the first one listed below is used.)

    end=*FLOATLIST*

        List of category endpoints. If the data extrema are not included, they are added to the list.

p=*FLOATLIST*
> Category endpoints are the quantiles corresponding to the list of probabilities *p*. If 0 and 1 are not included, they are added to the list.

n=*INT*
> The categories are *n* equal-length intervals.

c   The default: each distinct value defines a separate category.

quart The categories are the four quartiles of the data.

Lists of category endpoints are sorted in ascending order and any duplicate values are removed.

OUTPUTS
Output datasets, if requested, are created only for the first specified column of *x*.

count@
> Cell counts.

freq@ Cell frequencies.

cum@ Cumulative cell counts.

end@ Endpoints of the cell categories (or values if the {**c**} category option is in effect).

EXAMPLES
Discoveries is the number of 'great' discoveries and inventions per year, from 1860-1959, according to *The 1975 World Almanac and Book of Facts*. To calculate a frequency distribution of *discoveries*:

```
 . freq {all} discoveries

  discoveries, N = 100
```

Value	0	1	2	3	4	5	6	7
Count	9	12	26	20	12	7	6	4
Freq	.0900	.1200	.2600	.2000	.1200	.0700	.0600	.0400
Cum	9	21	47	67	79	86	92	96

Value	8	9	10	12
Count	1	1	1	1
Freq	.0100	.0100	.0100	.0100
Cum	97	98	99	100

SEE ALSO
xtab(*BRM*), for cross-tabulation (that is, bivariate frequency distributions).

fromunix — How to execute BLSS commands directly from UNIX

DESCRIPTION

Regular BLSS commands may be invoked directly from the UNIX shells. However, the syntax is different due to differences in argument handling and special characters. The general syntax for a regular BLSS command from the BLSS shell is:

> *blss-command* [*tag@*] *indata* ... {*opt;opt=val*} ... > [*tag@*] *outdata* ... [>> *file*]

The corresponding general syntax from the UNIX shells is:

> *blss-command* [*@tag*] *indata* ... −*opt* −*opt=val* ... \> [*@tag*] *outdata* ... [>> *file*]

The following table summarizes the differences.

Grammatical Function	BLSS Syntax	UNIX Syntax
Untagged dataset name	*name*	*name*
Tagged dataset name	*tag@ name*	*@tag name*
Option	{*opt*}	−*opt*
Option with value	{*opt=val*}	−*opt=val*
Output dataset separator	>	\>
Output text redirection	>> *file*	> *file* or >> *file*

Notes

Backslashes are necessary to escape characters which have special meaning to the UNIX shells, such as >. If the UNIX kill character is @, it too must be escaped (or reset; see *stty*(1)).

BLSS commands are normally stored in the directory ~*blss/bin*, although on some systems this is different. In any case, the pathname of the BLSS *bin* directory may be discovered from the BPATH environment string within BLSS. Either refer to BLSS commands by full pathname or add the BLSS *bin* directory to your PATH environment string.

Unless BLSS datasets are specified by pathname, they are assumed to be in the current directory. Similarly, output datasets are created in the current directory unless a pathname is specified.

BLSS commands accessed from UNIX shells may be run in background or batch in the usual fashion.

Let Commands

To run **let** commands from the UNIX shells, it is easiest to quote the entire series of let-expressions in order to escape the UNIX meaning of let-characters:

> **let** '*expression*; *expression*; ...'

Macros

The remarks above apply to regular BLSS commands and **let** commands only. To run BLSS macros from UNIX, one must either rewrite the macro as a shell script using UNIX syntax or else invoke BLSS as the macro interpreter. There are three such ways to invoke BLSS.

First, BLSS can run the macro in the usual protected BLSS environment (the work area). In this case, the macro must **load** and **save** files just as the user would in a regular BLSS session. The syntax is:

> **blss −m** *macroname* [*arguments in UNIX syntax*]

Second, BLSS runs in the current directory when invoked with the −s option. Hence a macro need not **load** or **save** files, and can access datasets in the current directory, when invoked as:

> **blss −s** *macroname* [*arguments in UNIX syntax*]

Third, the BLSS shell may be invoked directly on the macro, although, depending on particulars of BLSS installation at your site, the environment string BLSS_SYS may need to be set (see *blssf(BRM)*). The syntax is:

> **blssh −m** *macroname* [*arguments in UNIX syntax*]

As in the previous case, the process runs in the current directory.

EXAMPLES

Here is a typical BLSS command:

```
. regress iris.se w@ w.se {x=1,2;y=3} > iris.e rstat@ r
```

To run the same command from UNIX:

```
% ~blss/bin/regress iris.se @w w.se -x=1,2 -y=3 \> iris.e @rstat r
```

To produce a 10-by-100 matrix x of Normal(5,2) pseudo-random numbers from UNIX:

```
% ~blss/bin/rgau -dims=10,100 \> x
% ~blss/bin/let 'x=x*sqrt(2)+5'
```

SEE ALSO

BUG Section 10.8, for an introduction to macros.
intro(BRM), for BLSS command syntax.
blssf(BRM), for BLSS startup and use of UNIX directories.
unix(BRM), for executing UNIX commands from within BLSS.
Appendix A: Basic UNIX Command Summary.

---

help, man, doc, news — Access on-line documentation

SYNOPSIS

help *topics* ... { **key; list; lpr** =; **lpr;** ... }
man *topics* ... { **key; list; lpr** =; **lpr;** ... }
doc *topics* ... { **key; list; lpr** =; **lpr;** ... }
news

DESCRIPTION

Help gives access to on-line BLSS documentation: both help files that are available on-line only and entries in the *BLSS Reference Manual.* It searches the help, man, and doc areas for text files with the requested topic name(s) and informs you if it fails to find anything with a given name.

Man works like **help**, except that in case of a name conflict (both a help file and a manual entry exist with the same name) **help** shows the help file whereas **man** shows the manual entry. In addition, **man** searches the manetc and manlib areas. Thus it gives on-line access to entries in the BLSS support and library manuals.

Doc gives on-line access to longer BLSS documentation. It looks first in the doc area for the named topics and then in the other areas searched by **help**.

Typing 'help' by itself is the same as 'help blss'.

Typing 'news' by itself is the same as 'help news'.

Type 'help index' for a description of available BLSS system manual entries.

Type 'help {list}' for a complete list of all available manual entries and help topics, including your own.

Help automatically uses the *more* mechanism to pause after each screenful of text—see *BUG* Section 4.8. It normally invokes *more*(1) with the **–dfs** flags; however, if the environment string MORE is set, only the **–fs** flags are specified. (See *set*(*BRM*) for information on setting environment strings.)

OPTIONS

key Keyword search: list the names of manual entries which appear to be related to the specified *topics*.

list List all available on-line manual entries and help files.

lpr = *STRING*
lpr Send help files to the lineprinter instead of the terminal. See *options*(*BRM*) for details.

bin, class, data, doc, help, home, sys, text
dirs = *dirlist,* **tree** = *treelist,* **area** = *arealist*
These options specify alternate areas (or directories) in which to search for help files. For a full description see *area*(*BRM*). The default values are:

dirs   (none)
tree   trees on your BPATH environment string
area   **help:man:doc**, if invoked as **help**;
           doc:man:help, if invoked as **doc**;
           man:manetc:manlib:doc:help, if invoked as **man**.

EXAMPLES

To see the on-line manual entries about the **stat** and **scat** commands, type:

```
. help stat scat
```

To find out which manual entries appear to be related to the topic *regress*, type:

```
. help {key} regress
```

Your Own Help Files

You can write your own help files using the {**help**} option of the **addtext** or **edit** commands. For example, to document your own BLSS dataset called *pineapple*:

```
. addtext {help} pineapple
Creating text file "pineapple" in your help area.
Type your text; finish with RETURN and CTRL-D.
Pineapple is a BLSS dataset that contains data on per capita
pineapple consumption in the United States from 1920 to 1980.
Control-D
```

or:

```
. edit {help} pineapple        # invokes the UNIX text editor
```

To see this help file, type:

```
. help pineapple
```

As stated above, **help** normally looks for files of a given name in your own help area before looking for BLSS system help files of the same name. Thus, if you have your own help file named *scat,* the command:

```
. help scat
```

will show your file—not the BLSS manual entry *scat(BRM).* To see the BLSS manual entry, type:

```
. man scat
```

NOTE

To see *intro(BLR)* or *intro(BML)* on-line, you must type 'man intro_l' or 'man intro_m'.

SEE ALSO

BUG Section 4.11.
addtext(BRM), area(BRM), edit(BRM).

---

history, redo — Show command history; repeat or edit a command

---

SYNOPSIS
history [*n*] [–] [– –]
redo [*n*]

DESCRIPTION
History displays the most recent commands typed in the BLSS session. If a numeric argument *n* is given, the most recent *n* commands are displayed, up to a maximum of 100. The default display is 23 commands (one standard screenful). The option '–' suppresses the command numbers and leading tabs. It is useful when the display will be edited for insertion into a command file. The option '– –' suppresses the command numbers only. The options and numeric argument(s) may be given in any order. The final numeric argument given supercedes any others.

Redo invokes a one-line open mode text editor on a command in the history. By default, the most recent command is edited. If *n* is a positive number, then **redo** *n* invokes the editor on command number *n* in the history, and **redo** –*n* invokes the editor on the *n*-th most recent command.

Exiting **redo** with a carriage return causes the result to be executed as the next BLSS command. Exiting with a **q** or with an interrupt (which is most often control-C) causes no action to be taken. If you seem to be stuck, an interrupt should save you.

History and **redo** are internal commands to the BLSS shell.

EDITING COMMANDS
The following one-line open mode editor commands are implemented. Except as noted by a dagger †, commands and features of this editor are a subset of those of the UNIX editor *vi*. If an editor command is unrecognized, or cannot be executed, the editor beeps at you.

All commands are single characters. In some cases, alternate characters (separated by a comma in the list below) provide the same command.

In this editor, a *word* is any space- or tab-separated string of characters. †

Moving the Cursor
l, <SPACE>
     Move forward (right) one character.
L   Move forward ten characters. †
h, <BACKSPACE>
     Move backward (left) one character.
H   Move backward ten characters. †
b   Move to the beginning of the current (or previous) word.
e   Move to the end of the current (or next) word.
w   Move to the beginning of the next word.
0, ^   Move to the beginning of the line.

$ Move to the end of the line.
f*c*   Move forward to (find) the next occurrence of the character *c*.
F*c*   Move backward to the previous occurrence of the character *c*.
;   Repeat the last **f** or **F** command.
,   Repeat the last **f** or **F** command, but in the opposite direction.

Making Changes

x   Delete (x out) the character the cursor is over.
d   Delete the rest of the word, from the cursor to the next space. †
D   Delete the rest of the line, from the cursor to the end.
a   Append characters to the right of the cursor.
A   Append characters to the end of the line.
i   Insert characters to the left of the cursor.
I   Insert characters at the beginning of the line.
c   Change the rest of the word, from the cursor to the next space. †
C   Change the rest of the line, from the cursor to the end.
r*c*   Replace the character the cursor is over with the character *c*.
R   Replace characters by typing over them.
s   Substitute text for the character the cursor is on top of.

The **a**, **A**, **c**, **C**, **i**, **I**, **R**, and **s** commands accept new text until you type ESCAPE or control-D. You can also finish these commands by typing RETURN, which is equivalent to typing ESCAPE and then RETURN. †

~   Change the character the cursor is on top of from lowercase to uppercase—or vice versa—and move forward to the next character.
u   Undo the last change.

Exiting, Interrupting, and Redrawing

<RETURN>
        Finish editing and execute the edited string as a new command. †
q, <INTERRUPT>
        Quit the editor, with no action taken. †
^L, **^R**
        (Control-L or control-R.) Redisplay the line being edited.

EXAMPLES

Command 27 in the following history contains errors:

```
. history -3
26      stat plover
27      stat plover {{x=1:4,quiet} > fred
28      history
. redo 27
stat plover {{x=1:4:quiet} > fred        # now edit this
```

To fix the '{{', type 'f{' to find the first '{', followed by 'x' to delete it. To change the second ':' to ';', type 'f:' followed by ';', followed by 'r;'. To change the name of the output dataset from *fred* to *joe*, type 'ww' to move to the start of the word, then 'c' followed by 'joe', and then SMALL CAPS ESCAPE. The result-ing command:

```
stat plover {x=1:4;quiet} > joe
```

is executed when you hit RETURN.

In order to redo and edit the command, we could also have typed:

```
. redo -2
```

SEE ALSO
BUG Sections 10.1, 10.6, and 10.8.
set(*BRM*).

---

ht — Hypothesis tests (z-tests, t-tests) for means and differences of means

---

SYNOPSIS
ht x [y c] {**alpha**=;**b**;**ht**=;**nx**=;**ny**=;**p**;**pool**;**sdx**=;**sdy**=;**x**=;**y**=;**z**; ...} [> *mean* ...]

DESCRIPTION
Ht performs hypothesis tests for a population mean, probability of success or proportion of successes, and the difference between two population means, probabilities of success, or proportions of successes. It and the **ci** commands are closely related; many of their inputs, options, and outputs are identical.

One-Sample Hypothesis Tests
If x is specified but y is not, **ht** performs hypothesis tests for the mean of the population from which the sample x is drawn. If {**nx**=} and {**sdx**=} are specified, x contains sample means instead of data. If x contains several columns, **ht** performs hypothesis tests separately for each column. If c is specified, it contains the population mean under the null hypothesis. Other-wise, the population mean under the null hypothesis is 0.

Two-Sample Hypothesis Tests
If both x and y are specified, **ht** performs hypothesis tests for the difference of the means of the populations from which the data x and y are drawn. If x and y contain several columns, **ht** performs hypothesis tests for the difference between the means of corresponding columns. If c is given, it contains the difference between the population means under the null hypothesis. Other-wise, the difference under the null hypothesis is 0.

z-Test or t-Test?
If the population SDs are known and specified via the {**sdx**=} and {**sdy**=} options, **ht** performs a z-test. Otherwise, **ht** estimates the population SDs and performs a t-test by default; or a z-test if the {**z**} option is specified.

Hypothesis Tests for Proportions
A random variable which is 1 with probability p and 0 with probability $1-p$ is called *Bernoulli* with parameter p. The sum of n Bernoulli random variables

with parameter p is called *binomial* with parameters n and p. If the {**b**} option is specified, x is a binomial observation. If the {**p**} option is specified, x is the observed proportion of 1's in a Bernoulli sample of size n. In either case, n must be specified using the {**nx**=} option. If {**b**} or {**p**} is specified and y is given, it contains a second sum or proportion; the Bernoulli sample size for y must be specified by {**ny**=}.

If c is given and y is not given, c contains the value of the parameter p under the null hypothesis. The default value of p under the null hypothesis is .5. If x and y are both given, the null hypothesis is that the true probabilities are equal; c may not be given.

The Output Display

The output display shows: the number of observations; the mean (or mean difference) under the null hypothesis; the sample mean (or mean difference); the sample or known SE; the degrees of freedom (df) in the SE, if appropriate; the test statistic; the P-value; the value(s) of the significance level α; the critical value of the test for each value of α (for two-tail tests, only the positive critical value is shown); and, if the {**long**} option is given, whether the null hypothesis is accepted or rejected.

INPUTS

x@   Sample(s) or sample mean(s) from a population. If {**nx**=} is specified and x is a column vector, it is treated as a row vector. Otherwise, if x is a row vector or a sheet vector, it is treated as a column vector.

y@   Sample(s) or sample mean(s) from a population. If {**ny**=} is specified and y is a column vector, it is treated as a row vector. Otherwise, if y is a row vector or a sheet vector, it is treated as a column vector.

c@   The population mean(s), or difference between population mean(s), under the null hypothesis. If c contains only one value, it is assumed to be the null mean(s) (or difference(s) of means) for each column (or element) of x. The default value of c is 0 (or .5 for a one-sample test under the {**b**} and {**p**} options).

OPTIONS

alpha=*FLOATLIST*
a=*FLOATLIST*
level=*FLOATLIST*

Ht performs a hypothesis test at each of the specified significance levels α. Default values are {**alpha**= .05, .01, .001} or, equivalently, {**a**= .05, .01, .001} or {**level**= .05, .01, .001}.

Note: For hypothesis tests, the significance level is α; whereas for confidence intervals, the confidence level is $1-\alpha$.

b, p   The input x (and y, if given) is the sum (under the {**b**} option) or the average (under the {**p**} option) of n Bernoulli observations. Under the {**b**} option, if x or y is between 0 and n but is not an integer, it is rounded to the nearest integer. Unless specified otherwise (by the {cc} or the {exact} option), the usual Normal approximation (with no continuity correction) is used to calculate the z statistic and the P-value as follows. For one sample:

$$z = (\hat{p} - c) / \sqrt{c(1-c)/n_x}$$

where $\hat{p} = x/n_x$ under the {b} option or x under the {p} option, and c is the value of c@ defined above. For two samples,

$$z = (\hat{p}_1 - \hat{p}_2) / \sqrt{\hat{p}(1-\hat{p})/n_x + \hat{p}(1-\hat{p})/n_y}$$

where $\hat{p}_1 = x/n_x$ and $\hat{p}_2 = y/n_y$ under the {b} option or $\hat{p}_1 = x$ and $\hat{p}_2 = y$ under the {p} option, $\hat{p} = (n_x\hat{p}_1 + n_y\hat{p}_2)/(n_x+n_y)$, and the P-value is calculated as for the one-sample case.

cc   For one-sample hypothesis tests of the probability p of Bernoulli data (under the {b} or {p} option), use the continuity correction for approximating the binomial distribution by the Normal. Let p_0 be the hypothesized probability of success denoted by the c@ input dataset and let x be the number of successes in n observations. If x is the integer closest to np_0, then x is changed to np_0 (that is, the test statistic is 0 and the P-value is 1). Otherwise, x is changed as follows:

Type of Test	Option	New Value of x
two-tail	{ht=2}	$x-.5$, if $x > np_0$; $x+.5$, if $x < np_0$
lower-tail	{ht=L}	$x+.5$
upper-tail	{ht=U}	$x-.5$

exact For one-sample hypothesis tests of the probability p of Bernoulli data (under the {b} or {p} option), report the exact P-value using the binomial distribution instead of an approximation using a z statistic and the Normal distribution. For a two-tail test, the P-value is:

$$P = \begin{cases} I_{p_0}(x, n-x+1) + I_{(1-p_0)}(x, n-x+1) & \text{for } x > np_0 \\ I_{p_0}(n-x, x+1) + I_{(1-p_0)}(n-x, x+1) & \text{for } x < np_0 \\ 1 & \text{for } x = np_0 \end{cases}$$

where x is the number of 1's, n is the number of observations, p_0 is the probability of success under the null hypothesis, and $I_x(a,b)$ denotes the incomplete beta function (see *BUG* Table 7.1). For a one-tail test, the P-value is:

$$P = \begin{cases} 1 - I_{p_0}(x+1, n-x) & \text{(lower-tail test)} \\ I_{p_0}(x, n-x+1) & \text{(upper-tail test)} \end{cases}$$

No z statistic is calculated or displayed. This option overrides {cc}.

ht=*CHAR*
    Specifies the type of hypothesis test:

Type of Test	*Option*	*Alternative Hypothesis*
two-tail	{**ht**=2}	$\mu \neq c$ or $\mu_x - \mu_y \neq c$
lower-tail	{**ht**=L}	$\mu < c$ or $\mu_x - \mu_y < c$
upper-tail	{**ht**=U}	$\mu > c$ or $\mu_x - \mu_y > c$

In place of **L**, the value **l** may be used; in place of **U**, the values **u**, **R**, or **r** may be used. (You can think of **L** and **U** as referring to lower-tail and upper-tail, or **L** and **R** as referring to left-tail and right-tail.)

nx=*INTLIST*
    The number of observations if *x* contains means; the number of Bernoulli observations if {**p**} is specified; or the *n* parameter of the binomial distribution if {**b**} is specified. In all cases, one value of *nx* may be specified for each element of *x*. If only one value of *nx* is given, it is used for each element of *x*.

ny=*INTLIST*
    If *y* contains means, or {**b**} or {**p**} is specified, *ny* has a meaning for *y* analogous to that of *nx* for *x*.

long   The output display also states whether the null hypothesis is accepted ('acc') or rejected ('rej') at each significance level α. (Technically, it suffices to set **stalk** > 2. See *options(BRM)*.)

terse   Only the P-value is shown, not the significance level(s) or critical value(s). (Technically, it suffices to set **stalk** < 2.)

sdx=*FLOATLIST*
sdy=*FLOATLIST*
    SDs of the populations from which the columns of *x* and *y* are drawn: one for each column of *x* or *y*. Normally, {**sdx**=} and {**sdy**=} are taken to be the true population SDs, and a z-test is performed. However, if the {**nx**=} and {**ny**=} options are also given and the {**z**} option is *not* given, then {**sdx**=} and {**sdy**=} are taken to be estimates based on the given number of observations and the unbiased variance estimator (divide by $N-1$), and a t-test is performed.

    One SD may be given for each column of *x* or *y*. If only one value is specified, every column of the corresponding dataset is assumed to have that SD.

pool   If both *x* and *y* are specified and their population SDs are not, use a pooled estimate of the population SDs (which assumes that they are equal). If {**pool**} is not specified, the degrees of freedom for the t statistic are computed using Welch's approximation, as described in *ci(BRM)*.

z   Perform a z-test even when the true population SDs are unknown. This is appropriate only when the number of observations is large.

x= Columns of *x* to use. Default is all columns.

y= Columns of *y* to use. Default is all columns.

ci Display the confidence interval information which corresponds to the hypothesis tests. See *ci(BRM)* for details. Note that the column labeled *level* is the confidence level, or $1-\alpha$.

OUTPUTS

mean@
If only *x* is specified, the estimated population mean. If both *x* and *y* are specified, the estimated difference of the population means.

se@ Sample or known SE. If both *x* and *y* are specified, the SE of the difference of the population means.

df@ Degrees of freedom for each t statistic (when {**sdx**=;**sdy**=} or {**z**} are not specified).

t@ The test statistics, one for each test.

p@ The P-values, one for each test.

cv@ Critical values. If the test is two-tailed ({**ht=2**}), only the positive critical values are given. If a z-test is performed, each critical value corresponds to one value of *alpha*. Otherwise, *cv* is an array with dimensions (number of columns of *x*) by (number of values of *alpha*). Array element *cv[i,j]* contains the critical value for the test on data column *i* at significance level *alpha[j]*.

h@ 0 or 1 according to whether the null hypothesis is accepted or rejected. This is an array with dimensions (number of columns of *x*)-by-(number of values of *alpha*). The array element *h[i,j]* corresponds to the test statistic computed from column *i* of the data when compared to the critical value for the test with significance level *alpha[j]*.

ci@ Endpoints of the corresponding confidence intervals. See *ci(BRM)* for details.

EXAMPLES
Column 1 of the library dataset *ucprofs* contains the salary of 27 full-rank professors at UC Berkeley in spring 1984. Column 1 of *statprofs* contains the salary of all full-rank professors at the UCB Department of Statistics in spring 1984. To test the null hypothesis that the mean salary of all full-rank professors equals the mean salary of the full-rank Statistics professors against the alternative that it does not at the default significance levels:

```
. mean statprofs[1] > m
. ht {long} ucprofs[1] c@ m

Hypothesis Test: one-sample, two-tail t.  Data: ucprofs[1].  Null mean: m.

Col   N      null     sample    sample   df     test    P     alpha  crit  ?
             mean     mean       SE            stat  value         value
 1    27   4.751e+04 4.267e+04 1342.6  26.0   -3.60  .0013 .0500  2.06 rej
                                                           .0100  2.78 rej
                                                           .0010  3.71 acc
```

Each column of the library dataset *pigs* contains the average tooth length of guinea pigs fed ascorbic acid in varying forms and amounts. To test the null hypothesis that there is no difference between mean tooth length of guinea pigs fed 1 milligram and 2 milligram doses of ascorbic acid in orange juice (columns 5 and 6) against the alternative hypothesis that there is a difference at the significance levels α = .05 and .1:

        . ht pigs[5] pigs[6] {alpha=.05,.1}

SEE ALSO

BUG Sections 6.3, 6.4, 7.4.4, 7.4.5, and 9.5.2.

cdf(BRM), for commands to use for generating P-values and critical values.

ci(BRM), for the corresponding confidence intervals.

---

imprsk — Compute impulse response functions from transfer functions

---

SYNOPSIS

imprsk *gain phase* {x=;**conv**} [> *ar ai*]

DESCRIPTION

Imprsk computes impulse response functions from the transfer function arrays *gain* and *phase* produced by **trnfrk**.

INPUTS

gain@

> The gain output dataset produced by **trnfrk**.

phase@

> The phase output dataset produced by **trnfrk**. Missing values are not allowed.

If the input datasets are three-dimensional, so are the output datasets and the dimensions are the same. The *j*-th element of each sheet contains the real or imaginary part of the impulse response function for the *j*-th transfer function.

Otherwise, the output datasets are two-dimensional arrays. The *k*-th columns contain the impulse response function for the *k*-th transfer function, unless the {x=} option is specified, in which case they contain the impulse response function specified by the *k*-th element of {x=}.

OPTIONS

x=*INTLIST*

> If the *gain* and *phase* input datasets are two-dimensional, specifies the columns for which to compute the impulse response functions. Default is all columns.

conv   Tukey-Hamming convergence factors are used in computing the impulse response functions.

OUTPUTS

ar@   Real parts of the impulse response functions.

ai@   Imaginary parts of the impulse response functions.

If names for the output datasets are not specified, they are called *gain*.**ar** and *gain*.**ai** respectively, where *gain* is the name of the first input dataset.

REFERENCES

David R. Brillinger (1975). *Time Series: Data Analysis and Theory.* Holt, New York. (Expanded ed., 1981, Holden-Day, San Francisco.) Chapter 6.

Keith A. Haycock and David R. Brillinger (1985). *LIBDRB: A subroutine library for elementary time series analysis.* Technical Report No. 48, Department of Statistics, University of California, Berkeley.

SEE ALSO

trnfrk(BRM).

---

indic — Construct indicator variables

SYNOPSIS

indic *x* [*c*] {c=;cen;x=} [> *indic*]

DESCRIPTION

Indic constructs an indicator dataset from the categorical variable *x*. The output dataset, *indic*, contains a complete set of indicator variables which may be used for regressions on the categorical variable.

The indicator dataset consists of 0's and 1's and is constructed as follows. If there are *v* distinct categories, the indicator dataset contains *v* columns. The *j*-th column contains 1 or 0 in its *i*-th row according to whether or not the *i*-th element of *x* is in the *j*-th category.

If the {c=} option or the optional **c@** input dataset are given, they determine the possible categories. Otherwise, the possible categorical values are those of *x*.

INPUTS

x@   The categorical dataset for which to construct an indicator dataset. It must be a vector, or a single column of a dataset specified via the {x=} option.

c@   Optional dataset which contains the possible categorical values.

OPTIONS

c=*FLOATLIST*
Possible categorical values. Overrides the **c@** input dataset.

cen   The indicator dataset is centered—that is, it is constructed as described above, and then the mean of each column is subtracted off. In this case, the indicator dataset consists of $v-1$ columns: the column for the last category is omitted.

The purpose of the {**cen**} option is to construct indicator variables which are linearly independent of the constant vector. Thus, the corresponding regression coefficients are uncorrelated with the constant term in the regression.

x=*INT*

Construct an indicator dataset for the specified column of the input dataset.

OUTPUTS

indic@

The indicator dataset. If no output dataset is specified, the indicator dataset is put in *x*.**indic**, where *x* is the name of the input dataset.

SEE ALSO

BUG Section 8.6. The last example presents an alternative to **indic** which uses the equality operator '= =' and dimension-expansion in the case that *x* is a column vector and *c* is a row vector.

---

lag — Lag a time series

SYNOPSIS

lag *x* {**lag**=;**nona**} [> *lag*]

DESCRIPTION

Lag lags each column of *x*. By default, the lag interval is 1: $x[i,]$ is replaced by $x[i-1,]$.

INPUTS

x@   Dataset whose columns are the time series. Missing values are not allowed.

OPTIONS

lag=*INT*

Specifies the lag: if $lag = l$, $x[i,]$ is replaced by $x[i-l,]$. The lag value l may be negative. The default value is 1.

nona   Missing values at the start or end of the lagged series are omitted.

OUTPUTS

lag@   The lagged series.

If no output dataset is specified, the lagged series is put in *x*.**lag**, where *x* is the name of the input dataset.

SEE ALSO

diff(*BRM*), *tsop*(*BRM*).

---

let — Mathematical functions and matrix operations

SYNOPSIS
   [**let**] *expression* ; [**let**] *expression* ; . . .

DESCRIPTION
   Let performs arithmetic on BLSS datasets. The syntax for expressions is similar to that of most computer languages. For example:

   . *let c = sqrt(a^2 + b^2); let logodds = log(p/(1-p))*

   The word **let** may be omitted:

   . *cov = trn(x-ex) #* (x-ex)*

   Let may be accessed implicitly, and the result passed on to a regular BLSS command, by placing parentheses around an argument:

   . *scat x (sin(x) / x)*

   The outer parentheses may be omitted if all spaces in the implicit **let** expression are contained within parentheses or brackets. Thus:

   . *show   sin(x)/x   exp(-x ^ 2)*
   . *stat   y[1 4]*

   But:

   . *show (a*x^3 + b*x^2 + c*x)*

   For a detailed description of **let** see *BUG* Chapter 8. What follows is a list of functions available within **let** and a brief summary of other capabilities.

Elementwise Operators and Functions

+ – * /	Elementwise addition, subtraction, multiplication, division.
^ **	Elementwise exponentiation (alternate notations).
$x!$	x factorial.
abs(x)	Absolute value of x.
ceil(x)	Smallest integer not less than x.
exp(x)	Exponential function, e^x.
fac(x)	Factorial, $x!$.
floor(x)	Largest integer not greater than x.
gam(x)	Gamma function, $\Gamma(x)$.
int(x)	Integer part of x; rounds toward zero.
lfac(x)	Natural logarithm of factorial.
lgam(x)	Natural logarithm of absolute value of gamma function, $\ln|\Gamma(x)|$.
ln(x), **log**(x)	Natural logarithm (alternate names).
log10(x)	Base 10 logarithm.
neg(x)	Negative part of x: $-x$ if $x<0$; 0 otherwise.
pos(x)	Positive part of x: x if $x>0$; 0 otherwise.
pgau(x)	Standard Gaussian (Normal) cumulative distribution function (cdf), $\Phi(x)$. See also **pgau** and **qgau** in *cdf*(*BRM*).

qgau(p)	Inverse standard Gaussian (Normal) cdf, $\Phi^{-1}(p)$.
sin(x), **cos**(x), **tan**(x), **asin**(x), **acos**(x), **atan**(x)	
	Trigonometric and arc-trigonometric functions.
sinh(x), **cosh**(x), **tanh**(x), **atanh**(x)	
	Hyperbolic and arc-hyperbolic functions.
sgn(x)	Signum of x: -1 if $x<0$; 0 if $x=0$; 1 if $x>0$.
sqrt(x)	Square root of x.
BesJ0(x), **BesJ1**(x), **BesY0**(x), **BesY1**(x)	
	Bessel functions.
NA	Missing value.
PI, E	Values of π and e to seven decimal places.

Let also recognizes the elementwise logical operators listed at the end of this section and described in *BUG* Section 8.6.

Matrix Subscripting

In the forms below, the subscripts i, j, k, l, and m represent explicit integers.

vec[i]	i-th element of row vector *vec*.
vec[i,]	i-th element of column vector *vec*.
vec[$i\ j\ k\ l$]	Subvector consisting of elements i, j, k, and l of *vec*.
vec[$i{:}j$]	i-th through j-th elements of row vector *vec*.
vec[$i{:}j{:}k$]	i-th through j-th elements of row vector *vec*, with subscripts incremented by k.
mat[i,j]	[i,j]-th element of matrix *mat*.
mat[j] *mat*[$,j$] *mat*[*,j]	
	j-th column of matrix *mat* (alternate notations).
mat[$i,$] *mat*[$i,*$]	
	i-th row of matrix *mat* (alternate notations).
mat[$i{:}j, k\ l\ m$]	Submatrix consisting of rows $i, \ldots, j$ and columns k, l, and m.

See *BUG* Sections 8.3 and 8.4 for more information about subscripting, including three-way arrays.

Matrix Operations and Functions

a'	Matrix (or vector) transpose.
a #* b	Matrix multiplication.
a #^ n	Matrix a raised to the power n.
a,b	Columnwise matrix catenation: columns of matrix a followed by columns of matrix b.
$a,,b$	Rowwise matrix catenation: rows of matrix a followed by rows of matrix b.
a #, b	Sheetwise matrix catenation: sheets of array a followed by sheets of array b.
$a{:}b{:}c$	Sequence of numbers a through b incremented by c, returned as a row vector. The operands a, b and c must be scalars. If $:c$ is omitted, the increment is 1.
$'a$	Row-unravel operator (see *BUG* Section 8.3.3).
$''a$	Column-unravel operator (see *BUG* Section 8.3.3).

cumprod(*mat*)	Cumulative product of elements, in *row-major order* (row index varies fastest).
cumsum(*mat*)	Cumulative sum of elements in the array, in row-major order.
det(*mat*)	Determinant of the square matrix *mat*.
diag(*a*)	Construct a diagonal matrix from vector *a*; extract the main diagonal from matrix *a*.
inv(*mat*)	Inverse of the square matrix *mat*.
iota(*mat*)	Array containing the indices of element positions, in row-major order.
max(*mat*)	Maximum element of a matrix.
ncols(*mat*)	Number of columns in a matrix.
norm(*mat*)	2-Norm of matrix (square root of sum of squares of elements).
nrows(*mat*)	Number of rows in a matrix.
nsheets(*x*)	Number of sheets (third dimension) in BLSS dataset.
trace(*mat*)	Trace of a square matrix.

Operator Summary

Here is a summary of all **let** operators, listed in order of decreasing precedence. This table also appears in *BUG* Section 8.8, with definitions of terms and section references for each operator.

Operator	Meaning	Associativity
' "	Row-unravel (prefix '); Column-unravel	–
'	Matrix transposition (postfix ')	–
!	Factorial (postfix !)	–
^ #^	Exponentiation; Matrix exponentiation	Right
– !	Arithmetic negation (unary –); Logical negation (prefix !)	–
* / #*	Multiplication; Division; Matrix multiplication	Left
+ –	Addition; Subtraction (binary –)	Left
: ::	Sequence operators (binary and ternary)	–
,	Columnwise matrix catenation	Left
"	Rowwise matrix catenation	Left
#,	Sheetwise array catenation	Left
== != < <= > >=	Equality and order relationships	Left
&&	Logical and	Left
||	Logical or	Left
?:	Logical conditional operator (ternary)	Right
=	Assignment	Right

SEE ALSO

BUG Chapter 2, Sections 5.1 and 5.6.3, Chapter 8.

arrop(*BRM*), for commands that perform simple array operations.

cdf(*BRM*), for continuous distribution functions and their inverses.

count(*BRM*) and *select*(*BRM*), to count or select rows of a matrix that meet a specified condition.

redim(*BRM*), to change the dimensions of a dataset.

submat(*BRM*), to choose submatrices using datasets as subscripts.

vecop(*BRM*), for operations (sum, max, ...) applied to rows, columns, sheet vectors, or entire datasets.

---

list — List contents of work, data, text, ... areas

SYNOPSIS
> **list** [*files* ...] {**all;always;never;date;terse;** ... }

DESCRIPTION
> **List** describes each of the named *files* (or the entire contents, if no *file* arguments are given) in each of the specified areas (or the work area, if none is specified). If a file is a BLSS dataset, **list** reports its dimensions and describes its shape ('scalar', 'column vector', etc.). If a file is an ASCII text file, **list** reports its size in bytes, the number of lines it contains, and the width in characters of its widest line (taking into account any possible tab and backspace characters). Otherwise **list** indicates the file's type and size.
>
> UNIX pathnames are valid file names.

OPTIONS
> **bin, class, data, doc, help, home, sys, text**
> **dirs**=*dirlist,* **tree**=*treelist,* **area**=*arealist*
>> These options specify alternate areas (or directories) in which to list files. For full information, see *area(BRM)*. The default values are:
>>> **dirs**   your work area
>>> **tree**   your *blss* tree (normally *~/blss*)
>>> **area**   the data area
>
> **all**   List files whose names begin with '.' or '_' characters. By default, such files are listed only when specified as *file* arguments.
>
> **always**
> **never** By default, **list** reports on files it cannot access only when one area is being listed. These options cause **list** to always or never report on such files, respectively.
>
> **date**   Show the date of areas being listed.
>
> **terse** Omit the line count and width of widest line for text files (this causes **list** to run noticeably faster for large text files) and, if only one area is being listed, the area name at the head of the list.

EXAMPLES
> List the files *zeta* and *eta* in the work area:
>> `. list zeta eta`
>
> List the contents of the work area:
>> `. list`
>
> List the contents of the data area:
>> `. list {data}`
>
> List all files named *mu* and *nu* in the data, doc, and text areas:
>> `. list {data;doc;text} mu nu`
>
> or:
>> `. list {area=data,doc,text} mu nu`

List the BLSS datasets *births* and *deaths* in the BLSS system data area:

```
. list {sys} births deaths
```

LIMITATIONS

When deciding whether a file is ASCII text, **list** looks at its first few characters only, so it can be fooled.

SEE ALSO

BUG Sections 4.1, 4.5, 5.1.

area(BRM).

---

load — Retrieve saved datasets and text files

SYNOPSIS

load [*files* ...] {**ask;clobber;link;** ... }

DESCRIPTION

Load copies the named BLSS datasets or text files into the active work area. By default, **load** searches for the named files in the data and text areas corresponding to the bin areas on your BPATH. Unless your BPATH has been changed, this means that **load** searches first your own data and text areas and then the BLSS system data area (or *library*).

UNIX pathnames are valid file names.

If no file name is given, **load** asks you for one.

If you attempt to load a file with the same name as one which already exists in your work area, **load** asks whether you want to overwrite it. If yes, type 'y' or 'Y'.

OPTIONS

bin, class, data, doc, help, home, sys, text

dirs=*dirlist,* **tree**=*treelist,* **area**=*arealist*

These options specify alternate areas (or directories) for **load** to search. For a full description see *area(BRM).* The default values are:

dirs	(none)
tree	trees on your BPATH environment string
area	data:text

ask   **Load** asks you about each file in your data area—or some other area, if you have used one of the options listed above. The file will be loaded if your answer begins with a 'y' or a 'Y'.

clobber

Do not request permission before overwriting files which already exist in the work area.

link   Instead of copying files into the work area, *link* them. (On BSD UNIX, a symbolic link is made. On other UNIX systems, a hard link is made if possible; otherwise, the file is copied and a warning is issued.) This is useful when loading large files or datasets if disk space is limited. A *link* is an alternate file name for the original file. Thus any changes made to the file in the active work area are simultaneously made to the file in the permanent storage area from which it was loaded.

EXAMPLES

Load the file *kappa*, searching first your own data and text areas and then the BLSS system data library:

    . load kappa

Load *kappa*, searching the system data library only:

    . load {sys} kappa

Load *kappa*, searching text areas only:

    . load {text} kappa

Be asked for files to load from your data area:

    . load {ask}

Be asked for files to load from your text area:

    . load {ask;text}

SEE ALSO

BUG Section 4.6.
area(BRM), *data(BDL)*.

---

mdrace — Optimal transformations for dimensionality reduction

---

SYNOPSIS

  mdrace *x* [*w phi*] {**x=;k=;type=;** ... } [> *phi alpha cor l rsq load* ...]

DESCRIPTION

Mdrace uses the *alternating conditional expectation* (ACE) technique of Breiman and Friedman (1982, 1985) to determine transformations ϕ^* of multivariate data x which maximize R^2, the sum of the k largest eigenvalues of cov$[\phi^*(x)]$, over all transformations ϕ which satisfy:

$$\left.\begin{array}{l} E[\phi_i(x)] = 0 \\ E[\phi_i(x)]^2 = 1 \end{array}\right\} \text{ for all } i.$$

The final value of R^2 is displayed. Certain output datasets are intended for use in diagnostic plots.

INPUTS

  x@ *n*-by-*p* array whose columns are the observations on the variables.

  w@ Optional *n*-vector of nonnegative weights for the observations.

  phi@ Optional *n*-by-*p* matrix which contains initial values of $\phi(x)$. Default initial transform is the identity.

Missing values are not allowed in the inputs.

OPTIONS

x=*INTLIST*

> The columns of x to use as the variables. Default is all columns.

k=*INT*

> The number of factors (variables in the transformation) desired. The default is $p-1$.

type=*INTLIST*

> p integers with values as below. The i-th element specifies the value type for the i-th variable.

> *Type Meaning*
> 1   Variable type is unrestricted (default).
> 2   Variable assumes circular (periodic) values in the range [0., 1.] with period 1.
> 3   Variable assumes categorical values.

> If the {**x**=} option is in effect, then the i-th predictor variable is the variable specified by the i-th element of the {**x**=} array.

maxit=*INT*

> Maximum number of outer loop iterations. No greater than 20 is recommended. The default is 10.

wnd=*FLOAT*

> Size of the smoother window for Friedman's *super smoother* used in the inner loop. *Wnd* must be in the range [0, .8]. The default is $\min(3/\sqrt{n}, .8)$, which experimentally has given good results over a wide range of datasets. *Wnd*=0 gives variable window size, which may increase the running time considerably.

vloop=*INT*

> Validation loop checking:

> *Vloop Behavior*
> −1   Perform validation loop checking only upon exit from the outer loop (default).
> 0   No validation loop checking.
> n Perform validation loop checking every n outer loop iterations starting with the first and also upon exit from the outer loop.

OUTPUTS

phi@   Estimates of $\phi^*(x)$, the transformed values of x.

alpha@

> p-by-k matrix whose j-th column contains the component loadings corresponding to the $(p-j+1)$-st largest eigenvalue of *cor*.

cor@ p-by-p covariance matrix of the transformed variables. (Also called the correlation matrix because the variances are 1.)

l@   The p eigenvalues of *cor*.

rsq@   The values of R^2 for each iteration of the outer loop.

load@

Data to make a loading plot for variables: *p*-by-2 array whose columns are the loadings *alpha* ∗ (the eigenvectors corresponding to the two largest eigenvalues of *cor*).

cluster@

Data to make a variable clustering plot: *p*-by-2 array whose columns are the first and second principal components with signs aligned so as to minimize the sum of the squared Euclidean distances between the plotted points. The average of the squared Euclidean distances is normally displayed at the terminal when **cluster@** is requested.

factor@

Factor scores for cases: *n*-by-2 array whose columns are the first and second principal components.

REFERENCES

Robert A. Koyak (1985). *Optimal transformations for multivariate linear reduction analysis.* Ph.D. dissertation, Department of Statistics, University of California, Berkeley.

Leo Breiman and Jerome H. Friedman (1982). *Estimating optimal transformations for multiple regression and correlation.* Technical Report No. 9, Department of Statistics, University of California, Berkeley.

Leo Breiman and Jerome H. Friedman (1985). *Estimating optimal transformations for multiple regression and correlation.* **JASA**, **80**, 580-598.

SEE ALSO

ace(BRM).

---

medmad — Median and median absolute deviation

SYNOPSIS

medmad *x* {x=} [> *mm*]

DESCRIPTION

For each column of *x*, **medmad** displays the following:

Col	The column index.
N	Number of non-missing values.
Med	The median.
MAD	The median absolute deviation from the median.

INPUTS

x@   The input dataset. Missing values in *x* are ignored.

OPTIONS

x=*INTLIST*

Columns for which to compute the median and MAD. Default is all columns.

OUTPUTS

mm@ One row for each column processed, containing the four statistics N, Col, Med, and MAD. If the {quiet} option is in effect, the **mm@** output dataset is always created. If no output dataset name is specified, its name is *x*.**mm**, where *x* is the name of the input dataset.

SEE ALSO

biweight(BRM), stat(BRM), vecop(BRM).

---

medpolish — Median polish of two-way tables

SYNOPSIS

medpolish *x* {**eps**=} [> *res dia*]

DESCRIPTION

Medpolish performs a quick exploratory two-way additive fit to the two-dimensional table *x*. The row medians are subtracted off, and then the column medians. This is repeated until the relative change in the median absolute residual is less than *eps*. The overall center and spread (median absolute residual) and row and column effects are displayed at the terminal.

Missing values in *x* are ignored in computations.

OPTIONS

eps=*FLOAT*

As explained above. Default is *eps*=.01

long In addition to the usual display, show the median absolute residual at each iteration.

OUTPUTS

res@ Same shape as *x*; contains the residuals.

dia@ (*nr*∗*nc*)-by-3 file of diagnostic data:

Column	Contains
1	$a_i + b_j$
2	$a_i * b_j / t$
3	residual

where a_i is the row effect, b_j is the column effect, and *t* is the overall center. The diagnostic data are arranged so that the column effect index changes faster than the row effect index. These diagnostics may be used to help assess whether the model is appropriate.

EXAMPLE

Illit is a two-way table of median illiteracy rates for nine different regions of the U.S. (rows) and six different years (columns): 1900, 1920, 1930, 1950, 1960, and 1970. The scatterplot of the **dia@** output variable and the coded display of the residuals show a highly structured, nonlinear pattern:

```
load illit
```

```
. medpolish illit > z d
. code z
. scat d {x=1,2; y=3}
```

REFERENCES
Donald R. McNeil (1977). *Interactive Data Analysis.* Wiley, New York.
Chapter 5.
John W. Tukey (1977). *Exploratory Data Analysis.* Addison-Wesley, Reading,
Mass. Chapter 11.

SEE ALSO
code(BRM), twoway(BRM).

---

oneway — Least squares and robust one-way anova

SYNOPSIS
oneway *x* {**x**=;**c**=;**eps**=} [> *res cef w anova*]

DESCRIPTION
Oneway performs one-way analysis of variance on the dataset *x*. The columns
of *x* should contain the different groups to be compared. If the groups have
differing numbers of observations, the columns should be filled out with missing values so that they all have the same length. Missing values in *x* are not
used in computations.

Oneway displays the overall center and spread, and the column effects. In
robust anova, the degrees of freedom are estimated by the sum of the final
weights for the observations.

Oneway also displays an anova table which gives the degrees of freedom and
the weighted mean square error for each source of variation (total, columns,
and error), the F-statistic for the column effect, and the P-value of the F-
statistic.

By default, **oneway** obtains the unweighted least squares fit. If the {**c**=}
option is specified, **oneway** performs an iteratively reweighted analysis of variance which is robust against thick-tailed error distributions. At each iteration,
the bisquared weight function (see *biweight(BRM)*) is used: *z* is taken to be the
deviation of an observation from the current estimate of the group center and
s is taken to be the current weighted pooled SD of the residuals.

OPTIONS
x=*INTLIST*
Columns to be included in the analysis. Default is all.

c=*FLOAT*
Bisquare parameter. Default is 0, which gives least squares anova. For
robust estimation, values of *c* between about 4 and 10 are reasonable.
The larger *c* is, the closer the results are to those of least squares anova.

eps=*FLOAT*
Iterations stop when the absolute change in (1 − *SS~OLD~*/*SS~NEW~*) is less than *eps*, where *SS* is the weighted sum of squared residuals. Default is *eps*=.01. This stopping rule is almost equivalent to stopping when the weighted sum of residuals is zero.

terse   Display only the anova table.

long   In addition to the regular display, show the weighted sum of squared residuals for each iteration.

vlong   In addition to the display produced by the {**long**} option, show the weighted SDs and degrees of freedom of the column effects. Note, however, that these are computed in a nonstandard way.

OUTPUTS

res@   Same shape as *x*; contains the residuals.

cef@   One row for each column of *x*. Each row contains the estimated column effect, SD, and degrees of freedom.

w@   Same shape as *x*; contains the final weights for each observation.

anova@
The anova table. It can be subsequently displayed by the **anovapr** command—see *anovapr(BRM)* for a description of its contents.

EXAMPLES

Demopct records for each of 24 states the percentage of Democratic vote for the Presidential elections of 1960, 1964, 1968 and 1972. The boxplots generated by *compare(BRM)* show the overall pattern. The final boxplots help answer the question: In landslides (e.g., 1964), is variance between states reduced?

```
. load demopct
. compare demopct
. oneway demopct > res@ r cef@ c
. compare r
```

To make a scatterplot of the estimated SD against the estimated mean of each column:

```
. scat c
```

This plot helps assess whether a data transform is appropriate.

REFERENCE

Donald R. McNeil (1977). *Interactive Data Analysis*. Wiley, New York. Chapter 7, pp. 160-163, discusses robust one-way anova but gives an older version of the algorithm which used the mean absolute residual as *s* in the bisquared weight function and was less numerically stable.

SEE ALSO

biweight(BRM), *robust(BRM)*, *twoway(BRM)*.

---

options — Options understood by all regular BLSS commands

DESCRIPTION

This manual entry contains a list of options recognized by all regular BLSS commands (but not by **let** commands or BLSS shell internal commands such as **echo**, **history**, **alias**, and **set**). See the manual entry *intro(BRM)* for a general discussion of options, including their syntax and possible value types.

OPTIONS

usage   The correct usage of the command is displayed: Valid input tags, output tags, and option names are listed. The command then exits. See *BUG* Section 11.4 for more information.

more   Pipe output through the UNIX filter **more –ds** in order to pause after each screenful. If the environment string MORE is set, its contents replace **–ds** as the **more** options. See *BUG* Section 4.8 for more discussion.

lpr=*STRING*
lpr   Send output to the lineprinter instead of the terminal. *STRING,* if specified, is used as the lineprinter command. Otherwise, if the environment string LPRPROG is set, it is used as the lineprinter command. Otherwise, the default lineprinter command for your system is used. See *BUG* Section 4.3 or the manual entry *print(BRM)* for examples and more information.

filter=*STRING*
    Like the standard UNIX pipe-and-filter. For example, **show** *file* {**filter**=*STRING*} is analogous to **show** *file* | *STRING*. *STRING* may be any UNIX pipeline, although spaces must be protected by double-quote characters `""`.

Controlling the Output Display
talk=*INTLIST*
    This option controls the amount of output displayed by a command. The general form is {**talk**=*stalk,utalk*}. *Stalk* and *utalk* control, respectively, the amount of statistical and utility/diagnostic output. The values of *stalk* and *utalk* should be integers in the range 0 through 5; the default is 2. If {**talk**=} is given one value, it is used for both *stalk* and *utalk*.

The following options have conventional talk level meanings:

Option	Equivalent to	Meaning
quiet	**talk=0**	No output to the terminal, except error messages.
terse	**talk=1**	Less output than usual.
usual	**talk=2**	The usual (default) amount of output.
long	**talk=3**	More output than usual.
vlong	**talk=4**	Very long; all available statistical and utility outputs.
debug	**talk=5**	Debugging output.

The amount of output to the terminal controlled by the {**talk**=} options is cumulative: All output displayed by a given {**talk**=} level is also displayed by all greater {**talk**=} levels.

Note that the documentation for these options in individual manual entries may describe only a subset of the amount of printout control available under the {**talk**=} options.

The TALK environment string has the same effect as, and is overridden by, the {**talk**=} option.

The **show** command is an exception. It ignores the {**talk**=} option and all its variations described above, including the TALK environment string.

Controlling the Option Checks

Normally, if a BLSS command is invoked with incorrect option or tag names, BLSS displays the correct usage of the command and then exits. The following two options alter this behavior.

nocheck
> BLSS does not check the validity of option and tag names. This is sometimes useful with BLSS macros.

noerror
> Invalid option or tag names cause a warning message but are otherwise ignored.

EXAMPLES

The following examples illustrate the {**more**}, {**lpr**}, {**lpr**=}, and {**filter**=} options in conjunction with the **show** command.

To send output to the lineprinter instead of the terminal:

    . show {lpr} warpbreaks

To use the *more* mechanism so that **show** pauses at the end of each screenful:

    . show {more} plover

To pipe output through the UNIX filter **grep NA**:

    . show {filter="grep NA"} turbines

To further filter output through **wc** and write the result on *wcfile*:

    . show {filter="grep NA | wc > wcfile"} turbines

Note the double-quote characters "" which protect the spaces in the {**filter**=} option values.

SEE ALSO

area(BRM), for additional options recognized by all utility commands.
strings(BRM), for environment strings which affect many BLSS commands.
talk(BLR), *usage(BLR)*.

output — How to redirect text output to a text file

SYNOPSIS
command >> textfile

DESCRIPTION
Regular BLSS commands produce two distinct types of output: *output datasets*—datasets which contain numerical results; and *output text*—text which is normally displayed on the terminal. Output datasets are discussed in *intro(BRM)* and *BUG* Section 11.3. This manual entry concerns output text.

The output text from a command which is normally shown at the terminal (standard output, or 'stdout') may be redirected to a text file whose name follows the output text *redirection symbol* '>>', as shown above. This text file may be saved, edited, printed on a lineprinter, etc. Error messages (which are printed to standard error, or 'stderr') still appear on the user's terminal.

Three other output text redirection symbols may be used in place of '>>'. Here is the complete list of all four:

Symbol	Action
>!	Write output text (except for error output) to *textfile*.
>>	Append output text (except for error output) to *textfile*.
>&	Write all output text (including error output) to *textfile*.
>>&	Append all output text (including error output) to *textfile*.

To *write* to a file means to overwrite any preexisting contents it might have. To *append* to a file means to add new text at the end of the file if it already exists. Both actions cause the file to be created if it does not already exist.

Output text redirection works with all regular BLSS commands (including macros) and with most irregular commands. It does not work with **let** commands.

EXAMPLES
To append the summary statistics display from the **stat** command to the text file *final.smry*:

```
. stat final >> final.smry
```

To create a text file *plots* which contains three different scatterplots (and overwrite any previous contents of the text file it might have):

```
. scat plover {x=1;y=2} >! plots
. scat plover {x=1;y=3} >> plots
. scat plover {x=2;y=3} >> plots
```

SEE ALSO
BUG Sections 4.5, 4.10.1, 11.3, for more about output text redirection.
BUG Sections 5.4, 11.3, for more about output datasets.

---

pgrm1 — Periodogram of a univariate time series

SYNOPSIS
    pgrm1 *x* [> *pg*]

DESCRIPTION
    Pgrm1 computes the periodogram of a real-valued time series *x*. Periodograms of several series may be computed by placing the series in separate columns or separate sheets.

INPUTS
    x@   An *n*-vector which contains the time series. *N* should be reasonably composite. If *x* is a row vector or a sheet vector, it is treated as a column vector. Missing values are not allowed in *x*.

OUTPUTS
    pg@   The periodogram of *x*. If no output dataset is specified, the periodogram is put in *x*.**pg**, where *x* is the name of the input dataset.

REFERENCES
    David R. Brillinger (1975). *Time Series: Data Analysis and Theory.* Holt, New York. (Expanded ed., 1981, Holden-Day, San Francisco.) Chapter 5.
    Keith A. Haycock and David R. Brillinger (1985). *LIBDRB: A subroutine library for elementary time series analysis.* Technical Report No. 48, Department of Statistics, University of California, Berkeley.

SEE ALSO
    pgrmk(BRM), for *k*-variate periodograms.
    spec1(BRM) and *speck(BRM)*, for spectra.

---

pgrmk — Periodograms and cross-periodograms of *k*-variate time series

---

SYNOPSIS
    pgrmk *x* [> *pg*]

DESCRIPTION
    Pgrmk computes all periodograms and pairwise cross-periodograms of a *k*-variate series *x*.

INPUTS
    x@   An *n*-by-*k* array which contains the time series. The length *n* of the series should be reasonably composite. Missing values are not allowed.

OUTPUTS
    pg@   An *n*-by-*k*-by-*k* array which contains the periodograms. The *f*-th sheet, $pg[f, *, *]$, contains periodogram values at frequency $2\pi(f-1)/n$. Each sheet is arranged as follows:

For   pg[*, i, j] *Contains*
i = j   Periodogram of the *i*-th time series.
i < j   Cross-periodogram of the *i*-th and *j*-th series: Real part.
i > j   Cross-periodogram of the *i*-th and *j*-th series: Imaginary part.

If no output dataset is specified, **pg@** is put in *x*.**pg**, where *x* is the name of the input dataset.

REFERENCES
David R. Brillinger (1975). *Time Series: Data Analysis and Theory.* Holt, New York. (Expanded ed., 1981, Holden-Day, San Francisco.) Chapter 7.

Keith A. Haycock and David R. Brillinger (1985). *LIBDRB: A subroutine library for elementary time series analysis.* Technical Report No. 48, Department of Statistics, University of California, Berkeley.

SEE ALSO
pgrm1(BRM), for univariate periodograms.
spec1(BRM) and *speck(BRM)*, for spectra.

---

poisson — Poisson distribution probabilities and cdf

SYNOPSIS
poisson *n lambda* [> *pdf cdf*]

DESCRIPTION
Poisson computes the probabilities that a Poisson random variable with parameter λ assumes the values 0, 1, ..., *n*. It can also compute the cumulative distribution function (cdf).

INPUTS
n@   Maximum value for which to compute the Poisson probability and cdf.

lambda@
       The Poisson parameter λ.

OUTPUTS
pdf@ Row vector that contains the probabilities.

cdf@ Row vector that contains the cumulative distribution function.

NOTE
This command is implemented as a macro. It illustrates macro features described in *BUG* Section 10.8.4, and it uses methods described in Section 9.2.1 to compute the probabilities and the cdf.

---

polar, rect — Transform between polar and rectangular coordinates

---

SYNOPSIS

polar x [y] [> r th]
rect r [th] [> x y]

DESCRIPTION

Polar transforms from rectangular (Cartesian) coordinates (x,y) to polar coordinates (r,θ) without destructive overflow or underflow.

Rect transforms from polar coordinates (r,θ) to rectangular coordinates (x,y).

INPUTS

Inputs when transforming from rectangular coordinates:

> **x@**   X-coordinates.
> **y@**   Y-coordinates.

Inputs when transforming from polar coordinates:

> **r@**   Radii.
> **th@**   Angles (θ) in radians.

If there is only one input dataset and it contains exactly two columns, then the two columns contain the pairs of input coordinates. Otherwise, the input datasets must have the same dimensions unless they are both vectors, in which case they must have the same length. Missing values in either input dataset result in corresponding missing values in the output datasets.

OPTIONS

inv   Take the inverse transform: **polar** {inv} transforms to rectangular coordinates; **rect** {**inv**} transforms to polar coordinates.

OUTPUTS

Outputs when transforming to polar coordinates:

> **r@** $r = \mathrm{sqrt}(x^2 + y^2)$
> **th@** $\theta = \mathrm{atan}(y/x)$, in the range $-\pi$ to π.

Outputs when transforming to rectangular coordinates:

> **x@** $x = r\cos(\theta)$
> **y@** $y = r\sin(\theta)$

If the input consists of a single dataset with two columns, then so does the output unless two output datasets were explicitly requested. If no output datasets are specified, then output datasets are created with appropriate names.

EXAMPLES

To convert the coordinates of the library dataset *circle* from rectangular to polar coordinates:

```
. load circle
. polar circle > c.r c.th
```

Both these examples convert the coordinates of *circle* stored in polar coordinates back to rectangular coordinates:

```
. polar {inv} c.r c.th > c.x c.y
. rect c.r c.th > c.x c.y
```

---

print — How to send files to a printer

SYNOPSIS
show {**lpr;lpr**=;**ff**; ... } *file* ...
regular-command {**lpr**} {**lpr**=}

DESCRIPTION
The command **show** {**lpr**} *file* ... sends the named files to the lineprinter on your system. As always with the **show** command, the files named may be either text files or BLSS datasets. See *show*(*BRM*) for information on options particular to **show**—such as {**ff**}, which puts a formfeed between each file printed.

In general, the {**lpr**} and {**lpr**=} options can be used with any regular ISP command (although not with macros): the output text which would normally go to the terminal is sent to the lineprinter instead. However, it is often a more efficient use of paper to collect the output text from several commands and then print the text file(s) in a single print job using **show** {**lpr**} or **show** {**lpr;ff**}. See *output*(*BRM*) for information.

The {**lpr**=} option is used to invoke non-default printer commands; see below.

SPECIFYING A PRINTER
Large computer systems may have several lineprinters, each identified by its own name. Usually there is a default printer, but it may not be the one you want to use. How to choose a non-default printer depends not on BLSS, but on the version of UNIX at your site. Different versions of UNIX sites may specify alternate printers using one or more of the following methods.

Environment Strings
BSD (Berkeley) UNIX and its derivatives use the environment string PRINTER to specify a printer. System V UNIX and its derivatives use the environment string LPDEST. Environment strings may be set either in the UNIX shell before invoking BLSS, or within BLSS itself using the **set** command.

Printer Command Invocation
Some UNIX systems specify alternate printers via flags to the printer command (e.g., **–P***printer*) or by using different printer command names (for

example, **print** or **enscript**). The printer command name and flags may be specified in the LPRPROG environment string or in the {**lpr=**} option. Double-quote characters "" are necessary to protect spaces contained in an option value—see the **enscript** example below.

The specific method and list of printer names for your system must be obtained from your local documentation.

EXAMPLES

To make and print stem-and-leaf diagrams of every column in the *plover* dataset:

```
. stemleaf plover {lpr}
```

To make a printout of the entire *plover* dataset:

```
. show plover {lpr}
```

Some UNIX systems provide a (non-default) printing command **enscript –r** which prints output on a laser printer using a wide format. To use this command to print a copy of the *ozone* dataset on the laser printer named *lw*, give the command:

```
. show ozone {lpr="enscript -r -Plw"}
```

Choosing a Printer

Our computer center runs BSD UNIX. In order to use the printer named *Evans1* (it's in Evans Hall), we give the command:

```
. set PRINTER Evans1
```

and the *Evans1* printer is used for the remainder of the BLSS session or until we re-**set** PRINTER. On System V UNIX, we would instead give the command:

```
. set LPDEST Evans1
```

A non-default printer may be permanently established for BLSS by placing the appropriate command in your *.blss* file, or for UNIX by placing the appropriate command in your *.login* or *.profile* file (depending on whether you use the C shell or Bourne shell).

SEE ALSO

BUG Sections 4.5, 10.4, 10.5.
options(*BRM*), *output*(*BRM*), *set*(*BRM*), *show*(*BRM*).

---

qdata — Quantiles of a dataset

SYNOPSIS
  qdata x p {**p**=;**x**=} > q

DESCRIPTION
  Qdata computes the quantiles corresponding to the values in p for the columns of the dataset x and places the result in q.

  Quantiles are computed as follows. Let y denote the data vector (that is, a column of the input dataset x) with all missing values removed; let $y_{(i)}$ denote the i-th order statistic of y; and let n denote the number of elements in y. If $m = p * n$, the quantile q corresponding to the value p in the range $[0, 1]$ is:

$$q(p) = \begin{cases} y_{(1)}, & \text{if } p = 0; \\ y_{(n)}, & \text{if } p = 1; \\ (y_{(m)} + y_{(m+1)})/2, & \text{if } m \text{ is an integer and } 0 < m < n; \\ y_{(ceil(m))}, & \text{otherwise,} \end{cases}$$

  where ceil(m) is the smallest integer not less than m. This definition of q corresponds to evaluating, at the point p, the inverse empirical cdf of the data y (that is, the cdf which puts mass $1/n$ at each point of y). On intervals where the empirical cdf is flat (other than at the ends), the inverse value is taken to be the midpoint of the interval.

  If any value of p is out of the range $[0, 1]$, the corresponding value of q is missing.

INPUTS
  x@   Data for which to compute quantiles.

  p@   Vector of value(s) at which to compute quantiles.

  If either x or p is a row vector or a sheet vector, it is treated as a column vector.

OPTIONS
  p=*FLOATLIST*
       This option may be used in place of the **p@** input dataset to specify the values of p.

  x=*INTLIST*
       Columns of x for which to compute quantiles. Default is all columns.

OUTPUTS
  q@   The quantiles. The output dataset q has dimensions np-by-nc, where np is the number of values in the vector p and nc is the number of columns for which quantiles are computed.

  If no output dataset is specified, the quantiles are put in x.**q**, where x is the name of the input dataset.

EXAMPLES

The following command sequence creates an artificial dataset x which contains the values 1, 2, 3, ..., 20 and a data vector p which contains the values 0, .1, .2, ..., 1; it then computes the deciles (including the minimum and maximum) of x and places them in q:

```
. x = 1:20
. p = (0:10)/10
. qdata x p > q
```

This next command computes the 0.25 and 0.75 quantiles (that is, the 1st and 3rd quartiles) of x and places them in qq:

```
. qdata x {p=0.25,0.75} > qq
```

SEE ALSO

BUG Section 9.5.1.

qdata(BML).

---

qr, rank — QR decomposition and rank of a matrix

SYNOPSIS

   qr x {x=;**full**;**pivot**=;**pivot**;**tol**= } [> *q r e rank rcond*]
   rank x {x=;**full**;**pivot**=;**pivot**;**tol**= } [> *rank rcond*]

DESCRIPTION

Qr and **rank** are separate entries to the same command.

Qr computes the Gram-Schmidt QR decomposition of the *nr*-by-*nc* matrix **X** (contained in the dataset x) and places the results in the following datasets:

q Matrix **Q** with orthonormal columns (**Q′Q** = **I**) which span the same space as the columns of **X**. Its dimensions are *nr*-by-*nc* if $nr \geq nc$ (the usual case) and *nr*-by-*nr* otherwise.

r *nc*-by-*nc* upper triangular matrix **R** such that **X** = **QR**; or, when pivoting is in effect, **XE** = **QR**. **R** has nonnegative diagonal elements. Thus, if **X** has full rank, the decomposition is unique.

e *nc*-by-*nc* permutation matrix **E** when pivoting is in effect; otherwise, the identity matrix.

If **X** is rank deficient, **Q** nonetheless contains *nc* orthonormal columns. Thus, it contains extra columns which are orthogonal to the space spanned by **X**. In this case, **R** is determined correctly so that the equation **X** = **QR** (or **XE** = **QR**) still holds.

Note that **R′R** = **X′X**. Thus **R** is a Cholesky square root of **X′X** (which, when divided by $nr-1$, is the covariance of **X** if it has zero mean).

If invoked as **rank**, the command computes the rank of the matrix **X** and places the result in *rank*.

Algorithm. This command uses the LINPACK subroutines *sqrdc* and *sqrsl*.

INPUTS

x@   Dataset which contains the input matrix **X**. If it contains multiple sheets, each sheet is processed separately. Missing values are not allowed.

OPTIONS

x=*INTLIST*

The submatrix of *x* consisting of the specified columns is decomposed. Default is all columns.

full   A full, orthogonal *nr*-by-*nr* **Q** matrix is produced, not just the first *nc* columns. In this case, the **R** matrix has dimension *nr*-by-*nc* (the last *nr*−*nc* rows are zeroes) so that the equation **X** = **QR** still holds. If *nr* < *nc*, this option has no effect: **Q** is *nr*-by-*nr* and **R** is *nr*-by-*nc* regardless.

long   Display the rank and estimated reciprocal condition number.

pivot=*INTLIST*

pivot   Enable column pivoting: the columns of **X** are permuted in order to obtain the most numerically stable decomposition. In this case, the decomposition satisfies **XE** = **QR**, where **E** is the permutation matrix and **XE** is **X** with its columns permuted.

If *INTLIST* is specified, it controls pivoting as follows. Columns of **X** corresponding to positive elements of *INTLIST* are moved to the leading part of **X** before the decomposition begins and are frozen in place. Columns of **X** corresponding to negative elements are moved to the trailing part of **X** before the decomposition begins and frozen in place. Columns of **X** corresponding to zero elements are allowed to move freely during the decomposition. If *INTLIST* is not specified, all rows and columns are allowed to move freely. See the *LINPACK Users' Guide*, pages 9.3-9.4, for details.

The **rank** command always uses pivoting.

tol=*FLOAT*

Affects the computation of the rank as explained below. The default value is *tol* = 1.0e−7.

OUTPUTS

q@   The *nr*-by-min(*nr, nc*) matrix **Q**.

r@   The *nc*-by-*nc* matrix **R**.

e@   The *nc*-by-*nc* matrix **E**.

The **q@**, **r@**, and **e@** output datasets are possible only when the command is

rank@

The rank of **X**, defined as the number of diagonal elements of **R** which in absolute value exceed *tol* * sqrt(*nr*) * *rmax*, where *rmax* is the maximum absolute value of all diagonal elements of **R** and *tol* is explained above. *Warning:* the rank may be computed incorrectly if pivoting is not in effect. The **rank** command always uses pivoting; with the **qr** command the user must request pivoting.

rcond@

An estimate of the reciprocal condition number of **X**: the ratio *rmin*/*rmax*, where *rmin* and *rmax* are the minimum and maximum absolute diagonal elements of **R**. *Warning:* this estimate depends on the column scaling of **X** and on whether pivoting is in effect. See the reference.

If no output datasets are specified, the action taken depends on how the command was invoked. **Qr** puts **Q** and **R** in the datasets *x*.**q** and *x*.**r** where *x* is the name of the input matrix. If pivoting is in effect and no outputs are specified, **E** is put in *x*.**e**. **Rank** displays the rank and estimated reciprocal condition number of **X** (that is, the {**long**} option is automatically invoked).

EXAMPLES

Compute the QR decomposition with pivoting and then verify the equations given above:

```
. qr x {pivot} > q r e
. show (x#*e - q#*r)
```

To scale the columns of *x* so that each has the same absolute maximum value before computing the decomposition:

```
. colamax x > s      # see vecop(BRM).
. y = x/s            # use dimension-expansion; see BUG Section 8.3.5.
. qr y
```

REFERENCE

J. J. Dongarra et al. (1979). *LINPACK Users' Guide.* SIAM, Philadelphia. Chapter 9.

SEE ALSO

chol(*BRM*), *svd*(*BRM*), *linpack*(*BML*).

---

random — Random numbers from continuous probability distributions

---

SYNOPSIS

rbeta [*d*] {a=;b=;dims=;eps=} > *r*	# *beta random numbers*
rcauchy [*d*] {a=;b=;dims=} > *r*	# *Cauchy random numbers*
rchisq [*d*] {df=;dims=;eps=} > *r*	# χ^2 *(chi-square) random numbers*
rexp [*d*] {l=;b=;dims=} > *r*	# *exponential random numbers*
rf [*d*] {df=;dims=;eps=} > *r*	# *Fisher's F random numbers*
rgamma [*d*] {a=;l=;b=;dims=;eps=} > *r*	# *gamma random numbers*
rgau [*d*] {m=;s=;v=;dims=} > *r*	# *Gaussian (Normal) random numbers*
rlogis [*d*] {a=;b=;dims=} > *r*	# *logistic random numbers*
rt [*d*] {df=;dims=;eps=} > *r*	# *Student's t random numbers*
runi [*d*] {a=;b=;dims=} > *r*	# *uniform random numbers*

DESCRIPTION

Random generates pseudorandom numbers from continuous probability distribution functions (cdf's). Each type of random number generator has its own command name, of the form r*xxx*.

INPUTS

d@   The dimensions of the output dataset are the dimensions of *d*. The contents of the output dataset are not affected by the contents of *d*. If the {**dims=**} option is given, the **d@** input is ignored.

OPTIONS

a=*FLOAT*, **b**=*FLOAT*, **d**=*FLOAT*, **df**=*FLOATLIST*, **l**=*FLOAT*,
m=*FLOAT*, **s**=*FLOAT*, **v**=*FLOAT*

Parameter values of the cdf's. For their meanings, refer to *BUG* Table 7.1 or to the on-line table in 'help cdftable'. For the F distribution, the {**df=**} option takes two values to specify the two degrees of freedom parameters. All other parameter options take a single value.

eps=*FLOAT*

The required relative accuracy of the underlying inverse cdf. See *cdf(BRM)* for details. The default is *eps* = 1.0e−8. This option is not available for the **rcauchy**, **rexp**, **rgau**, **rlogis**, or **runi** functions.

dims=*INTLIST*

INTLIST is a list of up to three integers, separated by commas, which specify the number of sheets, rows, and columns in the output dataset. If only one or two integers are given, 1's are assumed to the left. For example, {**dims=5**} is equivalent to {**dims=1,1,5**}; both produce a row vector of length 5.

OUTPUTS

r@   The generated pseudorandom numbers.

EXAMPLES

To generate a dataset *uni.r* of dimensions $(20, 5)$ which contains uniform $[0, 10]$ random numbers:

```
. runi {b=10} {dims=20,5} > uni.r
```

To generate a dataset *samp* of dimensions $(10, 7)$ which contains standard Gaussian random numbers:

```
. rgau {dims=10,7} > samp
```

To generate a random sample *x.r* from a Gaussian distribution with mean=2 and SD=4, whose dimensions are those of another dataset *x*:

```
. rgau {m=2;sd=4} x > x.r
```

SEE ALSO

BUG Sections 7.3 and 9.3.

cdf(*BRM*), *cdf*(*BML*).

rans(*BML*), for information on the underlying uniform pseudorandom number generator.

rbinorm(*BRM*), *rmultinorm*(*BRM*), for bi- and multivariate random Normals.

seed(*BRM*), for how to reset the random number generator seeds.

---

rbinorm — Generate bivariate Normal random numbers

SYNOPSIS

rbinorm *r* {**n**=} [> *x*]

DESCRIPTION

Rbinorm generates *n* observations from the bivariate Normal distribution with mean 0, standard deviation 1, and correlation *r*.

INPUTS

r@   The correlation. It must be specified.

OPTIONS

n=*INT*

How many observations to generate. If not specified, 100 are generated.

OUTPUTS

x@   Matrix of dimensions $(n, 2)$ that contains the generated numbers.

NOTE

This command is implemented as a macro. It illustrates the macro features described in *BUG* Section 10.8.4; it uses methods described in Section 9.3.1 to generate the numbers.

SEE ALSO

rmultinorm(*BRM*), for multivariate Normal random numbers.

---

read — Enter data into a BLSS dataset

SYNOPSIS

　　read [*textfile*] {**autodims;count=;dims=;col;row;na=;clobber;** ... } [> *dataset*]

DESCRIPTION

　　Read converts an ASCII text file (such as those produced by **edit** or **addtext**) into a BLSS dataset. *Textfile* is the ASCII text file to be read. *Dataset* specifies the name of the BLSS dataset to be created.

　　The input text file is assumed to be in the work area unless specified otherwise, either by using an option or by using a UNIX pathname.

　　If the input file name is omitted, **read** asks you to enter data from the keyboard. If the number of values to read can be inferred from the {**dims=**} or {**count=**} options, data are read until enough items have been entered. Otherwise, data are read until you type a control-D at the beginning of a line. In either case, **read** tells you what it expects and when it is finished.

　　If the output dataset name is omitted and the input text file comes from another area (directory), then the output dataset has the same name as the input file. Otherwise, **read** asks you for the output dataset name.

　　If the work area already contains a file with the specified output dataset name, **read** asks whether you want to overwrite it. If yes, type 'y' or 'Y'. Any other reply causes **read** to abort.

　　Because the output dataset is put in the work area, it must explicitly be saved using the **save** command if you want it for future use. It can then be retrieved using the **load** command.

Input Text Format

　　Read assumes that the input text contains data in the following order:

sheet 1:	row 1, row 2, ..., last row of sheet 1
sheet 2:	row 1, row 2, ..., last row of sheet 2
...	
last sheet:	row 1, row 2, ..., last row of last sheet

　　Data values may be separated by any amount of white space (spaces, tabs, or linefeeds) or by commas, colons, or semicolons.

　　Missing values may be entered using 'NA'. Multiple commas, colons, or semicolons do not yield NA's; you must enter NA's explicitly.

　　Read understands E-notation (numbers such as 1.2e+3, etc.—see *BUG* Section 2.3 for an explanation). Any of 'e', 'E', 'd', or 'D' may be used. (However, BLSS stores all data in single precision regardless of 'd' or 'D'.)

　　Comments are allowed in the input text. All text to the right of a '/' or '#' in any line is regarded as a comment and ignored.

OPTIONS
Dimensioning Options
autodims

> **Read** infers the dimensions of the new dataset from the shape of the input text. The number of columns is the number of values in the first nonempty row of the input. Additional rows of data are read until none remain. If any (nonempty) rows contain a different number of items from the first row, they are truncated or padded with NA's as necessary and a warning is issued. If {**autodims**} is used, the command reminds you: 'Type your data one row per line.'

{**autodims**} is the default dimensioning option. However, it cannot be used to create datasets which contain more than one sheet.

count=*INT*

> *INT* is the largest number of elements to be read from the input text.

dims=*INTLIST*

> *INTLIST* is a list of up to three integers, separated by commas, which specify the number of sheets, rows, and columns in the output dataset. If less than three numbers are given, 1's are assumed to the left. Thus, if two numbers are given, **read** creates a matrix; if one number is given, **read** creates a row vector of that length.

If 0's are specified in any field, **read** reads in all data (except if the **count** specification would be exceeded) and deduces the correct count for that field. Missing values are inserted if necessary to fill out an incomplete row or sheet.

The {**dims=**} option overrides the {**autodims**} option. When {**dims=**} is used, any number of data may be entered on each input line.

col

> Force the new dataset to be a column vector, regardless of the shape of the input text. (Equivalent to {**dims=0,1**}; overrides {**autodims**}.)

row

> Force the new dataset to be a row vector, regardless of the shape of the input text. (Equivalent to {**dims=1,0**}; overrides {**autodims**}.)

Other Options
na=*FLOAT*

> Any values of *FLOAT* read in are converted to NA's.

clobber

> Do not ask for permission before overwriting a file that already exists.

bin, class, data, doc, help, home, sys, text
dirs=*dirname,* tree=*treename,* area=*areaname*

> These options may be used to specify an alternate area (or directory) from which to read the input text file. See *area(BRM)* for a full description. The default values are:

        dirs   your work area

        tree   your *blss* tree (normally *~/blss*)

        area   the text area

terse Fewer messages; in particular, no 'Type your data ...' message.

EXAMPLES

Create a 2-by-3 matrix from the keyboard using the default {**autodims**} dimensioning option:

```
. read > matrix
Type your data one row per line; finish with RETURN and CTRL-D.
1 2 3
4 5 6
Control-D
Read 6 values into "matrix"; dims=(2,3) (matrix).
```

Create a 2-by-2 matrix from the keyboard using the {**dims**=} option:

```
. read {dims=2,2}
Name for new BLSS dataset? twobytwo
Type your data; finish with RETURN.
1 2 3 4
Read 4 values into "twobytwo"; dims=(2,2) (matrix).
. show twobytwo

   1.00    2.00
   3.00    4.00
```

Create a column vector of unspecified length using the {**col**} option:

```
. read {col} > colvec
Type your data; finish with RETURN and CTRL-D.
7 6 5
4 NA 2 1
Control-D
Read 7 values into "colvec"; dims=(7,1) (column vector).
. show colvec

   7.00    6.00    5.00    4.00     NA    2.00    1.00
```

(Note that the **show** command normally shows column vectors as row vectors, to save space.)

Observe that in the first and third examples above, the user did not specify the number of values to enter so he ended his input with a control-D. In the second example, the command expected exactly four values and no control-D was necessary.

Read the text file *parsley* from the text area into the BLSS dataset *sage* in the work area:

```
. read {text} {dims=12,24} parsley > sage
```

Read an ASCII text file into a dataset; values of –1 in the text file are converted into NA's in the dataset:

```
. read rosemary {na= -1} > thyme
. show thyme
```

```
      1.00     NA     2.00     NA
```

Pathnames

As noted above, UNIX pathnames may be used as input file names. The BLSS shell recognizes '~' as an abbreviation for the pathname of your home directory, and '~jim' as an abbreviation for jim's home directory. Thus, both the following create the BLSS dataset *mydata* from the text file *datafile* in your home directory:

```
. read {home} datafile > mydata
. read ~/datafile > mydata
```

To read an input file from jim's home directory, type:

```
. read ~jim/datafile > jimdata
```

SEE ALSO

BUG Section 4.4, for more examples and discussion.

area(BRM).

intro(BRM), for more about UNIX pathnames and ~ expansion.

redim(BRM), which can be used to correct mistakes in the {**dims=**} option setting after entering a dataset.

---

redim — Redimension an array

SYNOPSIS

redim *x* [*d*] {**c;dims=;f=**} > *y*

DESCRIPTION

Redim generates a new dataset with the same data as *x* but new dimensions. Data elements are read from the input *x* and written to the output *y* in *row-major order* (that is, going along the rows). If the new dimensions call for fewer elements than *x* contains, the extra elements in *x* are not used. If the new dimensions call for more elements than *x* contains, then by default, NA's are used. This can be changed with the {**c**} or {**f=**} options.

INPUTS

x@   The original dataset.

d@   If the {**dims=**} option is not given, the dimensions of the output dataset are the dimensions of *d*. The contents of *d* are ignored.

OPTIONS

c   Cycle through the data x as often as necessary to fill all positions in y.

dims=*FLOATLIST*

The dimensions of the new dataset. If less than three numbers are given, 1's are assumed to the left. Thus, if two numbers are given, **redim** creates a matrix; if one number is given, **redim** creates a row vector of that length. If neither the {**dims**=} option nor the **d@** input is given, the dimensions (1,1,1) are used.

f=*FLOAT*

Fill extra elements in y with the specified value.

OUTPUTS

y@   The new dataset.

EXAMPLES

Suppose you use the **read** command to enter a dataset a by hand and discover afterward that you used the option {**dims**=3,15} but meant to use {**dims**=15,3}. You can correct this with the **redim** command:

    . redim a {dims=15,3} > b

This command is quite different from taking the transpose, because **redim** reads the data from x and enters it into y row by row.

If v is a vector of length 40, then to create a 10-by-4 matrix w which contains the same data:

    . redim v {dims=10,4} > w

This is the opposite of the row-unravel operation.

If x is a vector of length 10, then to create a matrix y that contains 5 rows, each of which are x:

    . redim x {c; dims=5,10} > y

To create a matrix z that contains 5 columns, each of which are x, give the same command and then take the transpose:

    . z = y'

SEE ALSO

BUG Section 8.3.4.

---

regress — Least squares multiple regression and weighted regression

---

SYNOPSIS

> **regress** x [y w] {**x**=;**y**=;**noint**;**rstat**;**anova**; ... } [> *fit res b se t p rstat* ...]

DESCRIPTION

Regress performs least squares multiple linear regression and least squares weighted linear regression. The default display shows: the number of observations and the number of parameters; the estimated parameters (that is, the intercept and coefficients), their SEs (standard errors), t-statistics, and P-values; the residual SD and variance; and the multiple R and R^2 of the regression. The display uses the column numbers of the **x@** input to refer to the independent variables.

The Design Matrix

Throughout this manual entry, the term *design matrix* refers to the matrix X of observations on the independent variables. If the regression includes an intercept, the design matrix includes an initial column of 1's, which is automatically added by **regress**.

Linear Dependencies in X

Regress automatically compensates for linearly dependent columns in the design matrix X. It drops as many columns as necessary to obtain a linearly independent subset which spans the same space. Regression coefficients for the dropped columns, and their SEs and t-statistics (which are meaningless), are reported as 0. The residuals, fitted values, and statistics which depend on them only (such as R^2 and the anova table) are uniquely determined. The regression coefficients are correct but not unique (because a different subset of columns could be dropped).

Multiple Dependent Variables

If the **y@** input contains more than one column, simultaneous regressions are run for each. In this case, the **fit@**, **res@**, **b@**, **se@**, **t@**, and **p@** outputs contain one column for each regression. The other outputs contain information for only the first regression.

INPUTS

x@   Dataset whose column(s) contain the independent variable(s).

y@   Dataset whose column(s) contain the dependent variable(s).

If the **y@** input dataset is not specified, then all columns but the last of the **x@** input dataset are taken to be the independent variables; the last column is taken to be the dependent variable. This default can be changed by options described below.

w@   Column vector that contains weights for a weighted regression. If observations have unequal variances, the weights should be inversely proportional to the variances. Observations with zero weights are ignored when estimating the regression model, but their fitted values and residuals can be computed.

Each row of the input datasets is considered to be one case, except if an input dataset is a row vector, it is treated as a column vector. Input datasets may not contain missing data. Use the **select** command if necessary to remove cases with missing observations.

OPTIONS

x=*INTLIST*
> Columns of the **x@** dataset to use as independent variables.

y=*INTLIST*
> Columns of the **y@** dataset (or **x@** dataset, if **y@** is not specified) to use as dependent variables.

w=*INT*
> Column of the **w@** dataset (or **x@** dataset, if **w@** is not specified) to use as weights.

noint No intercept is included in the model.

rstat By default, only the usual (centered) value of the R^2 statistic is displayed:

$$\text{(centered) } R^2 \;=\; 1 - \frac{\Sigma e_i^2}{\Sigma(y_i-\bar{y})^2} \;=\; \frac{\Sigma(\hat{y}_i-\bar{y})^2}{\Sigma(y_i-\bar{y})^2} \;=\; \frac{\|\hat{\mathbf{y}}-\bar{\mathbf{y}}\|^2}{\|\mathbf{y}-\bar{\mathbf{y}}\|^2}$$

where y_i denotes the observations on the dependent variable, $\bar{y}$ denotes the average of the y_i, $\hat{y}_i$ denotes the fitted values, and $e_i = y_i - \hat{y}_i$. In addition to that, the {**rstat**} option displays: the (centered) adjusted R^2, sometimes called $\bar{R}^2$:

$$\text{(centered) adjusted } R^2 \;=\; \bar{R}^2 \;=\; 1 - \frac{\Sigma e_i^2/(n-k)}{\Sigma(y_i-\bar{y})^2/(n-1)}$$

which penalizes itself for each extra parameter in the model (n is the number of observations and k the number of parameters); the uncentered R^2:

$$\text{uncentered } R^2 \;=\; \frac{\Sigma\hat{y}_i^2}{\Sigma y_i^2} \;=\; \frac{\|\hat{\mathbf{y}}\|^2}{\|\mathbf{y}\|^2}$$

which should be used whenever comparing models with and without intercepts; and the autocorrelation of the residuals and the Durbin-Watson statistic, which should be noted whenever there is a time- or serial-dependence in the data and ignored otherwise.

anova Display the analysis of variance (anova) table.

bcov Display the estimated covariance matrix of the regression coefficients.

bcor   Display the estimated correlation matrix of the regression coefficients.

long   Equivalent to {**rstat;anova;bcov;bcor**}. See *BUG* Figure 5.5 for an example of the {**long**} output display.

forceout
        Force all possible output datasets to be created.

OUTPUTS
        fit@   The fitted values.

        res@   The residuals.

        b@   The estimated regression coefficients. If the regression includes an intercept, the first element of **b@** is its estimated value.

        se@   The estimated SEs of the regression coefficients.

        t@   The t-statistics.

        p@   The P-values of the t-statistics.

        rstat@
                The 'rstat' array; that is, those statistics displayed by the {**rstat**} option.

        anova@
                The anova array; that is, the table of numbers displayed by the {**anova**} option. It can be subsequently displayed by the **anovapr** command— see *anovapr(BRM)* for a description of its contents.

        bcov@
                The estimated covariance matrix of the regression coefficients.

        bcor@
                The estimated correlation matrix of the regression coefficients.

        c@   The matrix $(X'X)^{-1}$, where X is the design matrix.

EXAMPLES
        This set of commands gives the linear regression of petal width on petal length for *iris versicolor*, makes a simultaneous plot of the original data and the fitted values, and then makes a plot of the residuals (Y-axis) against the first column of *iris.ve* (X-axis).

```
. regress {x=1;y=2} iris.ve > fit.ve res.ve
. scat iris.ve[1] (iris.ve[2],fit.ve)        # data and fitted values
. scat iris.ve[1] res.ve                     # residual plot
```

        The following command performs a linear regression of the first column of *y* on the columns of *x* and saves the fitted values in *f* and the residuals in *r*. Its output display includes the **rstat** and **anova** statistics.

```
. regress {y=1; rstat;anova} x y > f r
```

        This command saves the **rstat** and **anova** statistics as the BLSS datasets *r* and *a*, but does not display them:

```
. regress x y > rstat@ r anova@ a
```

To perform a weighted least squares regression with the independent variables in *x*, the dependent variable(s) in *y*, and the weights in *w*:

> . `regress x y w`

To perform a weighted least squares regression with both the independent and dependent variables in *x* and the weights in *w*:

> . `regress x w@ w`

ALGORITHM

Regress computes the QR decomposition of the design matrix *X* with column scaling and pivoting; it then projects the dependent variable(s) *Y* onto the space spanned by *X*. **Regress** uses the LINPACK subroutines *sqrdc* and *sqrsl*. The following additional options affect the algorithm:

nopiv Disable column scaling and pivoting in the QR decomposition. Not allowed if the design matrix is singular.

tol=*FLOAT*
Same as the {**tol**=} option of *qr(BRM)*.

The following additional output datasets are possible only when pivoting is in effect:

pivot@
Pivot indices. The matrix actually decomposed consists of the columns $X[pivot[1]], \ldots, X[pivot[k]]$, where *X* is the design matrix and *k* is its rank. If the model includes an intercept, then the initial column of 1's in *X* is always the first pivot column—that is, *pivot*[1] is always 1.

scale@
Scale factors for the columns of the design matrix.

LIMITATIONS

In a weighted regression, the values of the residual autocorrelation and Durbin-Watson statistics are computed as if the regression were unweighted.

Sufficiently peculiar column scaling may cause an arithmetic exception. This seldom occurs in practice; if it does, rescale the columns.

REFERENCE

J. J. Dongarra et al. (1979). *LINPACK Users' Guide*. SIAM, Philadelphia. Chapter 8.

SEE ALSO

BUG Sections 5.5, 5.6, and 9.4, for longer examples and discussion of the output display.

demo(BRM), for the on-line demonstration macros **demo.reg1**, **demo.reg2**, and **demo.reg3**.

qr(BRM) and *svd(BRM)*, for numerical decompositions pertinent to least squares regression.

robust(BRM), for an alternate method of fitting lines to data.

xvalid(BRM), for cross-validated estimates of regression R^2 and *RSS*.

---

remove — Remove datasets and text files

SYNOPSIS
> **remove** [*files*] {**all;ask;everything;noask;** ... }

DESCRIPTION
> **Remove** removes the named BLSS datasets and text files. In other words, the named objects cease to exist. By default, **remove** acts upon files in your work area. Other areas may be specified using options.
>
> With no arguments, **remove** questions you about each file in your work area and removes only those for which you respond 'yes'—that is, any response beginning with 'y' or 'Y'.

OPTIONS
> **all**   If the {**ask**} option is in effect, **remove** asks about files whose names begin with '.' or '_' characters. By default, such files are removed only when specified as *file* arguments.
>
> The {**all**} option is implied by the {**everything**} option.
>
> **ask**   Always ask for permission before removing any file. This option overrides {**noask**}.
>
> **bin**, **class**, **data**, **doc**, **help**, **home**, **sys**, **text**
> **dirs**=*dirlist*, **tree**=*treelist*, **area**=*arealist*
>   These options specify alternate areas (or directories) from which to remove files. See *area(BRM)* for full information. The default values are:

dirs	your work area
tree	your *blss* tree (normally ~/*blss*)
area	the data area

> **everything**
>   If set, **remove** asks whether you want to remove every file in the work area (or the area you have specified). If you respond yes, then everything in that area is removed without further asking.
>
>   If both the {**everything**} option *and* the {**noask**} option are set, **remove** removes everything in the specified area with no questions whatsoever.
>
> **noask** If multiple areas have been specified, **remove** asks for permission before removing files. The {**noask**} option turns off this protective feature.

EXAMPLES
> Remove *anise* and *basil* from your work area:
>
> ```
> . remove anise basil
> ```
>
> Selectively remove files from your work area:
>
> ```
> . remove
> ```

Selectively remove files from your text area:

```
. remove {text}
```

To remove *cumin* and *dill* from your data area:

```
. remove {data} cumin dill
```

SEE ALSO
BUG Section 4.7, for more examples and discussion.
area(*BRM*), *list*(*BRM*).

---

rename — Rename a dataset or text file

SYNOPSIS
rename {**clobber;** ... } *oldname newname*

DESCRIPTION
Rename causes the BLSS dataset or text file *oldname* to be renamed *newname*. If a file with the name *newname* already exists, **rename** asks for permission to overwrite it. It does so only if the response begins with 'y' or 'Y'.

OPTIONS
bin, class, data, doc, help, home, sys, text
dirs=*dirname,* **tree**=*treename,* **area**=*areaname*
Normally, **rename** renames files in the work area. These options may be used to rename files in other areas (or directories). For a full description, see *area*(*BRM*). The default values are:

dirs	your work area
tree	your *blss* tree (normally ~/*blss*)
area	the data area

clobber
Overwrite the file *newname* without asking permission.

EXAMPLES
Rename *junk* as *precious*:

```
. rename junk precious
```

Rename a file saved in your data area:

```
. rename {data} smeagol gollum
```

SEE ALSO
BUG Section 4.7, for more discussion.
area(*BRM*).

---

rmultinorm — Generate multivariate Normal random numbers

SYNOPSIS
   rmultinorm *cov* {**n**=} [> *x*]

DESCRIPTION
   Rmultinorm generates *n* observations from the multivariate Normal distribution with mean 0 and covariance matrix *cov*.

INPUTS
   cov@   The covariance matrix. It must be symmetric and positive semidefinite.

OPTIONS
   n=*INT*
      How many observations to generate. If not specified, 100 are generated.

OUTPUTS
   x@   Matrix that contains the generated numbers. It has *n* rows and the same number of columns as the covariance matrix *cov*.

NOTE
   This command is implemented as a macro. It illustrates the macro features described in *BUG* Section 10.8.4; it uses methods described in Section 9.3.2 and *chol(BRM)* to generate the numbers.

SEE ALSO
   rbinorm(BRM), for bivariate Normal random numbers.

---

robust — Robust weighted least squares regression

SYNOPSIS
   robust *x* {**x**=;**y**=;**c**=;**eps**=;**noint**;**anova**} [> *fit res b t p rstat anova se w*]

DESCRIPTION
   Robust performs an iteratively reweighted least squares regression that is resistant to thick-tailed error distributions. At each iteration, the bisquared weight function (see *biweight(BRM)*) is used: z is taken to be the deviation of an observation from its current fitted value and s is taken to be the current weighted residual SD.

   The default output display shows the estimated regression coefficients, their SEs (standard errors) and t-statistics, the R^2 of the regression, and the residual SD.

INPUTS
   x@   Dataset whose columns contain the dependent and independent variables. Rows of x that contain missing values in the {**x**=} or {**y**=} columns are ignored.

OPTIONS

x=*INTLIST*

> Columns of the *x* dataset to use as the independent variables. Default is all but the last column.

y=*INT*

> Column of the *x* dataset to use as the dependent variable. Default is the last column.

c=*FLOAT*

> Bisquare parameter. Default is 6. For robust estimation, values of *c* between about 4 and 10 are reasonable. The larger *c* is, the closer the results are to those of least squares regression. Setting *c* to 0 gives least squares regression exactly.

eps=*FLOAT*

> Iterations stop when the absolute change in $(1 - SS_{OLD}/SS_{NEW})$ is less than *eps*, where *SS* is the weighted sum of squared residuals. Default is *eps* = .01.

noint No intercept is included in the regression.

anova Display the anova table.

long Equivalent to {**anova**}.

vlong Display the default output, the anova table, and the weighted sum of squared residuals for each iteration.

OUTPUTS

fit@ The fitted values.

res@ The residuals.

b@ The estimated regression coefficients. If the regression includes an intercept, the first element of **b@** is its estimated value.

t@ The t-statistics corresponding to the estimated regression coefficients.

p@ The P-values of the t-statistics.

rstat@

> R^2 of the regression and the residual SD.

anova@

> The anova array. It can be subsequently displayed by the **anovapr** command—see *anovapr(BRM)* for a description of its contents.

se@ The SEs of the regression coefficients.

w@ The final weights for each observation.

EXAMPLES

Cars contains the initial speed and stopping distance for each of fifty drivers. The following commands perform a robust weighted least squares regression of stopping distance on speed, and plot the residuals against the fitted values. The residual plot displays a mild nonlinearity:

```
. load cars
. robust cars > fit res
. scat fit res
```

This command performs the same regression and, in addition, saves the final weights in the dataset *w*:

```
. robust cars > fit res w@ w
```

REFERENCE

Donald R. McNeil (1977). *Interactive Data Analysis.* Wiley, New York. Chapter 7, pp. 163-71, gives an older version of the algorithm which used the median absolute residual as *s* in the bisquared weight function and was less numerically stable.

SEE ALSO

biweight(BRM), *regress(BRM)*.

---

sample — Random sampling with and without replacement

SYNOPSIS

sample [*x*] {**x**=;**hi**=;**lo**=;**n**=;**nsamp**=;**rep**;**norep**;**shind**} [> *sample*]

DESCRIPTION

Sample generates a random sample from a user-specified population *x*. If *x* is specified, rows of *x* are selected at random with equal probability. Otherwise, integers are sampled with equal probability from the sequence *lo* through *hi*; see the options. By default, sampling is with replacement.

INPUTS

x@   Optional dataset from which to sample rows.

OPTIONS

x=*INTLIST*

When sampling from a dataset *x*, only the specified columns of *x* appear in the sample.

hi=*INT*

Largest integer to sample from, if *x* is not specified.

lo=*INT*

Smallest integer to sample from, if *x* is not specified. Default is 1 if *hi* is positive.

n=*INT*

Sample size. The default is the number of rows of *x*, if specified; otherwise, $hi - lo + 1$. When sampling without replacement, the sample size may not exceed the population size.

nsamp=*INT*

> Number of (independent) samples to draw. If sampling from the integers or from a column vector, the successive samples are placed in successive columns of the output dataset; otherwise, they are placed in successive sheets. By default, *nsamp* = 1.

rep   Sample with replacement. This is the default.

norep Sample without replacement.

shind Sample sheets independently. If *x* contains several sheets, then by default the identical set of rows is selected from each sheet. This option causes the sheets to be sampled independently.

OUTPUTS
sample@

> The generated sample. Its number of rows is equal to the sample size.

If no output dataset is specified, the sample is put into *x*.**sample** if *x* is specified—or into **n.sample** if *x* is not specified.

EXAMPLES
Sample 100 numbers with replacement from the integers 1 through 100:

```
. sample {hi=100}
```

Sample 8 numbers without replacement from the integers –5 through 5:

```
. sample {n=8;norep;lo=-5;hi=5} > mysample
```

Generate a random permutation of the rows of the *boston* dataset:

```
. sample boston {norep}
```

Draw a sample with replacement from the *boston* dataset of the same size as *boston* itself, and place it in *boot.s*.

```
. sample boston > boot.s
```

Such a command is useful for bootstrap techniques.

SEE ALSO
BUG Section 7.2.
sort(*BRM*).

---

save — Save datasets and text files for future use

SYNOPSIS
save [*files* ...] {**ask;clobber;force;noask;** ... }

DESCRIPTION
Save causes copies of the named *files* to be saved from the active work area into one of your permanent storage areas. Because the entire contents of the work area disappear when you exit from BLSS, it is necessary to save any

BLSS datasets or text files you wish to use in later sessions. Use the **load** command to retrieve saved files.

By default, BLSS datasets are saved in the data area and ASCII text files are saved in the text area. Alternate areas for saving may be specified using options.

If no *files* are specified, **save** asks you about each file in the work area and saves only those for which you respond 'yes' (any response beginning with 'y' or 'Y').

OPTIONS

ask   Force asking even when *files* have been specified.

bin, class, data, doc, help, home, sys, text
dirs=*dirname,* **tree**=*treename,* **area**=*areaname*
> These options may be used to specify an alternate area (or directory) in which to save files. For a full description, see *area(BRM)*. The default values are:
>
> | | |
> |---|---|
> | **dirs** | (none) |
> | **tree** | your *blss* tree (normally ~/*blss*) |
> | **area** | the data area for datasets; the text area for text files |

clobber
> Normally, **save** asks for permission before overwriting a previously saved file. The {**clobber**} option disables this safety feature.

force   Normally, **save** refuses to save text files in the data area or BLSS datasets in the bin, doc, help, or text areas. The {**force**} option overrides this refusal.

noask   Disable asking except when a file has been previously saved under the same name.

EXAMPLES

Selectively save everything in the work area:

    . save

Save the files *alpha* and *beta*:

    . save alpha beta

Save the file *alpha* regardless of whether anything was previously saved under that name:

    . save {clobber} alpha

Save *mymacro* into the bin area:

    . save {bin} mymacro

SEE ALSO

BUG Sections 4.6, 4.7, 4.12, for more examples and discussion.
area(BRM), load(BRM), remove(BRM).

---

scat — Scatterplots

SYNOPSIS
scat *x* [*y*] {x=;y=;height=;width=;big;noscale;xmin=;xmax=;ymin=;ymax= }

DESCRIPTION
Scat makes two-dimensional scatterplots (printer plots) from the columns of *x* and *y*. If only *x* is specified, both the X and Y variables of the plot are taken from columns of the *x* input dataset. If both *x* and *y* are specified, the X variables are columns of *x* and the Y variables are columns of *y*. Each X variable specified is used for a different plot. All Y variables specified are plotted for each plot. No more than eight Y variables may be plotted.

When one plot is made on a pair of axes, single data points are plotted as '*'. Two through nine points at the same location are plotted as the digits '2' through '9'. Ten through nineteen points are plotted as '%'. Twenty or more points are plotted as '#'.

When several Y variables are plotted on a single pair of axes, single data points of the first Y variable are plotted as *a*, the second variable as *b*, and so forth. Multiple data points of the first Y variable at the same location are plotted as *A*, of the second Y variable as *B*, and so forth. Points from different Y variables at one location are plotted as '$'.

Missing data are not plotted.

INPUTS
x@   Dataset whose columns contain the X variables. If the **y@** input dataset is not specified, *x* also contains the Y variables. The default behavior depends on whether or not the **y@** input dataset is specified. If it is, then all columns of the **y@** input dataset are plotted against all columns of the **x@** input dataset. If not, the second column of the **x@** input dataset is plotted against the first.

y@   Dataset whose columns contain the Y variables.

The number of rows in *x* and *y* must be equal. If either *x* or *y* is a row vector or a sheet vector, it is treated as a column vector.

OPTIONS
x=*INTLIST*
    Columns to use as the X variables for successive plots.

y=*INTLIST*
    Columns to use as the Y variables for each plot.

height=*INT*
    Approximate height, in characters, of the plot. The default height is 20.

width=*INT*
    Approximate width, in characters, of the plot. Actual width will be less than or equal to this unless the {**noscale**} option is used. **Scat** tries to

find a 'nice' minimum and maximum which contain the data and then decides on a good width. The default width is 75.

big Equivalent to {**height**=**43**; **width**=**80**}. Makes a big scatterplot, with aspect ratio suited to a printer rather than a video terminal.

noscale
Inhibit the scaling routine so that the plot is exactly the size chosen, or the default size. Minimum and maximum values may not be 'nice.'

xmin=*FLOAT*, **xmax**=*FLOAT*
ymin=*FLOAT*, **ymax**=*FLOAT*
The exact boundaries to use if {**noscale**} is in effect; approximate boundaries otherwise. The default values are obtained from the data.

terse Do not display the names of the X and Y variables above the plot.

EXAMPLES
Make a scatterplot with column 1 of *pressure* (temperature in degrees Celsius) as the X variable and column 2 (vapor pressure of mercury in millimeters of mercury) as the Y variable.

```
. load pressure
. scat pressure {x=1;y=2}
```

Get a more detailed look at the lower left corner of the previous plot:

```
. scat pressure {x=1;y=2} {xmax=100;ymax=1}
```

The options {**x**=**1**;**y**=**2**} could have been omitted in these two examples because these values are the defaults when only one input dataset is given.

Regress *kappa*[2] on *kappa*[1]; then plot the original data (as *a*'s and *A*'s) and the fitted values (as *b*'s and *B*'s):

```
. load kappa
. regress kappa[1] kappa[2] > fit
. scat kappa[1] kappa[2],fit
```

Plots for Lineprinters
The default plot size was chosen for video terminals; it is unpleasing when printed. For printer plots specify a greater plot height in proportion to width or use the {**big**} option, which produces an (approximately) 8-inch square plot when printed. For example, to make a large plot and send it to the lineprinter:

```
. scat {big; lpr} iris >> outputfile
```

NOTE
Future releases of BLSS will have improved graphics capabilities. The specifications of this command may be revised accordingly.

SEE ALSO
BUG Sections 5.2, 5.5.1.
BUG Section 9.5.1, for information on making Q-Q plots.
scatter(BRM), for an interactive video scatterplotter.

---

scatter — Interactive video scatterplots

SYNOPSIS
   scatter x {**d** = }

DESCRIPTION
   Scatter is an interactive visual scatterplot program for use with video terminals. Commands are single letters optionally followed by other characters as
   described below.

General Commands

h	**Help**: show the menu of commands.
q	**Quit** the session.
s *plot*	**Save** a copy of the current plot in the text file *plot* in your work area.
x	**Exhibit** information on x.
?	Same as **h**.
!	Gives an interactive UNIX shell. Uses the environment string SHELL if it exists. The default is the C shell, */bin/csh.*
! *command*	Execute the given UNIX command.

Plotting Commands

a	Turn **axis** on or off. Default is off.
b *c r*	Adjust **box** to *c* columns by *r* rows. Default is 79 columns and 22 rows.
c	**Clear** screen.
o *x y*	Put axis **origin** at (x,y). Default is (0,0).
p *m n*	**Plot** column *m* against column *n*. *M* or *n* may be 0; column 0 is considered to be the integers 1, 2, . . . , (number of rows in x).
p	Clean copy of current plot.
r *i*	Set plotting **range** of column *i*.
w	Turn **window** boundary on or off. Default is off.

Labeling Points on the Plot

d *n*	Use column *n* as **depth**. By default, points are plotted using sym- bols from the character string '.,o+x*#'. Characters are assigned to increasing values of column *n* from left to right. Column *n* may contain either a continuous variable or a (numerically coded) categorical variable.
d *−n*	Use column *n* as depth. The user is asked for a symbol string.
d *−n STRING*	Use column *n* as depth and *STRING* as the symbol string.
f *i*	**Find** point *i* and label with '0'.
f *i a*	Find point *i* and label with character *a*.
f *n1:n2*	Find points *n1* through *n2* and label with '0'.
f *n1:n2 b*	Find points *n1* through *n2* and label with character *b*.
i	Enter **identify**-mode (i-mode) and label points by steering cursor.
u	**Unlabel**: remove all labels.

I-Mode Commands

In i-mode, all characters other than those listed below become labels for the current point identified by the cursor.

?	Show menu of i-mode commands.
,	Move left to next point.
\<SPACE\>	Move down to next point.
\<RETURN\>	Move right to next point.
.	Move up to next point.
*	Move to center of screen.
\<ESCAPE\>	Exit from i-mode.
^D	Exit from i-mode. (^D denotes control-D.)
^G	Turn bell on or off. (^G denotes control-G, etc.)
^H	Move left.
^J	Move down.
^K	Move up.
^L	Move right.

OPTIONS

d=*STRING*

Use *STRING* as the initial depth string instead of the default string '.,o+x*#'. Whether specified via this option or via the **d** command, the depth string is truncated to 23 characters.

NOTE

Scatter uses the BSD *libcurses.a* screen addressing library. Individual terminal capabilities are obtained from */etc/termcap*.

---

seed — Reset the seed for the random number generator

SYNOPSIS

seed [*x*]

DESCRIPTION

In normal use, BLSS produces a different sequence of random numbers every time you use any command that generates random numbers. But each time you give the **seed** command with a given value of *x*, BLSS produces the same sequence of random numbers for a given random number command, provided that you are on a sufficiently similar type of computer. The following paragraphs explain this in more detail.

By default, the underlying uniform random number generator uses a reasonably unpredictable time- and user-dependent pair of seeds. These are known as *cseed* and *tseed*, as explained in *rans(BML)*.

If the **seed** command is given with no argument, or with the {**long**} option, it displays the current values of the seeds in hexadecimal.

Given an argument *x*, the **seed** command resets the seeds based on (a nonobvious function of) the first element of the dataset x. (Elements beyond the first are ignored.) This function is hardware-dependent, but is the same for all computers which use 32-bit 2's complement integer arithmetic. (This is the most common type of integer arithmetic and includes VAXes, IBM RTs, and M680x0-based computers, such as SUNs and Macintoshes.) It is also possible, though not recommended, to set the value of *cseed* and *tseed* directly, as in the last example below.

The following BLSS commands base their random sequence on the value set by **seed**: **box**, **coin**, **confid**, **sample**, **xvalid**, and all r*xxx* commands documented in *random(BRM)*.

INPUTS
   x@   The first element of x determines the new values of the random seeds.

OPTIONS
   long   Display the current values of the random seeds.

EXAMPLES
```
. seed 17 {long}
cb293e17,18f17eed
```

The following command sets the values of the seeds to a17, 1c9 (hexadecimal). Note the backslash character '\' which must precede the comma ','.

```
. seed a17\,1c9 {long}
00000a17,000001c9
```

SEE ALSO
   rans(BLR), for information on the underlying uniform pseudorandom number generator.

---

select — Select rows (cases) which meet a specified condition

---

SYNOPSIS
   select *data* [*cond*] {**log;nm;x**=} [> *sel rej*]

DESCRIPTION
   Select performs conditional selection of rows (that is, cases) from a dataset. *Data* is the dataset from which the rows are to be selected. *Cond* is an optional vector with the same number of elements as *data* has rows. *Cond* is used to select the rows from *data* in conjunction with the {**log**} and {**nm**} options as described below.

   If only the *data* input dataset is given, **select** selects all rows from *data* which are completely non-missing.

   If *cond* is given, then by default all rows are selected for which the corresponding element of *cond* is non-zero and non-missing.

   If *cond* is given and the {**nm**} option is set, all rows are selected for which the corresponding element of *cond* is non-missing.

If *data* has several sheets and *cond* is given, the same rows from each sheet of *data* are selected. Otherwise, only rows from the first sheet are selected.

INPUTS

    data@

        The input array from which rows are to be selected. If *data* is a row vector, **select** treats it as a column vector.

    cond@

        The condition vector. If a row vector, **select** treats it as a column vector. *Cond* may have only one sheet.

OPTIONS

    log   Treat the *cond* vector as a logical vector. (Default.)

    nm   If the *cond* vector is present, select only those cases of *data* for which the corresponding value of *cond* is non-missing.

    x=*INTLIST*

        Columns of *data* from which to select cases. Default is all columns.

OUTPUTS

    sel@   The array of selected rows. Unless the {**x**=} option is set, *sel* has the same number of columns as *data*. If no output dataset is specified, the selected rows are put in *data*.**sel**, where *data* is the name of the **data@** input dataset.

    rej@   The array of rejected (i.e., not selected) rows. Unless the {**x**=} option is set, *rej* has the same number of columns as *data*.

EXAMPLES

To select all rows from *turbines* for which all values are non-missing:

```
. load turbines
. select turbines > turbs
```

To select all rows from *turbines* for which the third column is non-missing:

```
. select {nm} turbines turbines[3]
```

Logical condition datasets can be created with implicit or explicit **let** commands—see *let(BRM)*. For example, the *prostate* dataset contains the survival time of 129 patients with prostate cancer. The second column takes on the value 0 if the patient withdrew from the study, or 1 if the patient died during the study. It may be of interest to study the two groups of patients separately. To create a dataset made up only of those patients who died during the study and call it *died*:

```
. load prostate
. select prostate (prostate[2] == 1) > died
```

To create two datasets, one of patients who died during the study and the other of those who withdrew:

```
. select prostate (prostate[2] == 1) > died withdrew
```

Another way to select all rows from *turbines* for which the third column is non-missing is to use a *cond* vector that expresses the condition logically:

```
. select turbines (turbines[3] == NA)
```

SEE ALSO

BUG Sections 5.4, 8.6, 8.7.

count(*BRM*), to count the number of cases which would be selected without actually selecting them.

---

set, unset — Define (or remove) strings and environment strings

---

SYNOPSIS

set *name* [*value*]
set [+] [–]
unset *name1 name2 ...*

DESCRIPTION

Set and **unset** are internal commands to the BLSS shell which manipulate *strings* (that is, sequences of characters). The command **set** *name value* stores the given *value* in the string *name*. *Name* may be no longer than 20 characters, may contain only letters, digits, and underscore characters, and may not begin with a digit. *Value* is optional. It can be any sequence of characters (including spaces) although certain special characters (such as the double-quote character '"' and the semicolon ';') must be escaped using a backslash '\'.

Any later occurrence of the word *name* in the BLSS shell immediately following a '$' character will be replaced by *value*. This text replacement is called *string expansion*. Strings which are **set** but have no value serve as flags to the BLSS shell or to macros. More elaborate string expansion forms are available; see *BUG* Section 10.8.4. String expansion in the BLSS shell is similar but not identical to that in the UNIX C Shell.

The command **unset** *name1 name2 ...* eliminates the named strings.

Strings with names that begin with an uppercase letter automatically become *environment strings*. When **unset,** they are removed from the environment. Environment strings are just like regular strings, except that their value is available not only to you when you type commands but also to the commands themselves, internally.

Set with no arguments displays the values of all currently defined strings. The command 'set +' displays only those strings whose names begin with uppercase letters (and hence, by convention, are also environment strings). The command 'set –' displays only those strings whose names begin with lowercase letters (and hence, by convention, are not environment strings).

To permanently define certain strings, create a file named *.blss* in your home directory which contains the appropriate **set** commands. See *BUG* Section 10.5 for a discussion.

Note that strings are completely different from BLSS datasets.

EXAMPLES

Some BLSS dataset names are descriptive but inconveniently long for repeated use. In such cases, the **set** command can provide a temporary abbreviation. For example:

```
. sample {n=100;norep} boston > boston.sample
. set b boston.sample
. stat $b
. scat $b {x=1;y=2}
```

If you often use **scat** with the options {**xmin**=−1; **xmax**=1; **ymin**=−1; **ymax**=1}, a string whose value is these options is useful:

```
. set opt {xmin=-1; xmax=1; ymin=-1; ymax=1}
```

The following two commands are then equivalent:

```
. scat x $opt
. scat x {xmin=-1; xmax=1; ymin=-1; ymax=1}
```

Alternatively, you can use:

```
. set opt xmin=-1\; xmax=1\; ymin=-1\; ymax=1
. scat x {$opt}
```

Because double-quote characters suffice to protect semicolons from being interpreted by the BLSS shell, you can also type:

```
. set opt "xmin=-1; xmax=1; ymin=-1; ymax=1"
. scat x {$opt}
```

SEE ALSO

BUG Sections 10.3 (more examples and discussion), 10.4 (setting environment strings), 10.5 (explains the *.blss* file), 10.6.2 (interaction with the **source** command), 10.8 (using strings within macros).

alias(BRM).

strings(BRM), for a list of strings with special meanings.

---

show — Display or print datasets and text files

SYNOPSIS

show *files* ... {**cols**=;**ff**;**lpr**;**more**;**shape**;**width**=; **format**=;**f**=;**f**;**g**=;**g**;**i**=;**i**; ... }

DESCRIPTION

Show prints BLSS datasets and text files. If an argument is a BLSS dataset, it formats and displays the contents of the dataset. If an argument is a text file, it displays the text in the file.

The **show** command is used so often that BLSS allows you to omit the word **show**. If the first word in a BLSS command is not a command name, and the command is not a **let** command, then BLSS **show**s the named files. (It simply infers the word **show** at the front of the command.)

OPTIONS

cols=*INT*

c=*INT*

> Number of columns across the page. The default is 8 if other defaults are in effect.

dataonly

> Print BLSS datasets only. Text files produce an error message.

ff   Put a formfeed between each file printed. When sending files to a printer, this makes it advance to a new page for each file.

lpr   Send the files to a printer instead of the terminal. See *print(BRM)* for more information.

more   Use the *more* mechanism to view the display one screenful at a time. See *BUG* Section 4.8 or *options(BRM)* for more information.

names

> Show the name of each BLSS dataset. Text file names are not shown.

nastr=*STRING*

> Missing value string is *STRING*. Default is 'NA'.

nobl   Suppress the blank lines which are normally inserted when sending files to the lineprinter. Makes no difference unless {**lpr**} is also in effect.

nosh

shape By default, column vectors and sheet vectors are displayed as row vectors in order to save space. That is, an $(n,1)$ or $(n,1,1)$ array is printed as if it were $(1,n)$ or $(1,1,n)$. The {**shape**} option causes vectors to be printed in their true shape. This option may also be specified by setting the SHAPE environment string. If SHAPE is set, the option {**nosh**} overrides it.

unravel

> Treat the dataset as one long row vector.

width = *INT*

> Maximum line width. The actual width will be less than or equal to this number. Default is 79; if the {**lpr**} option is in effect, the default is reset to 132. The latter is appropriate for most lineprinters. The WIDTH environment string may also be used; the {**width**=} option overrides it.

x   Equivalent to:

> {**shape; cols** = <*number of columns in dataset*>; **width** = <*unlimited*>; **g**=7}

> This option is primarily used when creating text representations of datasets for transfer between computers with different binary representations of data. The implicit format option {**g**=7} can be overridden by the explicit format options listed below, but not by the FORMAT environment string.

Format Options

By default, data are displayed in a modified general format: to at least three places of precision when E-notation is used and at least four places of precision otherwise. The following options allow different format specifications.

e = *INT*

e   Use E-notation for all data, regardless of magnitude. *INT* specifies how many places of precision to show. {**e**} is equivalent to {**e**=6} (and to {**format**=%12.5e} [sic]).

f = *INT*

f   Show a fixed number of places beyond the decimal point. *INT* specifies the number. {**f**} is equivalent to {**f**=2} (and to {**format**=%9.2f}).

g = *INT*

g   Use a general format. *INT* specifies how many places of precision to show. {**g**} is equivalent to {**g**=6} (and to {**format**=%#12.6g} or, on older UNIX systems, **%12.6g**); it is useful for showing data to full working precision (six places).

i = *INT*

i   Use an integer format. *INT* specifies how many digits to allow space for. {**i**} is equivalent to {**i**=3} (and to {**format**=%3.0f}); it is useful for showing small integers (in the range –99 to 999).

format = *STRING*
f = *STRING*

> Output format, specified C-style. Acceptable formats are e, f, g, E, or G. For example: {**f**=%8.1f}, {**f**=%14.5E}, {**f**=%#10.4g}. Simple Fortran-style formats are also accepted—for example {**f**=f8.1} or {**f**=g10.4}. They are converted to C-style.

Note that the width of the printing format affects the number of columns printed across the page and that one additional space is printed between columns regardless of the format specification. The {**f**}, {**f**=}, {**i**}, and {**i**=}

options are the only options which align decimal points for numbers of all magnitudes; as a result, they are poor formats for displaying very large or very small numbers.

The environment string FORMAT may also be used to specify the format; its value is interpreted equivalently to the {**format**=} option *STRING*. The options listed above override it.

Area Options
bin, **class**, **data**, **doc**, **help**, **home**, **sys**, **text**
dirs=*dirname,* **tree**=*treename,* **area**=*areaname*
> Files are normally shown from the work area. An alternate area (or directory) may be specified with these options. For a full description, see *area*(*BRM*). The default values are:

dirs	your work area
tree	your *blss* tree (normally *~/blss*)
area	the data area

EXAMPLES
These examples illustrate the format options:

```
. show 10:15

   10.00      11.00      12.00      13.00      14.00      15.00

. show 10:15 {i}

 10   11   12   13   14   15

. show sqrt(2)

   1.414

. show sqrt(2) {g}

   1.41421

. show 1.234*10^(-3:2)' {shape}

 0.001234
 0.01234
 0.1234
 1.234
 12.34
 123.40

. show 1.234*10^(-3:2)' {shape} {f}

   0.00
   0.01
   0.12
   1.23
   12.34
   123.40
```

SEE ALSO

> *BUG* Sections 2.1, 2.3, 4.5, for more examples and discussion.
> *area(BRM)*.
> *output(BRM)*, for information about saving text output (such as generated by **show**) into a text file.
> *print(BRM)*, for information about sending files to a printer.

---

smooth — Tukey robust smoothers (running medians, etc.)

SYNOPSIS

> **smooth** *x* {**by=;x=**} [> *smooth*]

DESCRIPTION

> **Smooth** performs most of John Tukey's resistant smooths on the columns of *data*. Missing values in *x* are not used in computations.

OPTIONS

> **by=***STRING*
>> Specifies the smoother. Characters in *STRING* are read and interpreted left to right. They have the following meanings:
>>
>> **3, 5, 7, 9** Running medians of length 3, 5, 7, or 9.
>>
>> **2, 4, 6, 8** Running medians of length 2, 4, 6, or 8, followed by a pass of 2.
>>
>> **r** Repeat the preceding operation until no change. Valid only if preceded by an odd digit.
>>
>> **s** Splitting mesas followed by another **3r**. Valid only if preceded by a **3**.
>>
>> **h** Hanning (same as **22**): $x[i,] = .5x[i,] + .25(x[i-1,] + x[i+1,])$.
>>
>> **+** Reroughing: $a+b$ means do a on the data; then b on rough; then add back.
>>
>> **t** Twicing: a**t** means $a+a$.
>>
>> The default smoother is **3r**.

> **long** Trace the entire smoothing operation by printing current smooth being executed. For example, **3rssh** may generate a trace of '3r3rs3r3rs3rh'.

> **x=***INTLIST*
>> Columns of *x* to be smoothed. Default is all columns.

OUTPUTS

> **smooth@**
>> Smoothed columns of *x*. If no output dataset name is specified, the smoothed dataset is named *x*.**smooth**, where *x* is the name of the input dataset.

EXAMPLES

> The library dataset *sunspots* contains 30 years of monthly mean sunspot

numbers. Column 1 is a time index; column 2 is the data itself. The following commands plot first the raw series and then the smoothed series:

```
. load sunspots
. scat sunspots {x=1;y=2}
. smooth sunspots {x=2; by=453h} > s
. scat sunspots[1] s {width=80}
```

REFERENCES

Donald R. McNeil (1977). *Interactive Data Analysis.* Wiley, New York. Chapter 6.

John W. Tukey (1977). *Exploratory Data Analysis.* Addison-Wesley, Reading, Mass. Chapter 7.

SEE ALSO

filter(BRM), in order to smooth via moving averages, etc.

---

solve, inv, det — Solve systems of linear equations; inverse; determinant

---

SYNOPSIS

solve *a b* {**b**=} > *soln* [*inv det detvec*]
inv *a* > *inv* [*det detvec*]
det *a* [> *det detvec*]

DESCRIPTION

Solve, inv, and **det** are separate entries to the same command.

Solve solves the system of linear equations $AX = B$ for the unknown **X**. The right-hand side **B** may be a column vector or a matrix. The solution **X** contains one column for each column of **B**.

Inv computes the inverse of the matrix **A**; it optionally computes the determinant and reciprocal condition number. However, as the LINPACK manual notes: "Many calculations formulated in terms of matrix inverses can be reformulated in terms of the solution of sets of linear equations. The reformulated versions often require less time and produce more accurate results."

Det computes the determinant, log determinant, and reciprocal condition number of the matrix **A**.

Algorithm. This command uses the LINPACK subroutines *sgeco, sgefa, sgesl,* and *sgedi.*

INPUTS

a@   The coefficient matrix. Must be square and nonsingular.

b@   The right-hand side of the system of equations to be solved. Must have the same number of rows as the **A** matrix.

The input datasets may not contain missing values. If the inputs contain multiple sheets, each set of sheets is acted upon separately.

OPTIONS
    b=*INTLIST*
        The submatrix of **B** consisting of the specified columns is solved for.
        Default is all columns.

    long   Display the following information about **A**:
        det   Determinant. Overflow is indicated as 'HUGE', underflow as
                   'TINY'.
        lnadet Natural logarithm of the absolute value of the determinant.
                   This is much less susceptible to overflow or underflow than
                   the determinant itself.
        sgndet Signum of the determinant: 1 if *det* > 0; 0 if *det* = 0; −1 if
                   det < 0.
        rcond   The estimate of the reciprocal condition number of **A**
                   returned by the LINPACK subroutine *sgeco*.

OUTPUTS
    soln@
        The solution vector or matrix. Has the same dimensions as **B**. Possi-
        ble only when the command is invoked as **solve**.

    inv@   The inverse of **A**. Possible only when the command is invoked as **solve**
        or **inv**.

    det@   The determinant of **A**.

    detvec@
        The four numbers displayed by the {**long**} option.

If no output datasets are specified, the action taken depends on how the command was invoked. **Solve** puts the solution in the dataset *a*.**soln**, where *a* is
the name of the **a@** input matrix. **Inv** puts the inverse of *a* in the dataset
a.**inv**. **Det** displays the determinant and other information about **A**—that is,
it automatically invokes the {**long**} option.

REFERENCE
    J. J. Dongarra et al. (1979). *LINPACK Users' Guide*. SIAM, Philadelphia.
    Chapter 1.

SEE ALSO
    qr(*BRM*), *svd*(*BRM*), *linpack*(*BML*).

---

sort — Sort rows of a dataset

SYNOPSIS
   sort *x* [*i*] {**r**;**s**=;**x**=;**shind**} [> *sort o*]

DESCRIPTION
   Sort reorders the rows of the dataset *x* (but does not sort the elements within
   individual rows). By default, the rows are sorted so that column 1 of *x* is
   arranged in ascending order, column 2 breaks any ties in column 1, column 3
   breaks any remaining ties in column 2, etc. There are several other sorting
   options, described below.

   For the purpose of sorting, missing values (NA's) are considered to be greater
   than all other values.

INPUTS
   x@   The dataset to sort. If *x* is a row vector or a sheet vector, it is treated
           as a column vector.

   i@   Optional input sorting vector (permutation vector). If *i* is given, the
           rows of *x* are rearranged according to the elements of *i*: the *j*-th row of
           the output dataset is the *i*[*j*]-th row of the input dataset.

           If *i* is given, the {**r**}, {**s**=}, and {**shind**} options are ignored.

           If the input dataset contains several sheets, then if *i* contains only one
           sheet, every sheet in the input dataset is sorted according to that per-
           mutation. If *i* contains the same number of sheets as *x*, then each sheet
           of *x* is sorted according to the corresponding sheet of *i*.

OPTIONS
   r   Reverse sort: sort the data in descending order.

   s=*INTLIST*
           The column(s) on which to sort *x*. If more than one {**s**=} column is
           given, the second column given breaks ties in the first, the third column
           breaks ties in the second, etc.

           If the {**s**=} option is not given, the dataset is sorted on the {**x**=}
           columns, if given. If neither {**s**=} nor {**x**=} is given, the data is sorted
           on column 1, then column 2, etc., as described above.

   x=*INTLIST*
           The sorted dataset consists of the specified columns, in the specified
           order, of the original dataset. The default is all columns in the original
           order. If {**s**=} specifies any columns not also specified by {**x**=}, those
           columns are implicitly added to the end of the {**x**=} list.

   shind Sort sheets independently. If the input dataset contains several sheets,
           then by default sheets 2 and beyond are sorted according to the same
           permutation as sheet 1. This option causes each sheet to be sorted
           independently of the others.

OUTPUTS

　　sort@ The sorted dataset.

　　o@　Output sorting vector: the permutation vector according to which x was sorted. The j-th row of the sorted dataset is the $o[j]$-th row of the input dataset. This can be used as the **i@** input sorting vector in order to sort other datasets according to the same permutation.

　　If no output dataset is specified, the sorted array is put in x.**sort**, where x is the name of the input dataset.

EXAMPLES

　　Sort the four-column BLSS library dataset *turbines* so that the first column is arranged in ascending order, the second column breaks ties in the first, the third column breaks ties in the second, and the fourth column breaks ties in the third:

```
. load turbines
. sort turbines > t
```

Sort the rows so that column 2 is arranged in ascending order, and column 4 breaks ties in column 2:

```
. sort {s=2,4} turbines > t2
```

Same as the previous example, except the output dataset consists of columns 2 and 4 only:

```
. sort {x=2,4} turbines > t2a
```

The following two commands use the **i@** input and **o@** output index vectors in order to sort the dataset x and then sort the dataset y according to the same permutation as x (the datasets x and y must have the same length):

```
. sort x > xx o
. sort y o > yy
```

SEE ALSO

　　sample(*BRM*), for random permutations of the rows.

---

source, run — Execute a command file or macro

SYNOPSIS

　　source *cmdfile*
　　run *cmdfile* [*arguments* …]
　　cmdfile [*arguments* …]

DESCRIPTION

　　A *command file* is a text file that contains BLSS commands. The **source** command causes the BLSS shell to read and execute the commands in the named command file just as if they were typed in at the terminal. The **run** command runs the named command file as a macro; inputs, options, and outputs can be specified to the macro just as to regular BLSS commands. *BUG* Section 10.8 describes in detail the differences between **source** and **run**.

If *cmdfile* does not exist in the work area, then BLSS searches for it in the other areas on the BPATH. (Generally, this is your bin area followed by the BLSS system bin area.) UNIX pathnames may also be used.

You should normally keep command files in your bin area. The **save, show, edit**, and **addtext** commands can all access files in your bin area using the {**bin**} option.

Command files in the bin area (or anywhere on the BPATH) may also be run in the same way as regular BLSS commands—by simply typing their name. Command files in the work area cannot be run this way unless their names are prepended with '.'. Thus, a file in the work area can be run by either **run** *cmdfile* or *./cmdfile*. In general, when UNIX pathnames which begin with '/', './', '../', or '~' are used, the word **run** may be omitted.

If a file called *.blss* exists in your home directory, it is automatically **source**-d each time you enter BLSS. You can use the *.blss* file to establish permanent aliases (command abbreviations), string settings, etc. for yourself.

SEE ALSO
BUG Sections 10.5 through 10.8.
alias(BRM), *set(BRM)*, for more information about aliases and string settings.
strings(BRM), for more information about the BPATH environment string.

---

spec1 — Spectral analysis of univariate time series

SYNOPSIS
spec1 *x* [*pg*] {**window**=;**bw**=;**d**=} [> *spec pg se df*]

DESCRIPTION
Spec1 performs spectral analysis of the input time series *x* or periodogram *pg*. Spectra for several time series or periodograms may be obtained by placing each time series or periodogram in a separate column or sheet.

INPUTS
Only one of the following inputs may be specified. Missing values are not allowed.

x@ An *n*-vector which contains the time series. *N* should be reasonably composite.

pg@ An *n*-vector which contains the periodogram of the time series previously computed by **spec1** or **pgrm1**.

OPTIONS
bw=*FLOATLIST*
  A vector of *nbw* smoothing bandwidths, specified in cycles per unit time. One spectral estimate is produced for each bandwidth. If more

than one bandwidth is specified and x contains more than one series, then the series must be contained in separate sheets, not separate columns. The default bandwidth is $5/n$.

d=*INT*

Decimate the output to d points; that is, estimate the output at d equally spaced frequencies between 0 and π inclusive. The value specified may not exceed the integer part of $n/2 + 1$. If $d = 0$, no decimation takes place (that is, d is set equal to the maximum). The default is the minimum of 101 and the largest acceptable value.

window=*CHAR*

The shape of the window used to smooth the periodogram. *CHAR* is one of:

r   rectangular (default)

c   cosine

t   triangular

OUTPUTS

spec@

A d-by-*nbw* array whose i-th column contains the estimated $\log_{10}$ spectrum for the i-th bandwidth.

pg@   The periodogram of the **x@** input series.

se@   The approximate standard errors of the estimated $\log_{10}$ spectra, one for each bandwidth.

df@   The degrees of freedom of the χ^2 approximation to the distribution of the spectral estimates, one for each bandwidth. See the reference listed below.

Note that **spec@** contains the $\log_{10}$ spectrum, but **pg@** contains the untransformed periodogram.

The **spec@** and **se@** output datasets are always created. The **pg@** output dataset is created whenever the input dataset is the **x@** dataset. If the name for an output dataset to be created is not specified, the dataset is named *in.tag*, where *in* is the input dataset name and *tag* is the output dataset tag name.

REFERENCES

David R. Brillinger (1975). *Time Series: Data Analysis and Theory.* Holt, New York. (Expanded ed., 1981, Holden-Day, San Francisco.) Chapter 5.

Keith A. Haycock and David R. Brillinger (1985). *LIBDRB: A subroutine library for elementary time series analysis.* Technical Report No. 48, Department of Statistics, University of California, Berkeley.

SEE ALSO

pgrm1(BRM) and *pgrmk(BRM)*, for periodograms.

speck(BRM), for k-variate spectra.

---

speck — Spectral and cross-spectral analysis of *k*-variate time series

---

SYNOPSIS
    speck *x* {**window**=;**bw**=;**d**=} [> *spec coher*]

DESCRIPTION
    Speck performs spectral and pairwise cross-spectral analysis of all columns of the array *x*.

INPUTS
    x@   An array which contains either the time series or previously computed periodograms of the time series. Missing values are not allowed in *x*.

             If *x* is a single sheet with dimension *n*-by-*k*, it is assumed to contain *k* times series of length *n*. The length *n* should be reasonably composite.

             If *x* has dimension *n*-by-*k*-by-*k*, it is assumed to contain *k*-variate periodograms as produced by *pgrmk*(*BRM*).

OPTIONS
    bw=*FLOAT*
        The smoothing bandwidth, specified in cycles per unit time. The default bandwidth is $5/n$.

    d=*INT*
        Decimate the output to *d* points; that is, estimate the output at *d* equally spaced frequencies between 0 and π inclusive. The value specified may not exceed the integer part of $n/2 + 1$. If $d = 0$, no decimation takes place; that is, *d* is set equal to the maximum. The default is the minimum of the largest acceptable value and 101.

        The default value for large *n*, $d = 101$, produces estimates at the frequencies $0, \pi/100, 2\pi/100, \ldots, \pi$. Note, however, that if the results are to be transformed back to the time domain (for example, via **trnfrk** followed by **imprsk**), 101 is a bad number. 101 is prime and the fft (fast Fourier transform) routines cannot factorize it. In this case, use a number such as 99 or 100.

    window=*CHAR*
        The shape of the window used to smooth the periodogram. *CHAR* is one of:
        r   rectangular (default)
        c   cosine
        t   triangular

OUTPUTS
    spec@
        A *d*-by-*k*-by-*k* array which contains the estimated spectra at *d* equally spaced frequencies. That is, the *f*-th sheet, *spec*[*f*,*,*], contains

spectrum values at frequency $\pi(f-1)/(d-1)$. Each sheet is arranged as follows:

For	*spec*$[*,i,j]$ *Contains*
$i = j$	Spectrum of the i-th time series.
$i < j$	Cross-spectrum of the i-th and j-th series: Real part.
$i > j$	Cross-spectrum of the i-th and j-th series: Imaginary part.

If no output dataset is specified, **spec@** is put in *x*.**spec**, where x is the name of the input dataset.

coher@

A d-by-k-by-k array which contains the estimated coherences between the components of x at d equally spaced frequencies. The i,j-th element of each sheet contains the coherence between the i-th and j-th components.

NOTES

Spec1 estimates the $\log_{10}$ spectrum; **speck** estimates the untransformed spectrum.

To estimate k-variate spectra using several bandwidths, it is most efficient to first call **pgrmk** to obtain the k-variate periodogram and then call **speck** with the periodogram as input once for each bandwidth of interest.

REFERENCES

David R. Brillinger (1975). *Time Series: Data Analysis and Theory.* Holt, New York. (Expanded ed., 1981, Holden-Day, San Francisco.) Chapter 7.

Keith A. Haycock and David R. Brillinger (1985). *LIBDRB: A subroutine library for elementary time series analysis.* Technical Report No. 48, Department of Statistics, University of California, Berkeley.

SEE ALSO

pgrm1(BRM) and *pgrmk(BRM)*, for periodograms.

spec1(BRM), for univariate spectra.

---

stat — Summary statistics: mean, SD, correlation, covariance, quartiles, etc.

SYNOPSIS

    stat x {**cor;cov;casex;pairx;x=;dn=;** ... } [> *stat all cor cov ncov cen stan*]

DESCRIPTION

    Stat calculates descriptive statistics for the columns of x. Each row is considered to be one case, except that if x is a row vector or a sheet vector it is treated as a column vector.

    Stat can compute covariances, correlations, and the following univariate statistics:

Col	The column index.
N	Number of non-missing values.
Mean	Arithmetic mean $\bar{x}$.
SD	Standard deviation = sqrt(variance).
Var	Variance = $\Sigma(x_i - \bar{x})^2/(N+dn)$, where dn is by default -1.
Skew	Skewness = $\Sigma z_i^3/N$ where $z_i = (x_i - \bar{x})/[\Sigma(x_i - \bar{x})^2/N]^{1/2}$.
Kurt	Kurtosis = $[\Sigma z_i^4/N] - 3$.
Min	Minimum.
25%	25th percentile (1st quartile).
50%	50th percentile (median, or 2nd quartile).
75%	75th percentile (3rd quartile).
Max	Maximum.

    If a quartile falls between two observations, its value is the average of the two observations. Correlations of constants (that is, columns with zero SD) are always reported as zero.

    Stat can also center and standardize the data in the columns of x.

OPTIONS

Univariate Statistics

    The following options control the display of univariate statistics. The first three options make displays which are 79 columns wide—suitable for viewing on most terminals. The {**all**} option makes displays about 130 columns wide—suitable for sending to lineprinters.

stat   Print Col, N, Mean, SD, Min, the three quartiles, and Max for each column of x. These statistics are displayed to only three or four decimal places; this precision is sufficient for most purposes. This is the default ouput.

mom   Print Col, N, and the five moment statistics: Mean, SD, Var, Skew, and Kurt. Full (six-place) working precision is used.

quart   Print Col, N, Min, the three quartiles, and Max to full precision.

all   Print Col, N, and all ten univariate statistics in the order shown above—all to full working precision.

The above options may be combined: for example, {**mom**} and {**quart**} may be combined to print all moment statistics and quartiles in a 79-column format.

Bivariate Statistics

cor Print the correlation matrix in addition to the univariate statistics.

coronly
Print the correlation matrix only.

cov Print the covariance matrix in addition to the univariate statistics.

covonly
Print the covariance matrix only.

Missing Data Treatment
The following options control the treatment of missing data. If the data are free of missing values, these options have no effect.

colx Columnwise exclusion of missing data: Univariate statistics are computed using all non-missing values in a column. This is the default when no bivariate statistics (covariance or correlation) are requested. This option is not allowed if bivariate statistics are requested.

casex Casewise exclusion of missing data: If a case (row) contains missing values in any column, then the entire case is excluded when computing all statistics, both univariate and bivariate. Hence, N is equal for all columns. This is the default if any bivariate statistics are requested.

pairx Pairwise exclusion of missing data for bivariate statistics. Univariate statistics are computed as with {**colx**}.

Additional Options

x = *INTLIST*
Columns for which to compute statistics. Default is all columns.

dn = *INT*
By default, SDs, variances, and covariances are calculated using the unbiased formula (divide by $N-1$), where N is defined above. The {**dn** = } option causes variances and covariances to be calculated by dividing by $N + dn$ instead. (*Dn* is usually negative or zero.) The divisor used is printed above the display as a reminder.

long Equivalent to {**mom;quart;cov;cor**}.

OUTPUTS

stat@ The array of univariate statistics displayed by the (default) {**stat**} option. The *i*-th row of this dataset contains nine entries for the *i*-th column of *x*, in the same order as in the display: Col, N, Mean, SD, Min, 25%, 50%, 75%, and Max.

all@ The array of univariate statistics displayed by the {**all**} option. The *i*-th row of this dataset contains all twelve entries for the *i*-th column of *x*, in the same order as in the {**all**} display (and the same order shown at the beginning of this manual entry).

cor@   Correlation matrix.

cov@   Covariance matrix.

ncov@

> Square matrix whose elements are the number of pairs of observations used to compute the corresponding covariances and correlations.

Note that if *x* has missing values and the {**casex**} option is *not* used, then the correlation and covariance matrices will not necessarily be positive definite.

The following two output datasets have the same dimensions as the *x* input dataset (possibly modified by the {**x=**} option).

cen@   Centered data: each column has its mean subtracted off.

stan@

> Standardized data: each column is centered and then divided by its SD. If an SD is zero, the elements of that column are set to zero.

Note that if *x* has missing values and the {**casex**} option *is* used, centered data will not necessarily have zero mean and standardized data will not necessarily have zero mean or unit SD.

EXAMPLES

Calculate univariate descriptive statistics for *turbines*:

```
. stat turbines
```

Calculate the same statistics for *turbines*, but use only those cases with no missing observations:

```
. stat {casex} turbines
```

Display all univariate statistics to full precision:

```
. stat {mom;quart} turbines
```

The following command centers the non-missing values of *turbines* and places the centered values in *centurbs*. No display appears on the terminal. This is useful when centering data for input to another command such as **qr** or **svd**.

```
. stat {quiet} stackloss > cen@ censtack
```

For additional examples, run the on-line demonstration macros **demo.stat1** and **demo.stat2**—see *demo(BRM)*.

SEE ALSO

BUG Sections 3.1, 5.1.2, 5.3.

biweight(BRM), *medmad(BRM)*, *qdata(BRM)*, *vecop(BRM)*, for other commands which produce summary statistics.

---

stemleaf — Tukey stem-and-leaf diagrams

SYNOPSIS
    stemleaf *x* {x=;count;cum;depth;width=;ds=;ldu=;lldu=;nl=; ... }

DESCRIPTION
    Stemleaf displays John Tukey's *stem-and-leaf* diagram (for short, *stemleaf*) of a
    univariate dataset. This diagram represents a set of data as a single vertical
    stem and many *leaves* growing on horizontal *branches* off the stem. Each
    datum is represented by one leaf character—a digit—in the diagram. Charac-
    ters in the stem represent the most significant digit(s) (possibly rounded) in
    the data. Leaf characters are the next most significant digit in the data. See
    BUG Section 3.2 for a detailed explanation.

INPUTS
    x@   The input dataset. A stemleaf diagram is made for each column of *x*.
          Missing values in the data are ignored.

OPTIONS
    x=*INTLIST*
          The columns of *x* to be diagrammed. The default is all columns.

    terse Do not print the header which usually precedes the diagram itself.

    width=*INT*
          The page width in characters per line. The minimum value allowed is
          14, or that necessary to print 5 leaves per line, whichever is greater. The
          default width is 79, or 132 if the {**lpr**} option is given. The WIDTH
          environment string can also be used; the {**width**=} option overrides it.

    c=*FLOAT*
          Number of inter-quartile distances from the nearest quartile for
          defining outliers. Outliers are printed above or below the stemleaf
          diagram. By default, no values are printed as outliers.

    brsym=*CHARLIST*
          A list of five characters to append to the stem labels to help distinguish
          between adjacent branches when the number of possible leaf values on
          a line (*nl*) is 2 or 5. The default character is '|'.

    brsym
          Specifies an alternate list of symbols to use for labeling branches whose
          stem values are the same when *nl* is 2 or 5. If *nl* = 2, the symbols are
          '*' for leaf digits 0, 1; 'T' for 2, 3; 'F' for 4, 5; 'S' for 6, 7; and '.' for 8,
          9. If *nl* = 5, the symbols are '*' for 0 through 4; and '.' for 5 through 9.
          This style is used by Hoaglin and Velleman (1981).

    Counting Options
          The following options display counts to the left of the diagram. If more than
          one type of count is requested, they are shown in the order (from left to right):
          depth, cum, count.

count For each branch, print the number of data points on the branch.

cum   For each branch, print the cumulative count of data points (through and including that branch)—that is, the number of data points on that branch plus the number data points on all previous branches.

depth For each branch, print the 'depth' of the branch: the total number of data points from that branch (inclusive) to the nearest end of the diagram, unless the branch contains the median. For the branch which contains the median, the number of data points on that branch is shown, enclosed in parentheses.

Scaling Options

By default, **stemleaf** chooses an appropriate scale for each column of x separately. The following options adjust the scale for all columns of x. Scaling options not specified by the user are chosen to satisfy $ds = nl * ldu$.

ds=*FLOAT*
   Difference between adjacent stem label values (unscaled). Overrides {**nl**=}, {**ldu**=}, and {**lldu**=}.

ldu=*FLOAT*
   Leaf digit unit: the unit of one leaf digit. Overrides {**lldu**=}. The value of *ldu* is printed above the display.

lldu=*INT*
   $\log_{10}$(leaf digit unit).

nl=*INT*
   Number of possible leaf values permitted on a branch: 2, 5, or 10.

EXAMPLES

The dataset *islands* contains the sizes of the world's largest land masses. According to McNeil (page 11), the following stemleaf diagram demonstrates conclusively that Australia is a continent, not an island:

```
. load islands
. stemleaf log(islands)

N = 48,  min = 2.485,  25% = 2.994,  50% = 3.713,  75% = 5.212,  max = 9.74
Leaf digit unit (ldu) = 0.1  (1|2 represents 1.2)

 2|566666778889
 3|01234444
 3|556778889
 4|13444
 4|5
 5|224
 5|67
 6|
 6|7
 7|
 7|
 8|02
 8|68
 9|14
 9|7
```

The values of log(*islands*) range from 2.48 to 9.74. In this stemleaf diagram, digits on the same horizontal line to the right of the '|' represent data values with the same integer part. On a given line, the digit to the left of the '|' is the integer part of a data value and the digits to the right of the '|' are the fractional part. For example, the data values represented by 5|224 are 5.2, 5.2, and 5.4. (The '8.0' is Australia; greater values are the continents.)

The dataset *boston* contains observations on 506 census tracts in the Boston area. Column 2 is the per capita crime rate by town. The logarithm of crime rate is bimodal:

```
. load boston
. stemleaf log(boston[2]) {depth;count;cum}

log(boston[2]),  N = 506
min = -5.064,  25% = -2.501,  50% = -1.361,  75% = 1.302,  max = 4.488
Leaf digit unit (ldu) = 0.1  (1|2 represents 1.2)

    1     1    1   -5|1
    3     3    2   -4|75
   15    15   12   -4|333322222100
   33    33   18   -3|999988776665555555
   78    78   45   -3|4444444433333333333322221111111111111100000000
  134   134   56   -2|99999999999999988888888887777777777777766665555555555555555
  197   197   63   -2|4444444444444444433333333333222222222222221111111111100000000000+
  246   246   49   -1|99999999998888888888888877777777777666666665555555555
 (38)   284   38   -1|44444444433333332222222211111111110000000
  222   313   29   -0|99998888777776666666665555555555
  193   332   19   -0|4444333322222221000
  174   351   19    0|01111222222233334444
  155   368   17    0|55566788888899999
  138   389   21    1|000012223333333334444
  117   422   33    1|5555556666666777777778888999999999
   84   460   38    2|000000111111112222222222333333333444444
   46   487   27    2|55555666666777777777778889999
   19   498   11    3|00011222234
    8   503    5    3|66789
    3   505    2    4|23
    1   506    1    4|5
```

The depth of each branch is shown in the first column, the cumulative count in the second column, and the count in the third column. The stem whose depth is indicated in parentheses, (38), has 38 data points and contains the median, −1.361, displayed as −1|4.

The '+' character at the end of the longest branch in the display indicates that the branch is too long to fit in 79 columns and that additional leaf digits were not printed. We can use the {**width**=} option to increase the width of the display and show all the data, as in:

```
. stemleaf log(boston[2]) {width=90} {depth;count;cum}
```

but for large enough datasets, increasing the width may not suffice. For large datasets we can make a *scaled-down* stemleaf diagram by sorting the data and plotting every other value of the sorted dataset. For example, with the *boston* crime rate:

```
. sort boston {x=2} > b2
. stemleaf log(b2[1:506:2,])

log(b2[1:506:2,]),  N = 253
min = -5.064,  25% = -2.501,  50% = -1.371,  75% = 1.301,  max = 4.298
Leaf digit unit (ldu) = 0.1   (1|2 represents 1.2)

 -5|1
 -4|5
 -4|332210
 -3|998765555
 -3|44433333322211111110000
 -2|99999998888877777777665555555
 -2|444444443333332222221111110000000
 -1|999988888877777666655555
 -1|4444333222211111000
 -0|998877766665555
 -0|443322210
  0|0112223344
  0|56788899
  1|00123333344
  1|5556667777889999
  2|0001111222223333444
  2|55566677778899
  3|01223
  3|679
  4|3
```

Note that the data must be sorted before choosing every other data point. Otherwise, **stemleaf** would display a haphazard sample (depending on how the data were ordered in the array) of half the data—the resulting stemleaf might not have the same shape as the entire stemleaf. To select every other value of the sorted dataset, we use the **let** sequence operator to specify the proper indices. (We could instead scale it down by any integer factor *n* by choosing every *n*-th data point.) See *BUG* Section 8.3.2 for more information on the colon operator.

REFERENCES

Donald R. McNeil (1977). *Interactive Data Analysis*. Wiley, New York. Chapter 1.

John W. Tukey (1977). *Exploratory Data Analysis*. Addison-Wesley, Reading, Mass. Chapter 1.

Paul F. Velleman and David C. Hoaglin (1981). *Applications, Basics, and Computing of Exploratory Data Analysis*. Duxbury, Boston. Chapter 1.

SEE ALSO

BUG Sections 3.2 and 3.3, for additional discussion and examples.

---

strings — Strings and environment strings with special meanings

---

DESCRIPTION

Here is a list of strings and environment strings which have special meanings
to BLSS. You may wish to set or change some of them to tailor BLSS to your
own liking. See *set*(*BRM*) for how to do so.

Strings

argv   Arguments to the BLSS shell.

blssh   If set, specifies an alternate BLSS shell which is used to execute macros.
(Intended primarily for testing new BLSS shells.)

cmderr

Determines the action taken by the BLSS shell when a command aborts
due to an error. Its three components correspond to the errors: non-
zero termination code, non-zero exit code, and output dataset(s) not
created. Permissible values for each component are **ignore**, **warning**,
and **error**; only the first letter (**i**, **w**, or **e**) is significant.

coredump

If set, commands which abort are permitted to make core dumps. Normally, core dumps are prohibited.

dolerr   Determines the action taken by the BLSS shell when a *dollar error*
(loosely speaking, a string expansion error) occurs. Permissible values
for each component are **ignore**, **warning**, and **error**; only the first letter
(**i**, **w**, or **e**) is significant.

echo   If set, the BLSS shell echoes commands before executing them. The
echo string may take an optional value which contains one or more of
the following characters:

Char Meaning

n   A blank line (newline) precedes each echoed command. The
value may contain repeated **n**'s for repeated effect.

.   A BLSS prompt '.' precedes each echoed command.

s   Echo the command before any expansions occur. This is the
default if the values **a**, **d**, **h** are not set. (The letter **s** is
mnemonic for 'simple'.)

a   Echo the command immediately after alias expansion.

d   Echo the command immediately after string (dollar) expansion.

h   Echo the command after ~ (home directory) expansion.

eofexit

If set, interactive BLSS shells exit upon end-of-file (control-D). Normally they do not.

pid   The process id number of the current BLSS shell. Because each process
on a given UNIX system at a given time has a unique id number, this is
useful for generating unique temporary names.

prompt

The prompt string for the BLSS shell. The default is a dot character '.' followed by a space.

in, in_*xxx*
opt, opt_*xxx*
out, out_*xxx*

(where *xxx* represents any legal sequence of characters). These strings are used by the macro facility. See the references listed below.

status   Status word returned for the most recent command by the UNIX *wait*(2) system call. BLSS shell internal commands have no effect on its value.

Environment Strings

BCLASS

This environment string is normally set by an instructor in the context of a class. It contains the pathname of a directory to be treated as a class-wide *blss* tree: for example, *~s20/blss* if the instructor's account name is *s20*.

If BCLASS is set in the student's environment before BLSS is invoked (the instructor usually arranges for this in the student account *.login* or *.profile* files), then when BLSS starts up: it displays the class message-of-the-day file (*$BCLASS/motd*) after the BLSS system message-of-the-day (*$BLSS_SYS/motd*); it sources the class startup file (*$BCLASS/startup*) after the BLSS system startup file (*$BLSS_SYS/startup*); and it prepends the directory *$BCLASS/bin* to the environment string BPATH (see below) so that students can access class-wide datasets, help files, and command files.

BLSS, BLSS1, BLSS2

If these environment strings exist, their values are used as locations for extra startup files to be read after the BLSS system startup file and the user's *.blss* startup file.

BLSS_SYS

The pathname of the BLSS system tree. Do not change BLSS_SYS unless you know what you are doing.

BPATH

Specifies the BLSS searchpath. The components of this path are areas (directories) which are searched, in order, for commands and macros to execute. BLSS executes the first command or macro it finds with the specified name. The work area (current directory) is always searched last. BPATH also determines where **source** and **run** searches for command files to execute (except that the work area is searched first by **source** and **run**); where **load** searches for files to load (it searches the data area and text area corresponding to each bin area); and where **help** searches for on-line documentation (it searches the help area and man area corresponding to each bin area). The default BPATH contains the components:

../bin	(your bin area)
~blss/bin	(BLSS system bin area)

If the environment string BCLASS exists when BLSS is invoked, then the following is added as the first BPATH component:

$BCLASS/bin   (class bin area)

EDITOR

If set, overrides the default editor used by the **edit** and **editdata** commands.

FORMAT

If set, overrides the default format used by the **show** command.

HOME

User's home directory. Used to locate the *.blss* file. *$HOME/blss* (equivalently, *~/blss*) is the usual base of the user's *blss* tree; see *blssf(BRM)* for details.

ISHOW

A programmer's debugging aid. See *imisc(BLR)*.

LPRPROG

If set, overrides the default lineprinter command used by the {**lpr**} option.

MORE

If set, specifies the options passed to the UNIX *more*(1) command by the {**more**} option in place of the –**d** option.

PATHNAMES

If set, messages displayed by the utility commands use UNIX pathnames (such as *../data*) instead of BLSS colloquial names (such as 'your data area'). See also *area(BRM)* and *cute(BLR)*.

SHAPE

If set, causes **show** to display column vectors and sheet vectors in their true shape. Normally they are shown as row vectors, in order to save space. See *show(BRM)* for details.

SHELL

If set, overrides the default shell used by the **unix** and **%** commands.

SHOW

Reserved for future use.

SHOWEXPAND

If set, BLSS displays the results of dimension-expansion every time it occurs. See *BUG* Section 8.3.5.

TALK Sets the verbosity level of BLSS commands, as does the {**talk**=} option. See *options(BRM)*.

TERM

Specifies the type of terminal being used.

WIDTH

If set, specifies the maximum width of displays made by a number of commands, including **boxplot**, **freq**, **show**, **stemleaf**, and **xtab**. Its value can be overridden by the {**width**=} option.

XSHOW

If set, error messages from commands include the name of the command. By default, this is set for macros but not for interactive shells.

SEE ALSO

BUG Sections 10.3, 10.4, 10.8.

options(BRM), *set(BRM)*.

environ(7) in BSD UNIX manuals; *environ*(5) in other UNIX manuals.

---

submat — Choose a submatrix from a matrix

SYNOPSIS

submat x [i j] [> *sub*]

DESCRIPTION

Normally, submatrices are specified using subscript notation. For example, the notation x[1 2, 3] specifies the submatrix of x consisting of rows 1 and 2 and column 3. See *let(BRM)* for more information. However, at present BLSS allows only explicit integers—not datasets—to be used in subscript notation.

The **submat** command selects the submatrix of x that consists of the rows whose indices are given by the dataset i and the columns whose indices are given by the dataset j.

If x is a row vector and the j input is not given, then the i input is understood to contain column indices, not row indices.

If x is a three-way array, the result is a three-way array for which the appropriate submatrix is chosen for each sheet.

Note that, as in all SYNOPSES, the square brackets in the SYNOPSIS above are not actually typed. They indicate that the i and j inputs are optional.

INPUTS

x@   The matrix from which to choose the submatrix.

i@   Array of row indices to choose. If omitted, all rows are chosen.

j@   Array of column indices to choose. If omitted, all columns are chosen.

The **i@** and **j@** arrays can have any shape; their elements are read in row-major order (that is, going along the rows).

OUTPUTS
   sub@   The submatrix.

EXAMPLES
   The following examples all use the ',' catenation operator to create vectors of indices. To choose elements 2 and 3 from the vector x (whether x is a row or column vector):

    . submat x (2,3) > y

To choose rows 2 and 3 from the matrix x, use the same command. To choose the submatrix which consists of the intersection of rows 2 and 3 with columns 4, 5, and 7:

    . submat x (2,3) (4,5,7) > y

To choose columns 7, 4, and 5 (in that order) from the matrix x:

    . submat x j@(7,5,4) > y

For the purpose of illustration, we used explicit integers in the examples above. However, the **submat** command is needed (in place of subscript notation) only when the i and j inputs are datasets whose values are the result of some calculation.

---

svd — Singular value decomposition of a matrix

SYNOPSIS
   svd x {x=} > $u\ d\ v$

DESCRIPTION
   Svd computes the singular value decomposition of the nr-by-nc matrix **X** (contained in the dataset x) and places the results in the following datasets:

   u nr-by-nc matrix **U** with orthonormal columns: $\mathbf{U'U} = \mathbf{I}$.
   d Row vector **d** of nc singular values, in descending order.
   v nc-by-nc orthogonal matrix **V**: $\mathbf{V'V} = \mathbf{VV'} = \mathbf{I}$.

These matrices satisfy $\mathbf{X} = \mathbf{UDV'}$ where $\mathbf{D} = diag(\mathbf{d})$ is the nc-by-nc diagonal matrix with entries **d**.

Rank Considerations
   The rank of **X** is the number of non-zero singular values in **d**. ('Non-zero' means up to machine accuracy, which is about six places because calculations are single precision.) Normally, $nr \geq nc$. If $nr < nc$, **U** and **d** contain the appropriate number of zero entries.

Eigenvalues and Eigenvectors
   The columns of **U** are the eigenvectors of **XX'**. The columns of **V** are the eigenvectors of **X'X**. The squares of the singular values in **d** are the non-zero eigenvalues of **XX'** and **X'X**.

Principal Components

If the columns of the data matrix $\mathbf{X}$ are centered, then the matrix $\mathbf{P} = \mathbf{XV} = \mathbf{UD}$ contains the principal component coordinates of the data. Thus, the sample covariance of $\mathbf{P}$ is the diagonal matrix $\mathbf{D}^2$ divided by nr or $nr-1$.

Algorithm. This command uses the LINPACK subroutine *ssvdc*.

INPUTS

x@   Dataset which contains the matrix $\mathbf{X}$ to be decomposed. If it contains multiple sheets, **svd** decomposes each sheet separately. Missing values are not allowed.

OPTIONS

x=*INTLIST*

Specifies a subset of columns for which to compute the decomposition. By default, the entire input matrix is decomposed.

OUTPUTS

u@   The *nr*-by-*nc* matrix $\mathbf{U}$.

d@   The row vector $\mathbf{d}$.

v@   The *nc*-by-*nc* matrix $\mathbf{V}$.

If just one or two output datasets are specified, only those datasets are created. If no output datasets are specified, all three output datasets are created: the names are *x*.**u**, *x*.**d**, and *x*.**v**, where *x* is the name of the matrix being decomposed.

EXAMPLE

The following commands compute principal components of the library data matrix *kappa*:

```
. load kappa
. stat kappa > cen@ k      # center the columns; put the result in k
. svd k > uk dk vk         # compute the svd; put results in uk, dk, vk
. let pk = uk #* diag(dk)   # pk contains the principal components
```

See *stat(BRM)* for more information on the *stat* command and *let(BRM)* for more information on matrix arithmetic.

REFERENCES

John Chambers (1977). *Computational Methods for Data Analysis.* Wiley, New York. Section 5.e, pp. 111-115.

J. J. Dongarra et al. (1979). *LINPACK Users' Guide.* SIAM, Philadelphia. Chapter 11.

Charles L. Lawson and Richard J. Hanson (1974). *Solving Least Squares Problems.* Prentice-Hall, Englewood Cliffs, N.J. Chapter 18.

SEE ALSO

eigen(BRM), *qr(BRM)*, *linpack(BML)*.

---

transfer — Compute the transfer function of a filter

SYNOPSIS
transfer *filter* {**d**= } [> *g p*]

DESCRIPTION
Transfer evaluates the transfer function of the filter whose coefficients are given in *filter* at equispaced frequencies spanning the interval [0, .5]. The results are in polar form.

This command is a macro which calls *dft(BRM)* and then *polar(BRM)*.

INPUTS
filter@
> A vector which contains the filter coefficients. Missing values are not allowed.

OPTIONS
d=*INT*
> The transfer function is evaluated at *d* equally spaced points. The default is the maximum of 101 and $n/2 + 1$, where n is the length of *filter*.

OUTPUTS
g@   The $\log_{10}$ gain of the filter. A column vector.

p@   The phase of the filter. A column vector.

If names for the output datasets are not specified, they are called *filter*.**g** and *filter*.**p** respectively, where *filter* is the name of input dataset.

SEE ALSO
dft(BRM), *polar(BRM)*.

---

trnfrk — Estimate pairwise transfer functions from *k*-variate spectra

SYNOPSIS
trnfrk *spec* {**x**=;**y**= } [> *gain phase*]

DESCRIPTION
Trnfrk estimates transfer functions from the spectral array *spec* produced by **speck**.

INPUTS
spec@
> The three-dimensional spectral array produced by **speck**. Missing values are not allowed.

OPTIONS

By default, all pairwise transfer functions are estimated. This is changed by the following options:

x=*INTLIST*

y=*INTLIST*

Lists of components of the original time series for which to estimate transfer functions. For k = 1, 2, ..., (number of elements in *INTLIST*), the gain and phase are calculated for the column of *spec* corresponding to the k-th element of {**y**=} on the column of *spec* corresponding to the k-th element of {**x**=}. If the {**x**=} and {**y**=} lists have different lengths, then the last element of the shorter list is paired repeatedly with all remaining values of the longer list to compute the gains and phases.

If either {**x**=} or {**y**=} is specified, both must be specified.

OUTPUTS

gain@

Gains of the estimated transfer functions.

phase@

Phases of the estimated transfer functions.

If names for the output datasets are not specified, they are called *spec*.**gain** and *spec*.**phase** respectively, where *spec* is the name of input dataset.

If {**x**=} and {**y**=} are not specified, the output datasets are three-dimensional arrays with the same dimensions as the input spectral array. The i,j-th element of each sheet contains the gain or phase of the transfer function which takes the i-th component of the original time series into the j-th component.

Otherwise, the output datasets are two-dimensional arrays; the k-th column contains the k-th gain or phase.

The f-th sheet (or row) of the output datasets corresponds to the f-th sheet of the input array.

REFERENCES

David R. Brillinger (1975). *Time Series: Data Analysis and Theory*. Holt, New York. (Expanded ed., 1981, Holden-Day, San Francisco.) Chapter 6.

Keith A. Haycock and David R. Brillinger (1985). *LIBDRB: A subroutine library for elementary time series analysis*. Technical Report No. 48, Department of Statistics, University of California, Berkeley.

SEE ALSO

speck(*BRM*).

---

tsop — Simple time series operations: remove means, trends, etc.

SYNOPSIS

    tsop *x* {**o**=;**ns**=;**freqrm**=;**taper**=;**taper**;**diff**=;**diff**;**lag**=;**lag**;**nona**} [> *adj b* ...]

DESCRIPTION

    Tsop performs various operations on each column of *x*, including: removing the mean, linear trend, seasonal means, and frequency components; tapering; differencing; and lagging. The operations to be performed are specified as options. They are performed in the order shown below. At least one option must be specified.

INPUTS

    x@   Array whose columns are the time series. Missing values are not allowed.

OPTIONS

    o=*INT*

        If *o* = 0, the mean of the series is computed and removed. If *o* = 1, the linear trend present in the series (as estimated by least squares) is also removed.

    ns=*INT*

        The number of seasonal means to compute and remove from the series.

    freqrm=*FLOATLIST*

        The frequencies (in radians per unit time) to remove from the series.

    taper=*FLOATLIST*

    taper The series is tapered using a raised cosine taper. The general form is {**taper**=*start*,*end*}. *Start* and *end* give the fraction of values to be tapered at the start and end of the series, respectively. If only one value is given, it is used for both *start* and *end*. The default value is .1 for both *start* and *end*.

    diff=*INTLIST*

    diff The series *x* is differenced repeatedly, once for each value in the list. The values give the spacing used: if a value is *l*, then $x[i,]$ is replaced by $x[i,] - x[i-l,]$, where *x* is the input series adjusted by any previous operations, including previous differencing. The default is 1.

    lag=*INT*

    lag The series is lagged: if *lag* = *l*, then $x[i,]$ is replaced by $x[i-l,]$, where *x* is the input series adjusted by any previous operations. The value *l* may be negative. The default is 1.

    nona When differencing or lagging, missing values at the start or end of the series are omitted.

OUTPUTS

    adj@ The adjusted series.

If no output dataset is specified, the adjusted series is put in *x*.**adj**, where *x* is the name of the input dataset.

The following output datasets are created whenever the corresponding option is specified. If no dataset name is specified, the name *x.tag* is used, where *x* is the name of the input series and *tag* is the output dataset tag name.

b@   The means removed (and slopes, if requested).

smeans@
The seasonal means removed.

REFERENCE
Keith A. Haycock and David R. Brillinger (1985). *LIBDRB: A subroutine library for elementary time series analysis.* Technical Report No. 48, Department of Statistics, University of California, Berkeley.

SEE ALSO
diff(BRM), *lag(BRM)*.

---

twoway — Least squares and robust two-way anova

SYNOPSIS
twoway *x* {**x=**;**c=**;**eps=**} [> *res dia ref cef w anova*]

DESCRIPTION
Twoway performs two-way analysis of variance on the *nr*-by-*nc* dataset *x*. X must be a two-way data matrix which contains one observation per cell. Missing values in *x* are ignored.

Twoway displays the overall center and spread, and the estimated row and column effects. In robust anova, the row (column) degrees of freedom are estimated by the sum of the final weights for the observations in each row (column).

Twoway also displays an anova table which gives the degrees of freedom and weighted mean square error for each source of variation (total, column, row, and error) the F-statistics for the column and row effects, and the P-values of the F-statistics.

By default, **twoway** obtains the unweighted least squares fit. If the {**c=**} option is specified, **twoway** performs an iterative analysis of variance which is robust against thick-tailed error distributions. At each iteration, the bisquared weight function (see *biweight(BRM)*) is used: *z* is taken to be the deviation of an observation from its current fitted value and *s* is taken to be the current weighted SD of the residuals.

Note. **Twoway** cannot handle interactions or multiple observations per cell. Because of these and other limitations, **twoway** will be replaced by another anova command in a future release of BLSS.

OPTIONS

x=*INTLIST*

Columns to be included in the analysis. Default is all.

c=*FLOAT*

Bisquare parameter. Default is 0, which gives least squares anova. For robust estimation, values of c between about 4 and 10 are reasonable. The larger c is, the closer the results are to those of least squares anova.

eps=*FLOAT*

Iterations stop when the absolute change in $(1 - SS_{OLD}/SS_{NEW})$ is less than *eps*, where SS is the weighted sum of squared residuals. Default is *eps*=.01. This stopping rule is almost equivalent to stopping when the weighted sum of residuals is zero.

terse Display only the anova table.

long In addition to the normal display, show the weighted sum of squared residuals for each iteration.

vlong In addition to the display produced by the {**long**} option, show the weighted SDs and degrees of freedom of the estimated row and column effects. Note, however, that these are computed in a nonstandard way.

OUTPUTS

res@ *nr*-by-*nc* matrix which contains the residuals.

dia@ (*nr**nc)-by-3 matrix of diagnostic data:

Column	Contains
1	$a_i + b_i$
2	$a_i * b_j / t$
3	residual

where a_i is the row effect, b_j is the column effect, and t is the overall center. The diagnostic data are arranged so that the column effect index changes faster than the row index. These diagnostics may be used to help assess whether the model is appropriate. See the example for details.

ref@ Fitted row effects. One row for each row of the data matrix. Each row contains: estimated row effect, SD, and degrees of freedom.

cef@ Fitted column effects. One row for each column of the data matrix. Each row contains: estimated column effect, SD, and degrees of freedom.

w@ *nr*-by-*nc* matrix which contains the final weights for each observation.

anova@

The anova table. It can be subsequently displayed by the **anovapr** command—see *anovapr(BRM)* for a description of its contents.

EXAMPLE
Illit contains the illiteracy rates for 1900, 1920, 1930, 1950, 1960, and 1970 (columns) for nine different regions of the country (rows). The boxplots generated by *compare(BRM)* show that there was a lot of interregional variation in illiteracy near the turn of the century, but that it dropped over time until all regions were almost equally literate. The scatterplot of the diagnostic data generated by **twoway** shows a distinct linear trend in the residuals plotted against the product of the factors, which suggests logarithmic reexpression. The scatterplot obtained after rerunning **twoway** with this transformation yields a much better fit of the data.

```
. load illit
. compare illit
. twoway illit > dia@ d
. scat d {x=1,2;y=3}
. let logill = log(illit)
. twoway logill > dia@ dd
. scat dd {x=1,2;y=3}
```

REFERENCE
Donald R. McNeil (1977). *Interactive Data Analysis.* Wiley, New York. Chapter 7, pp. 171-174, discusses robust two-way anova but gives an older algorithm which used the mean absolute residual as *s* in the bisquared weight function, did not handle missing values, and was less numerically stable.

SEE ALSO
biweight(BRM), medpolish(BRM), oneway(BRM), robust(BRM).

---

unix — How to execute UNIX commands from within BLSS

SYNOPSIS
unix [*unix-shell-command*]
% [*unix-shell-command*]
direct-unix-command

DESCRIPTION
This manual entry describes methods for executing UNIX commands from within BLSS.

Executing Commands via the UNIX Shell
The **unix** and **%** commands access the UNIX shell from within BLSS. Any arguments are passed directly to the UNIX shell, which interprets and executes them. The difference between **%** and **unix** is that **%** causes the entire remainder of the command line to be passed to the UNIX shell, whereas the **unix** command is terminated by semicolons ';' and keywords such as **ELSE** and **END** (see *BUG* Section 10.8.4). (In technical language, the **unix** command

is parsed like other BLSS commands, but **%** is not.) For example, both the following run the UNIX **date** and **who** commands:

```
. unix date ; unix who
. % date ; who
```

The first example invokes the UNIX shell twice. The second example invokes the UNIX shell just once—the UNIX shell interprets the ';' and knows to run two commands.

With no arguments, **unix** and **%** invoke a temporary interactive UNIX shell. To exit from this shell and return to BLSS, type control-D (or **exit**). The shell used—for both interactive shells and UNIX shell commands—is that in the environment string SHELL if it exists; otherwise, the default shell for your system.

Unix and **%** are internal commands to the BLSS shell.

Direct Execution of UNIX Commands
The **unix** and **%** commands execute commands via the UNIX shell. It is possible (and more efficient) to execute UNIX commands directly from the BLSS shell, bypassing the UNIX shell. There are two ways to do so. First, commands can be invoked by UNIX pathname beginning with '/', './', '../', or '~'. For example, the pathname of the UNIX command **ls** is */bin/ls*, so you can type:

```
. /bin/ls
```

Second, UNIX commands can be executed from the BLSS shell if the string BPATH includes the directories in which they are stored: */bin*, */usr/bin*, etc. Because these directories are already in the PATH environment string, you can reset BPATH with the command:

```
. set BPATH $BPATH $PATH
```

and then simply type UNIX command names. For example:

```
. ls
```

When directly invoking UNIX commands with options that begin with a '–' character, you must escape the '–' with a backslash '\' in order to inhibit its usual BLSS meaning (arithmetic negation):

```
. ls \-l
```

The backslash is not necessary with the **unix** or **%** commands.

Adding $PATH to the end of $BPATH makes BLSS somewhat less efficient, because it must search more directories for commands to execute. However, if you want to do this on a regular basis, add the following line to your *.blss* file:

```
IF ($#BPATH < $#PATH) set BPATH $BPATH $PATH
```

The **IF** test prevents repeated expansion if the *.blss* file is **source**-d more than once.

BSD UNIX Job Control

BSD UNIX systems provide an interactive *job control* facility which is more efficient than using the **unix** or **%** commands for running UNIX commands during an interactive BLSS session. (Job control cannot be embedded in command files.)

SEE ALSO

alias(BRM), for more about the *.blss* file.

fromunix(BRM), for information about executing BLSS commands directly from the UNIX command level.

Appendix A: Basic UNIX Command Summary.

---

vecop — Functions (sum, max, . . .) applied to vectors within datasets

---

SYNOPSIS

vecop *x* {**x=**;**dn=**} > *vecop-func@ output* [*vecop-func@ output* ...]

vecop-func x {**x=**;**dn=**} [> *output* [*vecop-func@ output* ...]]

DESCRIPTION

Vecop performs a variety of vector operations on the rows, columns, or sheet vectors of the BLSS dataset *x*. What operation to perform, and how to perform it, is specified by naming a *vecop-func*. The *vecop-func* names may be used as output tag names and also as BLSS command names, in which case the result of the named *vecop-func* is the default output dataset.

Here is the list of *vecop-func* names:

sum	**colsum**	**rowsum**	**sheetsum**
mean	**colmean**	**rowmean**	**sheetmean**
max	**colmax**	**rowmax**	**sheetmax**
min	**colmin**	**rowmin**	**sheetmin**
norm	**colnorm**	**rownorm**	**sheetnorm**
amax	**colamax**	**rowamax**	**sheetamax**
asum	**colasum**	**rowasum**	**sheetasum**
ss	**colss**	**rowss**	**sheetss**
sd	**colsd**	**rowsd**	**sheetsd**
var	**colvar**	**rowvar**	**sheetvar**
med	**colmed**	**rowmed**	**sheetmed**
mad	**colmad**	**rowmad**	**sheetmad**
prod	**colprod**	**rowprod**	**sheetprod**

The basic functions are:

sum	Sum of the elements.
mean	Mean of the elements.
max	Maximum of the elements.
min	Minimum of the elements.
norm	Euclidean norm (square root of the sum of squares).

amax	Absolute maximum (maximum of the absolute values).
asum	Absolute sum (sum of the absolute values).
ss	Sum of squares.
sd	Standard deviation (default divisor = $N-1$).
var	Variance (default divisor = $N-1$).
med	Median of the elements.
mad	Median absolute deviation from the median.
prod	Product of the elements.

The **norm** and **sd** functions are less susceptible to overflow or underflow than the **ss** and **var** functions. Products of long vectors are liable to overflow or underflow.

How a function is applied depends on its *vecop-func* name prefix. The unadorned *vecop-func* names (**sum, mean,** ...) act upon the entire array x as if it were a single vector; the result is a scalar. The **col** *vecop-func*'s (**colsum, colmean,** ...) act upon the columns of x; the result is a row vector. The **row** *vecop-func*'s (**rowsum, rowmean,** ...) act upon the rows of x; the result is a column vector. All three of these *vecop-func* types act sheetwise: if the input array x consists of multiple sheets the action is repeated for each sheet.

The **sheet** *vecop-func*'s (**sheetsum, sheetmean,** ...) act on the sheet vectors of x: the result is a single-sheet array in which each element contains the function result (sum, mean, ...) for the corresponding sheet vector—that is, the corresponding elements in all the sheets.

Missing Value Treatment
Elements of any *vecop-func* result which depend on missing values will themselves be missing.

INPUT
x@   The input dataset.

OPTIONS
x=*INTLIST*

Columns of the **x@** input dataset to use. Default is all.

dn=*INT*

The divisor used for computing variances and SDs is $N+dn$. The default divisor is $N-1$; that is, the default *dn* is -1.

OUTPUTS
vecop-func@

The result of the named *vecop-func*.

If the command is invoked using a *vecop-func* name and no output dataset corresponding to that *vecop-func* is specified, then the result of that *vecop-func* is put in the output dataset named *x.vecop-func*.

EXAMPLES
Suppose that x contains the following data:

```
. show {i} x

   2   7  12   5
   1   3  10   6
   8   9   4  11
```

Both the following commands compute the sums of the columns of x and place the result in the corresponding element of the row vector y:

```
. vecop x > colsum@ y
. colsum x > y
. show {i} y

  11  19  26  22
```

The following command puts the minimum of each row of x into the corresponding element of the column vector z:

```
. rowmin x > z
. show {shape; i} z

   2
   1
   4
```

If several *vecop-func*'s are needed, it is more efficient to compute them all with a single command. For example, to put the maximum of x in $x.max$ (a scalar) and the minimum of x in $x.min$ (also a scalar), use any of the following:

```
. vecop x >  max@ x.max  min@ x.min
. max x >  x.max  min@ x.min
. max x >  min@ x.min
. show {i} x.max x.min

  12

   1
```

LIMITATIONS
In order to allow sheetwise processing, all input and output datasets are open simultaneously. Hence the number of *vecop-func*'s which can be computed with one single command is limited by the maximum number of open files on UNIX.

SEE ALSO
arrop(BRM), for additional array operations.
let(BRM), for additional matrix, vector, and scalar functions.

---

xtab — Cross-tabulation

SYNOPSIS

    xtab *x* [*y*] {x=;y=;chisq;exp;freq;res;all; ... } [> *count exp res contr freq* ...]

DESCRIPTION

    Xtab cross-tabulates the columns of data *x* and *y*. By default, each value of *x* and *y* is treated as a separate category. Categories which are intervals may be specified by various options.

    A value is considered to lie in an interval with endpoints *a* and *b* if the value is at least *a* but less than *b*. The mathematical notation for such an interval is [*a*, *b*). This notation is used in the row and column labels of **xtab**.

    Missing data, if any, are tabulated in the last row or column.

INPUTS

    x@   Dataset from which to take the *X* variable: each row of the table is one *X* category.

    y@   Dataset from which to take the *Y* variable: each column of the table is one *Y* category. If not given, the *Y* variable is also taken from the **x@** input.

    The number of cases in *x* and *y* must be equal. If either *x* or *y* is a row vector or a sheet vector, it is treated as a column vector.

OPTIONS

    x=*INT*

        Column of **x@** to use as the *X* variable. Default is column 1.

    y=*INT*

        Column of **y@** (or **x@**, if **y@** is not given) to use as the *Y* variable. Default is column 1 if both **x@** and **y@** are given; column 2 if only **x@** is given.

    chisq Display the χ^2 (chi-square) statistic for the test of independence of *X* and *Y*, $\Sigma[(count - exp)^2 / exp]$, its degrees of freedom, and its P-value under the null hypothesis of independence.

    nobar Suppress the vertical bars which normally separate columns of cells.

    width=*INT*

        Maximum width of the output display. If the table exceeds this width, it is printed in blocks of columns. The WIDTH environment string can also be used; the {**width=**} option overrides it. Default width is 79, or 132 if the {**lpr**} option is set.

Cell Contents

    The following options control the contents of the cells in the table. They are listed in the order in which they appear in the cells. The upper left corner of the output display lists the cell contents in the order of appearance.

nocount
> Omit the cell counts from the table. By default, the cell counts (only) are tabulated.

exp Tabulate the expected cell counts, assuming that X and Y are independent.

res Tabulate the residual cell counts (*count* – *exp*).

contr Tabulate the contribution to the χ^2 statistic for each cell (*res*^{2} / *exp*).

freq Tabulate the cell frequencies (proportions).

xfreq Tabulate the frequencies within each X (row) category.

yfreq Tabulate the frequencies within each Y (column) category.

xcum Tabulate cumulative counts within each X (row) category.

ycum Tabulate cumulative counts within each Y (column) category.

all Tabulate and display all possible cell contents.

Determining the Categories
The following options specify the categories. They are listed in order of precedence, from high to low. (That is, if conflicting options are specified, the first one listed below is used.)

xend=*FLOATLIST*, **yend**=*FLOATLIST*
> Lists of X and Y category endpoints. If the data extrema are not included, they are added to the list.

xp=*FLOATLIST*, **yp**=*FLOATLIST*
> X and Y category endpoints are the quantiles corresponding to the lists of probabilities *xp* and *yp*. If 0 and 1 are not included, they are added to the list.

xn=*INT*, **yn**=*INT*
> The X and Y categories are *xn* and *yn* equal-length intervals.

xc The default: each distinct X value defines a separate category.

yc The default: each distinct Y value defines a separate category.

end=*FLOATLIST*
> List of X and Y category endpoints. If the data extrema are not included, they are added to the list.

p=*FLOATLIST*
> X and Y category endpoints are the quantiles corresponding to the list of probabilities *p*. If 0 and 1 are not included, they are added to the list.

n=*INT*
> The X and Y categories are *n* equal-length intervals.

quart The X and Y categories are the four data quartiles.

Lists of category endpoints are sorted in ascending order and any duplicate values are removed.

Suppressing NAs

The following options suppress the tabulation of NAs (missing data).

noxna
> Do not tabulate NAs in the X variable.

noyna Do not tabulate NAs in the Y variable.

nona Do not tabulate NAs in either the X or Y variable.

Note that although these options omit cases from tabulation when either an X or Y variable is missing, they do not omit cases from consideration when computing category endpoints from the data.

OUTPUTS
count@
> Cell counts. Includes both the X and Y margins.

exp@ Expected cell counts, assuming that X and Y are independent.

res@ Residual cell counts (*count – exp*).

contr@
> Contribution to the χ^2 statistic for each cell (*res^2 / exp*).

freq@ Cell frequencies (proportions). Includes both the X and Y margins.

xfreq@
> X (row) frequencies. Includes the X margin.

yfreq@
> Y (column) frequencies. Includes the Y margin.

xcum@
> X cumulative counts.

ycum@
> Y cumulative counts.

chisq@
> The χ^2 statistic, its degrees of freedom under the null hypothesis of independence of X and Y, and its P-value.

xend@
> Endpoints for the X data categories (intervals) used in the table.

yend@
> Endpoints for the Y data categories (intervals) used in the table.

EXAMPLES

Cross-tabulate the median value of owner-occupied homes (thousands of dollars) by the per capita crime rate for the Boston area, using the data quartiles as category endpoints. Note that the upper-left corner of the table lists the cell contents:

```
. load boston
. xtab boston {x=1;y=2;quart;freq} {nobar}

            X (rows): boston[1]    Y (cols): boston[2]

count   |[0.006, [0.082, [0.257, [3.68,
freq    | 0.082)  0.257)  3.68)    89]    Total
--------+------------------------------+-------
[    5, |    2      12      28      84 |   126
    17) | .0040   .0237   .0553   .1660 | .2490
        +------------------------------+-------
[   17, |   25      41      37      22 |   125
  21.2) | .0494   .0810   .0731   .0435 | .2470
        +------------------------------+-------
[ 21.2, |   40      47      26      10 |   123
    25) | .0791   .0929   .0514   .0198 | .2431
        +------------------------------+-------
[   25, |   59      27      35      11 |   132
    50] | .1166   .0534   .0692   .0217 | .2609
        +------------------------------+-------
Total   |  126     127     126     127 |   506
freq    | .2490   .2510   .2490   .2510 | 1.000
```

Ozone contains measurements on Los Angeles air pollution data in 1976. We cross-tabulate daily ozone measurements collected in Upland, California by the day of the week (1=Monday, ..., 7=Sunday). The **xfreq** entry shows the proportion of days of the week within each level of ozone concentration.

```
. load ozone
. xtab ozone {x=4;y=3; xend=0,10,20,30,40} {freq;xfreq;chisq}

count   |   X (rows): ozone[4]    Y (cols): ozone[3]
freq    |
xfreq   |    1       2       3       4       5       6       7     Total
--------+-------+-------+-------+-------+-------+-------+-------+-------
[    0, |   26 |   26 |   23 |   30 |   29 |   23 |   24 |   181
    10) | .0710 | .0710 | .0628 | .0820 | .0792 | .0628 | .0656 | .4945
        | .1436 | .1436 | .1271 | .1657 | .1602 | .1271 | .1326 |
        +-------+-------+-------+-------+-------+-------+-------+-------
[   10, |   16 |   12 |   19 |   16 |   15 |   22 |   20 |   120
    20) | .0437 | .0328 | .0519 | .0437 | .0410 | .0601 | .0546 | .3279
        | .1333 | .1000 | .1583 | .1333 | .1250 | .1833 | .1667 |
        +-------+-------+-------+-------+-------+-------+-------+-------
[   20, |    7 |   13 |    7 |    4 |    7 |    6 |    7 |    51
    30) | .0191 | .0355 | .0191 | .0109 | .0191 | .0164 | .0191 | .1393
        | .1373 | .2549 | .1373 | .0784 | .1373 | .1176 | .1373 |
        +-------+-------+-------+-------+-------+-------+-------+-------
[   30, |    2 |    1 |    3 |    0 |    2 |    1 |    0 |     9
    40] | .0055 | .0027 | .0082 | 0.000 | .0055 | .0027 | 0.000 | .0246
        | .2222 | .1111 | .3333 | 0.000 | .2222 | .1111 | 0.000 |
        +-------+-------+-------+-------+-------+-------+-------+-------
NA      |    1 |    0 |    0 |    3 |    0 |    0 |    1 |     5
        | .0027 | 0.000 | 0.000 | .0082 | 0.000 | 0.000 | .0027 | .0137
        | .2000 | 0.000 | 0.000 | .6000 | 0.000 | 0.000 | .2000 |
        +-------+-------+-------+-------+-------+-------+-------+-------
Total   |   52 |   52 |   52 |   53 |   53 |   52 |   52 |   366
freq    | .1421 | .1421 | .1421 | .1448 | .1448 | .1421 | .1421 | 1.000
```

Chi-square statistic 28.13 has 24 degrees of freedom and P-value 0.255.

Note that (because the {**nona**} and {**noxna**} options were not given) the NA's in the X variable were tabulated as a separate category.

SEE ALSO
chisq(BRM), for χ^2 statistics from already-tabulated data.
freq(BRM), for univariate tabulations (that is, frequency distributions).

---

xvalid — Cross-validated estimate of regression R^2 and *RSS*

SYNOPSIS
    xvalid x {**x=;y=;noint;i=**} [> b]

DESCRIPTION
Xvalid gives I-fold cross-validated estimates of the R^2, the *RSS* (residual sum of squares), and the *SSM* (sum of squares about the mean) for multiple linear regression. It is a useful tool when searching for the best prediction equation. If used several times with different models, the equation that produces the maximum cross-validated value of R^2 is a good candidate for use as a prediction equation in a regression command such as **regress**.

The usual multiple regression estimate of R^2 increases whenever more parameters are added to the prediction equation. This problem does not occur with the cross-validated R^2. Typically, the graph of the cross-validated R^2 estimate against the number of predictors is unimodal—it increases up to a point and then decreases.

Algorithm. The cases of x are divided into I random subsets of nearly equal size. For $i = 1, 2, \ldots, I$, the least squares fit and the mean are calculated for all cases but the i-th subset. The regression and the mean are each used to predict the observations in the i-th subset. The cross-validated estimate of *RSS* is the sum of the squared prediction errors (summed over all observations) using the regression model. The cross-validated estimate of *SSM* is the sum of the squared prediction errors using the mean. The estimate of R^2 is $1-RSS/SSM$.

Note that it is possible for the cross-validated estimate of R^2 to be negative, especially when overfitting, because the regression model is competing with the mean. This indicates that the mean is a better predictor than the regression. Note also that, because the subsets are selected randomly, the results contain some randomness. Therefore, it is a good idea to repeat the procedure several times and take the average.

INPUTS
    x@   Array whose columns contain the independent and dependent variables. Rows of x that contain missing values in the columns selected by the {**x=**} or {**y=**} options are ignored.

OPTIONS

x=*INTLIST*

Independent variables. Default is all but the last column of *x*.

y=*INT*

Dependent variable. Default is the last column of *x*.

noint No intercept is included in the regression.

long In addition to the default display, show the detailed cross-validation results: the residuals, the values of the dependent variable ('Y'), the fitted values ('Yhat'), the mean of the dependent variable ('Ybar') for each subset, and the contribution of each observation and each subset to the cross-validated *RSS* and *SSM*.

i=*INT* Number of subsets into which the data is divided. Default is 10.

OUTPUTS

b@ *I*-by-*k* matrix, where *k* is the number of parameters in the regression. The *i*-th row of **b@** contains the *i*-th set of estimated regression coefficients. If the regression includes an intercept, its estimated values are the first element in each row.

If the {**quiet**} option is in effect, the **b@** output dataset is always created. If no output dataset name is specified, its name is *x*.**b**, where *x* is the name of the input dataset.

EXAMPLE

The dataset *kappa* contains 30 cases. Because *I*=10 by default, these cases are split into 10 random subsets of size 3 each:

```
. load kappa
. xvalid kappa {x=1;y=2; long} > beta
```

Obs Subset	Residual	Y	Yhat	Ybar	Contribution to RSS	Contribution to SSM
11	-2.174	104.000	106.174	113.556	4.727	91.31
12	-0.569	96.000	96.569	113.556	0.3241	308.2
23	8.012	143.000	134.988	113.556	64.19	867.0
1					69.24	1266.
7	4.427	118.000	113.573	113.667	19.59	18.78
10	0.908	117.000	116.092	113.667	0.8245	11.11
21	-6.055	105.000	111.055	113.667	36.66	75.11
2					57.08	105.0
(results for subsets 3, . . . , 9 omitted)						
5	-6.941	104.000	110.941	113.444	48.18	89.20
15	-1.948	114.000	115.948	113.444	3.793	0.3086
24	12.052	128.000	115.948	113.444	145.3	211.9
10					197.2	301.4

X-Valid R-squared	Sample size	Residual Sum of Squares	Sum of Squares about Mean
0.8473	30	769.3	5038.

The {**long**} output shown here indicates, for example, that the first subset consisted of cases 11, 12, and 23. Case 11 contributed 4.727 to the *RSS* and 91.31 to the *SSM*; the cases in the first subset together contributed 69.24 to the *RSS* and 1266 to the *SSM*; etc.

The output dataset *beta* is a 10-by-2 matrix which contains the 10 sets of estimated coefficients.

Successive **xvalid** runs yield slightly different cross-validated estimates.

SEE ALSO
regress(BRM).

PART THREE

BLSS Data Library

Introduction to the Data Library

Part Three of this book is the *BLSS Data Library,* or *BDL.* It consists of about fifty short entries, each of which describes a dataset, or a group of closely related datasets, in the library. Individual entries are referred to by the form *entryname(BDL).*

The first entry, *data(BDL),* contains a complete list of the datasets. The remaining entries have names of the form *data.name(BDL),* where *name* is the dataset name, and are arranged in alphabetical order. They give the dimensions and contents of the dataset(s) and references where appropriate.

At Berkeley, the data library is used not only for examples in BLSS documentation, but also by instructors when creating their own examples or giving assignments. Datasets in the library fall into three main categories. The first is datasets which have appeared in the literature—some, such as *iris* and *sunspot,* are classics. The second category is datasets contributed by users, including Berkeley students who collected data for statistics projects. Often, the student-collected datasets are smaller in size and lighter in nature than datasets taken from the literature—thus they are particularly useful for introductory courses. Finally, for historical reasons the BLSS data library includes the datasets assembled by Donald R. McNeil for his book *Interactive Data Analysis.*

---

data — Introduction to BLSS system data library

SYNOPSIS
 load {**sys**} *dataset*

DESCRIPTION
 Here is a list of datasets in the BLSS data library, with dimensions and brief description of each. Detailed information for each dataset (or group of closely related datasets) is in the manual entry *data.name(BDL)*, where *name* is the dataset name. These manual entries may be seen on-line by typing **help data.***name*.

 Library datasets may be loaded into the work area with the **load** {**sys**} command as above.

DATASETS

Name	Dims.	Brief Description
acid †	(6,2)	Optical density of formaldehyde.
airmiles †	(24,2)	Passenger miles on U.S. airlines, 1937-1960.
alkaloid	(45,4)	Frequency of alkaloids in plants.
bees †	(8,8)	Consumption of sucrose solutions by bees.
births †	(8,4)	Time until conception.
books	(30,8)	Size and price of college textbooks.
boston	(506,14)	Harrison/Rubinfeld Boston housing tract data.
cars †	(50,2)	Speed and stopping distance of cars.
chickwts †	(14,6)	Weights of chicks given different feed supplements.
circle	(32,2)	Equally spaced points on the unit circle.
consumption †	(5,5)	Personal consumption expenditures for the U.S.
deaths †	(5,4)	Death rates in Virginia in 1940.
demopct †	(24,4)	Percent Democratic vote in U.S. Presidential elections.
die	(6,1)	Box model for a fair die: the numbers 1 through 6.
discoveries †	(100,1)	Number of 'great' discoveries per year, 1860-1959.
fake	(6,1)	Small 'fake' dataset with mean 0 and SD 2.
final	(63,1)	Alternate name for *final84*.
final84	(63,1)	Stat 2 final exam scores, summer 1984, UCB.
final85	(62,1)	Stat 2 final exam scores, summer 1985, UCB.
finalexam	(62,1)	Alternate name for *final85*.
heart	(184,4)	Stanford heart transplant data, 1967 to 1980.
hodgkin	(56,3)	Hodgkin's disease patients and DNCB.
illit †	(9,6)	Median illiteracy rates in the U.S.
insects †	(12,6)	Insect counts in agricultural experimental units.
iris	(150,5)	Anderson/Fisher iris data.
iris.se	(50,4)	*Iris* data, columns 1:4 for *setosa* only.
iris.ve	(50,4)	*Iris* data, columns 1:4 for *versicolor* only.
iris.vi	(50,4)	*Iris* data, columns 1:4 for *virginica* only.

Name	Dims.	Brief Description
islands †	(48,1)	Land area of the largest 48 islands and continents.
iudfailure	(430,4)	Time to IUD failure.
kappa	(30,4)	Height and weight of KKΓ Sorority members.
lobsters †	(8,2)	Percentages of lobsters recaptured.
micecon	(1080,2)	Mice in control group of survival experiment.
micerad	(1454,2)	Mice in treated (irradiated) group.
ozone	(366,13)	Los Angeles ozone pollution data, 1976.
phones †	(7,7)	Number of telephones, by continent.
pigs †	(10,6)	Tooth length of guinea pigs fed ascorbic acid.
plover	(68,5)	Snowy plover egg/chick data, 1985.
precip †	(69,1)	Average annual precipitation in the U.S.
presidents †	(120,1)	Approval ratings of U.S. presidents, 1945-74.
pressure †	(44,2)	Vapor pressure of mercury at various temperatures.
prostate	(129,4)	Prostate cancer survival.
rivers †	(141,1)	Lengths of 141 'major' rivers in North America.
stackloss	(21,4)	Loss of ammonia while oxidizing it to nitric acid.
statprofs	(16,2)	Salary and age of Statistics professors at UCB, 1984.
steam	(25,10)	Steam used at an industrial steam plant.
sunspot	(2820,1)	Monthly mean sunspot numbers, 1749–1983.
sunspotx	(360,2)	*Sunspot* excerpt, 1950–1979, with time index.
tractor	(107,1)	Tractor brake failure times.
tubercle	(65,2)	Survival times of guinea pigs infected with tubercle bacilli.
turbines	(57,4)	Crack growth rate for disk cracks in U.S. plants.
ucprofs	(27,2)	Salary and age of professors at UCB, 1984.
uspop †	(20,2)	U.S. population, census years 1790-1980.
viennaberlin	(2184,2)	Monthly mean temperatures of Vienna and Berlin.
warpbreaks	(9,6)	Number of warp breaks for different types of yarn.
women †	(15,2)	Average weight of American women aged 30-39.

NOTE

For historical reasons, the BLSS data library includes the datasets assembled by Donald R. McNeil for his book *Interactive Data Analysis* (1977), Wiley, New York. These datasets are marked by a dagger † in the list above.

---
data.acid — Optical density of formaldehyde

DATASETS
  acid   (6.2).

DESCRIPTION
  From a chemical experiment to prepare a standard curve for the determina-
  tion of formaldehyde by the addition of chromatropic acid and concentrated
  sulfuric acid. The first column is the amount of carbohydrate used in milli-
  liters; the second is the optical density.

  Source: Carl A. Bennett and Norman L. Franklin (1954), *Statistical Analysis
  in Chemistry and the Chemical Industry*, Wiley, New York, pp. 216-217.
  Used by McNeil, pp. 48-49.

---
data.airmiles — Passenger miles on U.S. airlines, 1937-1960

DATASETS
  airmiles (24,2).

DESCRIPTION
  Thousands of passenger miles on U.S. airlines for the years 1937-1960.
  Source: Robert Goodell Brown (1963), *Smoothing, Forecasting and Prediction
  of Discrete Time Series*, Prentice-Hall, Englewood Cliffs, N.J., p. 427. Used
  by McNeil, p. 62.

---
data.alkaloid — Frequency of alkaloids in plants

DATASETS
  alkaloid (45,4).

DESCRIPTION
  Frequency of alkaloids in plants. Each row represents a family. By column:
  number of species containing and number not containing alkaloids at high
  latitude; number of species containing and number not containing alkaloids at
  low latitude.

  Source: J. W. McCoy (1978), *Comments on the geographic distribution of
  alkaloids in angiosperms*, **American Naturalist, 112**, pp. 1126-1133.

---
data.bees — Consumption of sucrose solutions by bees

DATASETS
  bees   (8,8).

DESCRIPTION
  Individual cells of dry comb were filled with 8 different concentrations of lime

sulfur emulsion in sucrose solution. The concentrations ranged from 1/100 down to 1/1562500 in steps of ratio 1/5; the final concentration was zero. The cells were placed in a chamber with 100 honeybees for two hours. The response was the decrease in the amount of the solution in each cell, measured in milligrams. The cells were arranged in an 8 by 8 Latin square; however, the data are rearranged so that the 8 columns correspond to the 8 different concentrations, from the largest to zero.

Source: D. J. Finney (1947), *Probit Analysis*, Cambridge University Press, pp. 188-189. Used by McNeil, pp. 76-77.

---

data.births — Time until conception

DATASETS
  births   (8,4).

DESCRIPTION
  The number of months from the time of marriage or termination of contraception until conception. Column 1 is the number of months; columns 2-4 are the proportions of Hutterite, Taichung, and U.S. women who conceived in less than or equal to that number of months.

Source: S. N. Singh, K. C. Chakrabarty, and V. K. Singh (1976), *A modification of a continuous time model for first conception*, **Demography**, **13**, pp. 37-44. Used by McNeil, p. 105.

---

data.books — Size and price of college textbooks

DATASETS
  books   (30,8).

DESCRIPTION
  Can the price of a hardbound college textbook be predicted from its observable physical properties? To help answer this question, observations were made on 30 hardbound textbooks chosen at random from the shelves of the ASUC Textbook Store at UC Berkeley in spring 1983.

Column   Contains
  1   Price in dollars.
  2   Number of pages.
  3   Thickness in inches.
  4   Length of a page in inches.
  5   Width of a page in inches.
  6   Area (length * width) in square inches.
  7   Volume (length * width * thickness) in cubic inches.
  8   Percentage of special features (see below for definition).

Special features: 50 pages were chosen at random from each book. Out of these 50 pages, the number of pages with more than half the page consisting of diagrams, mathematical text, or foreign languages was noted; then this number was converted to a percentage of 50.

Sampling method: The store had 10 aisles. Each aisle was subdivided and within each subdivision were shelves containing columns of books. Each book was chosen as follows: choose (at random) an aisle, then a subdivision, then a shelf, then a column, then a book.

Data collected by Karen S. Louie of UC Berkeley in March 1983.

---

data.boston — Harrison/Rubinfeld Boston housing tract data

DATASETS
   boston   (506,14).

DESCRIPTION
   Boston housing tract data of Harrison and Rubinfeld. Each case is one U.S. Census tract in the Boston area.

Column	Code	Contains
1	MV	Median value of owner-occupied homes (in thousands of dollars).
2	CRIM	Crime rate (per capita) by town.
3	ZN	% Residentially zoned land for lots greater than 25,000 square feet, by town.
4	INDUS	% Industrial (nonretail) business acres by town.
5	CHAS	1 if tract on Charles River; 0 if not.
6	NOX	Nitrogen oxide concentration (parts per billion).
7	RM	Average number of rooms in owner units.
8	AGE	% of owner-occupied units built before 1940.
9	DIS	Weighted distance to five employment centers.
10	RAD	Index of accessibility to radial highways.
11	TAX	Full property tax rate ($ per $10,000).
12	PT	Pupil/teacher ratio by town school district.
13	BT	(% Blacks in population $- 63)^2$.
14	LSTAT	% Lower status population.

Data used by D. Harrison and D. L. Rubinfeld (1978), *Hedonic housing prices and the demand for clean air*, **Journal of Environmental Economics and Management**, **5**, pp. 81-102. Transformed data appear in David A. Belsley, Edwin Kuh, and Roy E. Welsch (1980), *Regression Diagnostics*, Wiley, New York, Appendix 4A.

---

data.cars — Speed and stopping distance of cars

DATASETS
  cars   (50,2).

DESCRIPTION
  Speed in miles per hour (column 1) and stopping distance in feet (column 2) for 50 motorists. Source: M. Ezekiel (1930), *Methods of Correlation Analysis*, Wiley, New York, Table 10. Used by McNeil, p. 52.

---

data.chickwts — Weights of chicks given different feed supplements

---

DATASETS
  chickwts (14,6).

DESCRIPTION
  Newly hatched chicks were randomly assigned to six different groups (columns). Each group was given a different protein supplement. The data are the weights of the chicks in grams after six weeks. Because some chicks died before six weeks, some columns are filled out with missing values.

  Source: Anonymous query (1948), **Biometrics**, **4**, pp. 213-214. Used by McNeil, p. 30.

---

data.circle — Equally spaced points on the unit circle

DATASETS
  circle   (32,2).

DESCRIPTION
  Circle contains the Cartesian coordinates (x, y) of 32 equispaced points on the unit circle. Such coordinates can be created for any number of points on the unit circle as follows. First, express the points in polar coordinates (r, θ), where $r = 1$ and $0 \leq \theta < 2\pi$. Then, convert the polar coordinates to Cartesian coordinates:

```
. n = 100                       # example for 100 points
. theta = 2*PI*(0:n-1)'/n       # θ = 0, 2π/n, 4π/n, ... , 2π(n-1)/n
. const 1 theta > r             # radius 1, same dimensions as theta
. polar {inv} r theta > x y     # transform to Cartesian coordinates
```

---

data.consumption — Personal consumption expenditures for the U.S.

DATASETS
consumption (5,5).

DESCRIPTION
Personal consumption expenditures for the U.S. in billions of dollars for five years (columns): 1940, '45, '50, '55, and '60; and five categories (rows): food and tobacco, household operation, medical and health, personal care, and private education, according to *The World Almanac and Book of Facts*, 1962, New York, p. 756. Used by McNeil, pp. 99-100.

---

data.deaths — Death rates in Virginia in 1940

DATASETS
deaths   (5,4).

DESCRIPTION
Death rates per 1000 in Virginia in 1940. The five rows are age groups: 50-54, 55-59, 60-64, 65-69, and 70-74. The four columns are rural males, rural females, urban males, and urban females.

Source: L. Moyneau, S. K. Gilliam, and L. C. Florant (1947), *Differences in Virginia death rates by color, sex, age, and rural or urban residence*, **American Sociological Review**, **12**, pp. 525-535. Used by McNeil, p. 94.

---

data.demopct — Percent Democratic vote in U.S. Presidential elections

---

DATASETS
demopct (24,4).

DESCRIPTION
The percent Democratic vote (among major party votes) for 24 northeastern and central states in the U.S. Presidential elections of 1960, '64, '68, and '72, according to *The World Almanac and Book of Facts*, 1975, New York, pp. 736-746. Used by McNeil, p. 67.

---

data.die — Box model for a fair die: the numbers 1 through 6

DATASETS
die   (6,1).

DESCRIPTION
Box model for a fair die: the numbers 1 through 6. For an explanation of box models, see *BUG* Section 7.1.2, or David Freedman, Robert Pisani, and Roger Purves (1978), *Statistics*, Norton, New York.

---

data.discoveries — Number of 'great' discoveries per year, 1860-1959

DATASETS
discoveries   (100,1).

DESCRIPTION
The number of 'great' discoveries and inventions per year, from 1860-1959, according to *The World Almanac and Book of Facts*, 1975, New York, pp. 315-318. Used by McNeil, p. 121.

---

data.fake — Small 'fake' dataset with mean 0 and SD 2

DATASETS
fake   (6,1).

DESCRIPTION
Fake is a small, artificial dataset with mean 0 and standard deviation 2. It contains the data: –3, –1, 0, 0, 1, 3.

---

data.final — Stat 2 final exam scores, UC Berkeley

DATASETS
final84   (63,1).
final85   (62,1).

DESCRIPTION
Final84 contains final exam scores in Statistics 2, summer 1984, at UC Berkeley. Also available on-line under the name *final*.

Final85 contains final exam scores in Statistics 2, summer 1985, at UC Berkeley. Also available on-line under the name *finalexam*.

Both exams were graded on the scale 0 through 100.

Source: Ani Adhikari, Department of Statistics, UC Berkeley.

---

data.heart — Stanford heart transplant data, 1967 to 1980

DATASETS
heart   (184,4).

DESCRIPTION
Stanford heart transplant data, October 1967 to February 1980.

Column   Contains
    1   Survival time in days, censored as in column 2.
    2   0 = alive; 1 = dead, by February 1980.
    3   Age in years at entry.
    4   T5 mismatch score.

Persons alive in February 1980 are considered censored. Source: Rupert Miller and Jerry Halpern (1982), *Regression with censored data*, **Biometrika**, **69(3)**, pp. 521-531.

---

data.hodgkin — Hodgkin's disease patients and DNCB

DATASETS
  hodgkin   (56,3).

DESCRIPTION
  Hodgkin's disease patients and the chemical DNCB, stage 1 of 4 only.

  Column   Contains
    1   Change-time to sensitize group (in days).
    2   Lifetime: observation time (in days).
    3   Censoring: 1 = recurrence at lifetime; 0 = censored.

  Source: Gail Gong (1980), *Do Hodgkin's disease patients with DNCB sensitivity survive longer?*, in **Biostatistics Casebook III**, Technical Report 57, Division of Biostatistics, Stanford University, pp. 70-100.

---

data.illit — Median illiteracy rates in the U.S.

DATASETS
  illit   (9,6).

DESCRIPTION
  Median illiteracy rates for nine regions of the U.S. (New England, Mid-Atlantic, East North Central, West North Central, South Atlantic, East South Central, West South Central, Mountain and Pacific) in six different years (1900, '20, '30, '50, '60, and '70).

  Source: *Statistical Abstract of the United States*, 1975, Bureau of the Census, Washington, D.C., p. 120. Used by McNeil, p. 42.

---

data.insects — Insect counts in agricultural experimental units

---

DATASETS
  insects   (12,6).

DESCRIPTION
  Counts of live tobacco hornworms found in agricultural experimental units treated with six different insecticides (columns). Six blocks were divided into twelve plots each; each insecticide was applied to two plots within each block. Rows 1 and 2 are from plots in block 1; . . . ; rows 11 and 12 are from plots in block 6. Source: Geoffrey Beall (1942), *The transformation of data from entomological field experiments so that the analysis of variance becomes applicable*, **Biometrika**, **32**, pp. 243-262. Used by McNeil, p. 12.

---

data.iris — Anderson/Fisher iris data

DATASETS
    iris   (150,5).
    iris.se (50,4).
    iris.ve (50,4).
    iris.vi (50,4).

DESCRIPTION
Anderson/Fisher iris data. Biometric data on 150 irises: 50 from each of 3 varieties. All measurements are in centimeters. *Iris* contains:

Column Contains
  1   Petal length.
  2   Petal width.
  3   Sepal length.
  4   Sepal width.
  5   Variety: 1 = *setosa*; 2 = *versicolor*; 3 = *virginica*.

iris.se is columns 1:4 of *iris* for *setosa* only.
iris.ve is columns 1:4 of *iris* for *versicolor* only.
iris.vi is columns 1:4 of *iris* for *virginica* only.

Data collected by Edgar Anderson (1935), *The irises of the Gaspé Peninsula*, **Bulletin of the American Iris Society**, **59**, pp. 2-5; reported in R. A. Fisher (1936), *The use of multiple measurements in taxonomic problems*, **Annals of Eugenics**, **7**, Part II, pp. 179-88.

---

data.islands — Land area of the 48 largest islands and continents

---

DATASETS
    islands (48,1).

DESCRIPTION
Land area, in thousands of square miles, of all islands and continents with area greater than ten thousand square miles. Source: *The World Almanac and Book of Facts*, 1975, New York, p. 406. Used by McNeil, p. 9.

---

data.iudfailure — Time to IUD failure in parous and nulliparous women

---

DATASETS
    iudfailure(430,4).

DESCRIPTION
4963 women took part in a study of a new type of IUD (birth control device). Each observation represents an event time. The data form two subsets: nulliparous and parous women.

Column   Contains
1   Event time: number of days from insertion of the IUD.
2   Number of IUD failures at the event time (due to expulsion or removal of the IUD, accidental pregnancy, etc.).
3   Number of observations censored at the event time (due to leaving the study area or IUD still in place as of the last time of observation, which is recorded as the event time).
4   Parity: 0 = nulliparous; 1 = parous.

Source: Sue Leurgans (1980), *Exploring the influence of several factors on a set of censored data*, in *Biostatistics Casebook*, Rupert Miller et al., eds., Wiley, New York, pp. 47-72.

---

data.kappa — Height and weight of Kappa Kappa Gamma Sorority members

DATASETS
kappa   (30,4).

DESCRIPTION
The first two columns contain the height and weight of 30 members of the Kappa Kappa Gamma Sorority at UC Berkeley; the third and fourth columns are the same for 30 members at UCLA. Data collected by Ian Yellin of UC Berkeley in March 1980.

---

data.lobsters — Percentages of lobsters recaptured from tagged populations

DATASETS
lobsters   (8,2).

DESCRIPTION
Percentages of lobsters recaptured from tagged populations released from eight locations off Long Island between April 1968 and June 1969. Column 1 contains the numbers of lobsters released from the eight locations; column 2 contains the percentages of lobsters recaptured before April 1970.

Source: R. A. Cooper and J. R. Uzmann, (January 1971), *Migrations and growth of deep-sea lobsters homarus americanus*, **Science**, Series 2, **171**, pp. 288-290. Used by McNeil, pp. 158-159.

data.mice — Survival experiment with mice

DATASETS
    micecon (1080,2).
    micerad (1454,2).

DESCRIPTION
Micecon and *micerad* contain, respectively, mice in the control group and treatment group of a survival experiment with serial sacrifice. The treated group was irradiated by $300R$ gamma radiation at age 70 days. Column 1 is age at death (or sacrifice) in days, minus 70. Column 2 is pathological state at death—an integer which is 0 (if none), or the sum of values which correspond to other states: 1 = thymic lymphoma; 2 = glomerulosclerosis; 4 = other diseases; 8 = sacrifice.

Source: B. Berlin, J. Brodsky, and P. Clifford (1979), *Testing disease dependence in survival experiments with serial sacrifice*, **JASA, 74,** pp. 5-14.

data.ozone — Los Angeles ozone pollution data, 1976

DATASETS
    ozone   (366,13).

DESCRIPTION
Los Angeles ozone pollution data, 1976. Each observation is one day.

Column Contains
    1   Month: 1 = January, . . . , 12 = December.
    2   Day of month.
    3   Day of week: 1 = Monday, . . . , 7 = Sunday.
    4   Daily maximum one-hour-average ozone reading (parts per million) at Upland, CA.
    5   500 millibar pressure height (m) measured at Vandenberg AFB.
    6   Wind speed (mph) at Los Angeles International Airport (LAX).
    7   Humidity (%) at LAX.
    8   Temperature (degrees C) measured at Sandburg, CA.
    9   Temperature (degrees C) measured at El Monte, CA.
    10   Inversion base height (feet) at LAX.
    11   Pressure gradient (mm Hg) from LAX to Daggett, CA.
    12   Inversion base temperature (degrees F) at LAX.
    13   Visibility (miles) measured at LAX.

Source: Leo Breiman, Department of Statistics, UC Berkeley. Data used in Leo Breiman and Jerome H. Friedman (1985), *Estimating optimal transformations for multiple regression and correlation*, **JASA, 80,** pp. 580-598.

---
data.phones — Number of telephones, by continent

DATASETS
   phones   (7,7).

DESCRIPTION
   The number of telephones on various continents, in thousands. The seven
   columns are North America, Europe, South America, Oceania, Africa, and
   Middle America; the seven rows are the years 1951, '56, '57, '58, '59, '60, and
   '61. Source: *The World's Telephones*, 1961, AT&T, New York, pp. 2-3. Used
   by McNeil, pp. 109-110.

---
data.pigs — Tooth length of guinea pigs fed ascorbic acid

DATASETS
   pigs   (10,6).

DESCRIPTION
   Average length of teeth in 10 guinea pigs fed ascorbic acid in doses of .5, 1,
   and 2 milligrams (columns 1-3) and orange juice containing the same levels of
   ascorbic acid (columns 4-6). The rows are different sets of guinea pigs.

   Source: C. I. Bliss (1952), *The Statistics of Bioassay*, Academic Press, New
   York, p. 500. Used by McNeil, p. 80.

---
data.plover — Snowy plover egg/chick data, 1985

DATASETS
   plover   (68,5).

DESCRIPTION
   Snowy plover egg/chick data. Each observation is an egg or the chick which
   hatched from it.

   Column   Contains
      1   Egg length, in millimeters.
      2   Egg breadth, in millimeters.
      3   Egg weight, in grams.
      4   Chick weight, in grams.
      5   Did the chick survive to fledgling? 0 = no; 1 = yes.

   These data are part of a much larger dataset collected on free-living snowy
   plovers (a shore bird) in the Monterey Bay area, 1985 season, by the Point
   Reyes Bird Observatory of Stinson Beach, California. Source: Gary Page,
   Point Reyes Bird Observatory.

---

data.precip — Average annual precipitation in the U.S.

DATASETS
    precip   (69,1).

DESCRIPTION
Average annual precipitation, in inches, for 69 weather stations in the U.S. Source: *Statistical Abstract of the United States*, 1975, Bureau of the Census, Washington, D.C. Used by McNeil, p. 3.

---

data.presidents — Approval ratings of U.S. presidents in polls, 1945-1974

DATASETS
    presidents   (120,1).

DESCRIPTION
Approval ratings of U.S. presidents in Gallup polls taken over the period 1945-1974. The lowest and highest ratings within each six-month period are given, in the order in which they occurred. (Polls were taken at a greater frequency, but irregularly.) Missing values indicate six-month periods over which fewer than two polls were taken. Used by McNeil, p. 126.

---

data.pressure — Vapor pressure of mercury at various temperatures

---

DATASETS
    pressure (44,2).

DESCRIPTION
Vapor pressure of mercury in millimeters of mercury (column 2) at $-30°$, $-20°$, ..., $400°$ Celsius (column 1). Source: *Handbook of Chemistry and Physics*, 1973, Robert C. Weast, ed., CRC Press, Cleveland, p. D-161. Used by McNeil, p. 57.

---

data.prostate — Prostate cancer survival

DATASETS
    prostate (129,4).

DESCRIPTION
Prostate cancer survival data.

Column	*Contains*
1	Survival time in months.
2	0 = withdrawal; 1 = death.
3	Number at risk at that time.
4	Number of ties.

'Death' denotes death from prostate cancer. 'Withdrawal' denotes censored observations: death from other causes or patient still alive at last recording.

Source: M. Hollander and F. Proschan (1979), *Testing to determine the underlying distribution using randomly censored data*, **Biometrics**, **35**, pp. 393-401.

---

data.rivers — Lengths of 141 'major' rivers in North America

DATASETS
rivers   (141,1).

DESCRIPTION
Length, in miles, of 141 'major' rivers in North America according to *The World Almanac and Book of Facts*, 1975, New York, p. 406. Used by McNeil, p. 14.

---

data.stackloss — Loss of ammonia while oxidizing it to nitric acid

---

DATASETS
stackloss (21,4).

DESCRIPTION
Measurements on the loss of ammonia in the process of oxidizing it to nitric acid. Observations were made over 21 consecutive days at a single plant.

Column   Contains
   1   Air flow—rate of operation (units not reported).
   2   Temperature of cooling water (units not reported).
   3   Concentration of nitric acid in the absorbing liquid (minus 50, times 10).
   4   Stack loss: Percentage of ingoing ammonia which escapes processing (times 10).

Source: K. A. Brownlee (1965), *Statistical Theory and Methodology in Science and Engineering*, 2nd ed., Wiley, New York, pp. 454-455. Also in: N. R. Draper and H. Smith (1966, 1981), *Applied Regression Analysis*, Wiley, New York, Chapter 6; Cuthbert Daniel and Fred S. Wood (1971, 1980), *Fitting Equations to Data*, Wiley, New York, Chapter 5.

---

data.statprofs — Salary and age of Statistics professors at UCB, 1984

DATASETS
statprofs (16,2).

DESCRIPTION
Annual gross salary in dollars (column 1) and age in years (column 2) of all full-rank professors in the Department of Statistics at UC Berkeley, spring 1984. Data assembled by Mark Thomas of UC Berkeley in spring 1984, from publicly available records. Compare with *ucprofs*.

---

data.steam — Steam used at an industrial steam plant

DATASETS
steam   (25,10).

DESCRIPTION
Column   Contains
  1   Pounds of steam used monthly at an industrial steam plant.
  2   Pounds of real fatty acid in storage per month.
  3   Pounds of crude glycerin made.
  4   Average wind velocity (mph).
  5   Calendar days per month.
  6   Operating days per month.
  7   Days below 32 degrees F.
  8   Average atmospheric temperature, degrees F.
  9   (Average wind velocity)^2.
 10   Number of startups.

Source: N. R. Draper and H. Smith (1966, 1981), *Applied Regression Analysis*, Wiley, New York, Appendix A.

---

data.sunspot — Monthly mean sunspot numbers

DATASETS
sunspot   (2820,1).
sunspotx (360,2).

DESCRIPTION
Sunspot contains monthly mean sunspot numbers, January 1749 through December 1983.

Sunspotx contains (in column 2) the excerpted series from January 1950 through December 1979 and (in column 1) a time index of the form *year + fraction*, where *fraction* assumes the values 1/24, 3/24, ..., 23/24 for the months January, February, ..., December. A corresponding time index for the entire series can be constructed by the command:

```
index = (1749+1/24 : 1984-1/24 : 1/12)'
```
Data provided by David R. Brillinger, Department of Statistics, UC Berkeley.
A longer description and brief bibliography appear in D. F. Andrews and A.
M. Herzberg (1985), *Data: A Collection of Problems from Many Fields for the
Student and Research Worker*, Springer-Verlag, New York, pp. 67-68.

---

data.tractor — Tractor brake failure times

DATASETS
   tractor   (107,1).

DESCRIPTION
   Tractor rear brake failure times in hours. Source: R. E. Barlow and R.
   Campo (1975), *Total time on test processes and applications to failure data
   analysis*, in *Reliability and Fault Tree Analysis*, R. E. Barlow et al., eds.,
   SIAM, Philadelphia.

---

data.tubercle — Survival times of guinea pigs treated with tubercle bacilli

---

DATASETS
   tubercle   (65,2).

DESCRIPTION
   Survival times, in days, of a control group of 107 guinea pigs (column 1) and
   a treatment group of 61 guinea pigs (column 2) which received a dose of
   tubercle bacilli. The data were truncated at 735 days.

   Source: T. Bjerkedal (1960), *Acquisition of resistance in guinea pigs infected with
   different doses of virulent tubercle bacilli*, **Amer. J. Hygiene**, **72**, pp. 246-253.

---

data.turbines — Crack growth rate for disk cracks in U.S. plants

---

DATASETS
   turbines   (57,4).

DESCRIPTION
   Apparent crack growth rate, obtained by dividing crack depth by rotor operating time, for disk cracks in U.S. plants (usually nuclear plants).

   Column   Contains
   1   Crack location:
            1 = bore; 2 = web face; 3 = keyway; 4 = rim attachment.
   2   Estimated disk temperature (degrees F).
   3   0.2% offset yield strength.
   4   Apparent crack growth rate.

   Source: Leo Breiman, Department of Statistics, UC Berkeley.

---

data.ucprofs — Salary and age of professors at UCB, 1984

DATASETS
  ucprofs   (27,2).

DESCRIPTION
  Annual gross salary in dollars (column 1) and age in years (column 2) of 27 professors at UC Berkeley, spring 1984. These 27 are a simple random sample from the population of 1100 full-rank, full-time professors. Data assembled by Mark Thomas of UC Berkeley in spring 1984, from publicly available records. Compare with *statprofs*.

---

data.uspop — U.S. population, census years 1790-1980

DATASETS
  uspop   (20,2).

DESCRIPTION
  Population, in millions, of the U.S. in the census years 1790-1980. Used by McNeil, pp. 64-66.

---

data.viennaberlin — Monthly mean temperatures of Vienna and Berlin

---

DATASETS
  viennaberlin (2184,2).

DESCRIPTION
  Monthly mean temperatures in degrees Celsius of Vienna (column 1) and Berlin (column 2), from January 1769 through December 1950. The first 72 Vienna temperatures are missing; Vienna observations begin January 1775.

  Data provided by David R. Brillinger, Department of Statistics, UC Berkeley; used in David R. Brillinger (1975), *Time Series: Data Analysis and Theory*, Holt, New York (expanded ed., 1981, Holden-Day, San Francisco).

---

data.warpbreaks — Number of warp breaks for different types of yarn

---

DATASETS
  warpbreaks   (9,6).

DESCRIPTION
  From a weaving experiment to compare the strengths of six different types of yarn. A *warp* is a quantity of yarn that goes into a loom as one unit. For each type of yarn, the number of breaks for each of nine warps was recorded and expressed as a rate of breaks per unit length of warp. More breaks mean

weaker yarn. The six yarn types were labeled, respectively, *AL*, *AM*, *AH*, *BL*, *BM*, and *BH*, where *A* and *B* are two different types of cotton (type *B* was more expensive) and *L*, *M*, and *H* refer to the number of twists (that is, turns per inch) in the yarn: low, medium, and high.

Source: L. H. C. Tippett (1950), *Technological Applications of Statistics*, Wiley, New York, pp. 105-106. Used by McNeil, p. 28.

---

data.women — Average weight of American women aged 30-39, by height

---

DATASETS
   women   (15,2).

DESCRIPTION
   Average weight (column 2) of American women in the age group 30-39 for various heights (column 1), according to *The World Almanac and Book of Facts*, 1975, New York. Used by McNeil, pp. 54-55.

Basic UNIX Command Summary

UNIX is the computer operating system on which BLSS runs. You do not need to know any UNIX in order to use BLSS, but a knowledge of UNIX will allow you to take advantage of other facilities it provides, such as file manipulation and electronic mail.

This appendix gives condensed descriptions of basic UNIX commands and features. It serves as an introduction to UNIX for the curious and as a quick reference for those who are already learning it. UNIX has many more commands than we show, and those which we do show have many more options and capabilities than we describe. For more information, see the on-line manual (explained in Section A.2) or one of the references we list at the end (in Section A.7). The contents of this appendix are:

At the time of this writing, there are two main versions of UNIX: BSD and System V. We note differences between them where appropriate. It appears that BSD and System V will grow closer together in the future, so these distinctions may become less important.

Note that, in general, uppercase and lowercase letters on UNIX are distinct. We will see several examples of this.

A.1   Files and Directories

Files. A *file* is the fundamental UNIX object for containing stored information. This term was discussed in the context of BLSS in *BUG* Section 1.5 and is used throughout the book. Recall that some files are *text files* which you can look at and edit, whereas other files—such as BLSS datasets, executable binaries, and directories (see below)—contain binary data which you cannot directly display.

Here are the basic UNIX commands for manipulating files. Note that options to UNIX commands are preceded by a '–' sign and that options to UNIX commands must (unlike options to BLSS commands) generally precede any arguments to the command.

cat *file1* [*file2 ...*]

Print text files to the terminal. 'Cat' is an abbreviation for catenate: if you **cat** more than one file, they run together. Long files will disappear off the top of the screen—see below, under *paging commands*, for a solution. If you **cat** a file that is not a text file, the binary data will be displayed on your terminal as garbage characters.

cp *oldfile newfile*

Make a copy of *oldfile*; the new file is called *newfile*.

ls [*options*] [*file1 file2 ...*]

List the specified files or directories. If none are specified, the contents of the current directory are listed. Here are a few of the options to **ls**. Note that several options can be specified together.

–a List all files, including those whose names begin with '.'. Normally, files whose names begin with '.' are listed only when explicitly specified.

–C Display the file names in several columns instead of one column. This is the default on BSD UNIX, but not on System V.[1]

–F Indicate each file's type next to its name using a single character. The '*' character denotes an executable file; '/' denotes a directory; on BSD UNIX only, '@' denotes a symbolic link (see below under **ln –s**).

–l Long listing: includes the file's permissions (see below under *permission settings*), owner, size in bytes, and date of creation or last modification.

–R Recursively display the contents of directories.

–s Show size of file in blocks.[2]

mv *oldname newname*

Rename (move) *oldname* to *newname*.

rm *file1* [*file2 ...*]

Remove (delete) the named files. The –r option removes files *recursively*: if a named file is a directory (see below), then it and all its contents are also removed. Without the –r option, **rm** cannot remove directories.

vi [*file1 file2 ...*]

Edit the named files using the **vi** editor. See Appendix B.

---

1. Because uppercase and lowercase letters are distinct, the –C, –F, and –R options have completely different meanings than the –c, –f, and –r options (which we don't discuss here).
2. A block is 1024 bytes (1 kilobyte) in BSD UNIX, 512 bytes in System V.

Paging commands allow you to display long text files one screenful (or *page*) at a time.

more *file1* [*file2* ...]

Display the named files one screenful at a time using the *more* mechanism described in *BUG* Section 4.8. Standard on BSD UNIX; available on some but not all System V UNIX systems.

pg *file1* [*file2* ...]

The standard System V UNIX paging command—similar to **more**.

Directories. A directory is a special type of file whose purpose is to contain other files, including other directories. Directories exist so that related files can be organized together. For example, all files that BLSS creates for you are contained within the directory *blss* in your account.

pwd

Print the name of the current ('working') directory.

chdir [*dir*]

Change to the directory *dir* (by default, your home directory); it becomes your current directory. **Cd** is a synonym for **chdir**.

mkdir *dir*

Create (make) the directory *dir*.

rmdir *dir*

Remove the empty directory *dir*. See also the **–r** option to **rm**.

The following file management commands operate on the contents of directories.

ls *dir*

For the **ls** command, any named file can also be a directory, in which case the contents of that directory are listed.

mv *file1* [*file2* ...] *dir*

Move the specified files (or directories) into the directory *dir*. Some versions of UNIX impose restrictions on moving directories.

cp *file1* [*file2* ...] *dir*

Copy the specified files into the directory *dir*.

du [*dir*]

Disk usage: Display the total size in blocks of files in the specified directories (if none are specified, the current directory) and, recursively, all directories contained within them.

ls –Rs

Show the size of all files in the current (or named) directory and, recursively, all directories contained within it.

File names. Names for files and directories should normally contain only letters, digits, and the dot '.' and underscore '_' characters. Other punctuation characters can also be used, but most punctuation characters other than '.' and '_' have special meanings to the UNIX shells.[3] We do not discuss here which other punctuation characters are safe to use in file names under what circumstances.

Uppercase and lowercase letters are distinct in file names. For example, the names *ab* and *Ab* refer to two distinct files.

On BSD UNIX, file names can be quite long (over 80 characters). On other versions of UNIX, including System V, only the first 14 characters of a file name are significant. For example, the names *AbraCaDabra1023* and *Abra-CaDabra1024* refer to two different files on BSD, but to the same file on System V.

Pathnames. To refer to files and directories outside the current directory, UNIX provides *pathnames*. There are two types. *Relative pathnames* specify the location relative to the current directory. For example: the name *a/b* refers to the file *b* contained in the directory *a*, which is in the current directory; the name *a/b/c* refers to the file *c* in the directory *a/b*.

Absolute pathnames specify the location relative to the *root* directory, */*, which ultimately contains all files and directories. For example: the name */a* refers to the file or directory *a* in the root directory; the name */a/b* refers to the file *b* in the directory */a*; the name */a/b/.../c/d* refers to the file named *d* in the directory */a/b/.../c*.

The 14 character limit on file names imposed by non-BSD versions of UNIX applies only to the individual components of a pathname (between the '/' characters)—not to full pathnames.

Here are some abbreviations for referring to directories:

.   The current directory.

..   The *parent* directory—that is, the directory which contains the current directory.

~   Your *home* directory—that is, where you first login to UNIX.

~user The home directory of the named user.

For example: *../help* is a file or directory contained in the parent directory; *~/blss* is the *blss* directory in your home directory; *~george/blss* is the *blss* directory in *george*'s home directory.

Relative and absolute pathnames and the names '.' and '..' can be used anywhere that plain filenames can be used. The ~ abbreviations can be used in many contexts (especially on BSD UNIX) but not everywhere. For example, the C shell ('%' prompt), **vi**, and BLSS all recognize the ~ abbreviations, but the Bourne shell ('$' prompt) does not.

---

3. When necessary, the special meanings of punctuation characters in the UNIX shells (see Section A.5) can be escaped by preceding them with a backslash character '\', just as in the BLSS shell (see *BUG* Section 4.9.1).

More file manipulation. Here are a few more commands for operating on files.

diff *file1 file2*
> Show all lines from *file1* and *file2* which differ. Lines marked with '<' come from *file1*; lines marked with '>' come from *file2*.

grep [*options*] *string file1* [*file2* ...]
> Print all lines from the listed file(s) which contain the specified string. Here are a few of its many options:
>
> –i   Treat uppercase and lowercase letters in the named files identically when searching for *string*. For example, 'grep –i blss file' prints lines which contains the words 'blss', 'Blss', 'BLSS', etc.
>
> –n   Show the line number in the file for each line printed.
>
> –v   Print all lines *except* those which contain the specified string.

ln [–s] *name1 name2*
> Create a *link* (that is, an alternate name; also called a *hard link*) for the file named *name1*; the alternate name is *name2*. (A file by the name *name2* cannot already exist.) Thereafter, *name1* and *name2* refer to the same file. If the –s option is given (BSD UNIX only), **ln** creates a *symbolic link* instead of the default hard link. Symbolic links differ from hard links in that they distinguish between the new name and the original name: the new name 'points to' the original. Symbolic links can be created to directories; hard links cannot be. **Ls –l** shows the name that a symbolic link points to.

sort [*file* ...]
> Sort the lines in a text file—by default, into alphabetical order—or sort and merge together the lines in several text files. This command has many options for specifying different sorting criteria, etc.

wc [*file* ...]
> Word count: report the number of lines, blank-separated words, and bytes in each specified file, and the total number of lines, words, and bytes in all specified files if more than one is given.

A.2   Information

Information about things. The following commands display information about things (as opposed to people).

cal *year*
cal *month year*
> The first usage displays the calendar for the specified four-digit year. The second usage displays the calendar for the requested month (1 = January, 2 = February, ...) in the given *year*. For example: 'cal 1988' shows the calendar for the year 1988; 'cal 5 1988' shows the calendar for May 1988. Be aware that specifying the year 88 produces a calendar for a year during the reign of the Roman emperor Domitian.

date
Display the current date and time (on the 24-hour clock).

help [*topic*]
Some sites provide a **help** command that displays *help files*—usually local information. However, the default BSD and System V **help** commands are not too helpful. See also the **man** command, below.

man *command*
Display the on-line UNIX manual entry for the specified command or topic.

man –k *keyword* ...
(BSD UNIX only.) List all manual entries with the specified keyword(s) in their title.

Information about people. The following commands are helpful when trying to locate or communicate with other people who use your system.

finger [*user1 user2 ...*]
(BSD UNIX only.) Display information about the named users—including their real names, if known. If no users are specified, display information about all users currently logged in to the system. A command of the form **finger** *user@hostname*[4] displays information about the named user on the named remote computer system; **finger** *@hostname* displays information about all users currently logged in to that system.

last [*–n*] [*user1 user2 ...*]
(BSD UNIX only.) Display the login and logout times for the specified users (by default, all users) back to the beginning of tracking. If *–n* is specified, only the first *n* lines of the display are given.

mail [*user1 user2 ...*]   # *BSD UNIX*
mailx [*user1 user2 ...*]   # *System V UNIX*
Read or send electronic mail. **Mailx** on System V corresponds to **mail** on BSD. **Mail** on System V is an older, more restricted command.

If no user name is specified, read your mail. **Mail** gives a numbered list of the messages you have received. To read a message, type its number. Type **?** for a synopsis of commands available within **mail**; type **q** to exit (quit) **mail**.

If user names are specified, send mail to them. Whatever you type becomes part of the message to be sent, until you type a control-D on a separate line. To edit what you have written using **vi**, type **~v** on a separate line. For a synopsis of other commands available when sending mail, type **~?** on a separate line. By default, the mail is sent to users on

---

4. A hostname is a word such as *bach* or *bach.berkeley.edu* which specifies the name of a computer system.

your system. To send mail to a user on another system, use the syntax *user@hostname.*[5]

who

Display the login names of users currently logged in to the system.

rwho

(BSD UNIX only.) Display the login names of users currently logged in to the local network of systems. On the SUN-OS version of UNIX, the command **rusers –l** performs a similar function.

A.3   Printing Files on a Lineprinter

The commands for printing files on a lineprinter are different for BSD UNIX and System V.

BSD lineprinter commands.

lpr [–P*printer*] *file ...*

Send the named text files to the lineprinter. The –**P** option is used to name a specific (non-default) printer. (See *BUG* Section 10.4.)

lpq [–P*printer*] [*user*]

List all jobs in the printer queue for the specified user (all users if none are given). One of the pieces of information is the job number, which is useful for removing a job from the printer queue.

lprm [–P*printer*] [*jobno*] [–]

Remove the specified job from the printer. If '–' is given instead of a job number, all your jobs are removed.

On BSD UNIX, the environment string PRINTER can also be used to specify a non-default printer.

System V lineprinter commands.

lp [–d*printer*] *file ...*

Send the named text files to the lineprinter. The –**d** option is used to name a specific printer. On System V UNIX, the environment string LPDEST can also be used to specify a non-default printer. (See *BUG* Section 10.4.)

lpstat [–p*printer ...*] [–u*user ...*]

List all jobs in the specified printer queues for the specified user (all users if none are given). One of the pieces of information given is the job number, which is useful for removing a job from the printer queue.

cancel [*jobno*] [*printer*]

Remove the specified job from the printer.

---

5. Other, older, address syntaxes also exist, but as of this writing it seems that they will gradually decrease in importance.

A.4   Security

Although UNIX security is not perfect—in the sense that a sufficiently determined and knowledgeable person can break it—the mechanisms described here can keep your account and its files reasonably secure.

Your password. Your account—and all it contains—can be accessed by everyone who knows your password. If you want to keep your account private, you need to keep your password private (and not easy to guess).

passwd
> Change your password. You will be asked to prove that you know the old password before you can change it.

Permission settings on your files and directories control who has access to them. There are three types of permission: read (**r**), write (**w**), and execute (**x**).[6] For the purpose of permissions, users are divided into three categories: the user who owns the file (**u**), the group (**g**),[7] and other users not in the group (**o**).

The command **ls –l** shows the permission settings on files and directories. The permissions are listed to the extreme left as a string of 10 characters. Characters 2 through 4 indicate read, write, and execute permission for the owner; characters 5 through 7 indicate permissions for the group; and characters 8 through 10 indicate permissions for others. The characters are **r**, **w**, or **x** if the corresponding permission is granted; – if the corresponding permission is denied. For example, the display for the file *.blss*:

```
% ls -l .blss
-rw-r--r--  1 fran           233 Dec 20 12:42 .blss
```

indicates that read and write permission is granted to the owner (in this case, *fran*); read permission is granted other users, both in the group and outside it; and all other permissions are denied. The **chmod** ('change mode') command is used to change permission settings:

chmod [*who*[+–]*access*] *file1 file2 ...*
> Grant (+) or deny (–) access to the named files or directories. *Who* may be any combination of **u**, **g**, and **o**; *access* may be any combination of **r**, **w**, and **x**. For example, to grant execution permission to everyone, type 'chmod ugo+x *files*'. To deny read permission to people outside your group, type 'chmod o–r *files*'. To deny all permissions for the current directory to everyone but yourself, type 'chmod go–rwx .'.

---

6. For a directory, you need read permission to **ls** it, write permission to create files in it, and execute permission to **cd** to it or access files within it.

7. Each user is assigned to one or more *groups*. Groups are created and administered by your system administrator. For example, when accounts are created for a course of instruction, usually all users in the class are put in the same group. The **groups** command (BSD UNIX only) shows the groups you belong to. On System V UNIX, **ls –l** shows the group that the file belongs to, as well as the owner. On BSD UNIX, use **ls –lg** to see the group and the owner.

A.5   UNIX Shells

The command interpreters on UNIX—that is, the programs which display the
'$' or '%' prompts and which interpret and execute your UNIX commands—
are called *shells* and UNIX, true to its democratic history, provides more than
one.

The different shells have different capabilities, strengths, and weaknesses.
You can get a new (and different) shell by invoking it as a UNIX command
from within your current shell.[8] On BSD UNIX systems and some others, you
can change your default shell (the shell you get when you log in) using the
chsh command. On systems with no **chsh** command, contact your system
administrator if you wish to change your default shell.

The Bourne shell (**sh**) is the standard shell on System V UNIX, and is available on all UNIX systems. Its prompt is the '$' symbol. Its startup file is
called *.profile* and is read only upon login—*not* upon each invocation.

The C shell (**csh**) is the standard shell on BSD UNIX and is available on many
System V systems. Its prompt is the '%' symbol. It provides string, alias, and
history features similar to those of BLSS. Its startup file is called *.cshrc*, and
is read upon each invocation. It reads a second startup file called *.login* only
upon login.

The Korn shell (**ksh**) is a comparatively new shell with a '$' prompt. It is an
extension of the Bourne shell which provides the best features of the C shell
and many other new features. It is not yet widely available.

Input/output redirection. Normally, UNIX commands read their input from
your terminal keyboard and write their output to your terminal screen. You
can cause them to take input from—or send output to—files instead. This is
called *I/O redirection* and is analogous to the BLSS *output text redirection*
facility described in *BUG* Sections 4.10, 11.3, and elsewhere. The following
I/O redirection symbols apply to all UNIX shells:

command < file
  Take input from the named *file* instead of the keyboard.

command > file
  Send output to the named *file* instead of to the terminal. If *file* already
  exists, overwrite its previous contents.

command >> file
  Send output to the named *file* instead of to the terminal. If *file* already
  exists, the new output is appended to its previous contents.

The above two forms redirect the standard output[9] of commands, but not

---

8. It is from this ability to put one shell inside another that the UNIX shells take their name.
9. In UNIX jargon, the regular input to commands which normally comes from the keyboard
is called *standard input* (or *stdin*) and the regular output from commands which normally
goes to the terminal is called *standard output* (or *stdout*). The error message output is called
standard error (or *stderr*).

error messages. Error messages can be redirected with standard output, but the syntax depends on the shell. In the C shell:

command >& file
> Redirect standard output and error messages from the terminal to the named *file*. If *file* already exists, overwrite its previous contents.

command >>& file
> Like *>&*, but append the output to the end of *file* if it is not empty.

To redirect standard output and error messages in the Bourne and Korn shells:

command > file **2>&1**
> Overwrite any previous contents of *file*.

command >> file **2>&1**
> Append to any previous contents of *file*.

Pipes allow the output from one command to be used directly as the input to another command.

command1 | command2
> Cause the output from *command1* to be the input to *command2*. (In jargon, *pipe* output from *command1* into *command2*.) For example, to count the number of lines in a file which contain the word 'blss', type 'grep blss *file* | wc'.

The pipe syntax shown above is equivalent to the command sequence:

> *command1 > tmpfile*
> *command2 < tmpfile*
> **rm** *tmpfile*

but it is more efficient.

A.6   Process Control

Loosely speaking, each separate UNIX command or job is a separate *process*. The system assigns to each process a *process id number* (or *pid* for short) with which you can identify and track it.

Processes can be run in *background* (as opposed to the default *foreground*) using the syntax:

> *command* **&**

When you do this, the shell displays the command's *pid* and immediately gives you a new prompt. You can continue to type other commands to the shell—even additional background commands—at the same time that the background command is running. Of course, each additional simultaneous job is an extra burden on your computer and slows it down that much more.

The background symbol **&** should follow any I/O redirection symbols.

jobs
(C shell only.) Display a list of jobs started in the current invocation of the C shell which are still active.

ps Process status: Display information on processes which are running, including their pids (under the heading PID) and how much CPU time they have consumed (under the heading TIME). This command has many options to control what processes are shown and how much information is displayed; the options are completely different on BSD and System V.

kill [*signal*] *pid*
Terminate the process which has the specified pid. The default **kill** command kills most processes. If a command does not die, use the –9 signal. To kill a login shell gracefully, use the –1 signal.[10]

The *job control* facility of the C shell allows you to temporarily suspend and subsequently resume jobs and to move jobs from foreground to background and vice versa.

The **nice** and (on BSD UNIX only) **renice** commands allow you to run your jobs at low priority. These command names refer to the fact that to do so is to be gracious to the people with whom you share your computer.

A.7 References

Peter Birns, Patrick Brown, and John C. C. Muster (1985). *UNIX for People.* Prentice-Hall, Englewood Cliffs, N.J. User-oriented introduction; extensive examples. Emphasis on BSD; no discussion of System V-only features; extensive discussion of **vi**.

S. R. Bourne (1983). *The UNIX System.* Addison-Wesley, Reading, Mass. A terse, but reasonably comprehensive, introduction to UNIX. Emphasis on System V; no discussion of BSD-only features; brief discussion of **vi**.

Brian W. Kernighan and Rob Pike (1984). *The UNIX Programming Environment.* Prentice-Hall, Englewood Cliffs, N.J. Excellent introduction to UNIX for programmers. Emphasis on System V; no discussion of BSD-only features; no discussion of **vi**.

UNIX Programmer's Manual. This is the UNIX reference manual. Each distinct version of UNIX has its own version of the manual. In recent versions of the manual (both BSD and System V), it is becoming more common to separate the material for users, programmers, and administrators into separate volumes.

---

10. On BSD UNIX, **ps** shows login shells with a leading '–' sign: for example, as '–csh' or '–sh'. On System V, **ps –f** shows login shells with a leading '–'.

Basic Vi Command Summary

A *text editor* is a program for interactively editing (that is, changing) the contents of text files. This appendix provides a concise overview of **vi** (pronounced *vee-eye*), a *visual* text editor which is available on BSD and System V UNIX.

General remarks. At any given time, **vi** displays one screenful of the file you are editing. If the file is empty or contains less than one screenful of text, or if you are at the bottom of the file, the left margin of the remainder of the screen contains '~' characters which denote that the file extends no further. Thus, any blank line on the screen which does not contain a '~' is actually a blank line in your file.

Most **vi** commands do something immediately. Instead of showing the command itself as you type it, **vi** shows the result of the command by updating the terminal screen.[1] Some **vi** commands are *bottom-line* commands,[2] which **vi** *does* show as you type—on the bottom line of the screen. The bottom-line commands must be terminated with the RETURN key before they can execute.

If **vi** doesn't like a command, it beeps at you. You can cancel an incomplete command with the ESCAPE key, or by giving an INTERRUPT.[3] You can verify that **vi** is ready to accept commands by pressing the ESCAPE key—**vi** should beep.

In this appendix, the notation ^{X} denotes a control-X character, where X is any letter.[4] However, a single '^' denotes the caret character itself (usually located on the '6' key), not the CTRL key.

Terminal type. In order to function properly, **vi** needs to know what type of terminal you are using. If **vi** behaves strangely—for example, it prints garbage characters, or commands do not behave as described here—then it probably has incorrect information about your terminal type. If necessary, check with a local expert to find out how to specify your terminal type.

---

1. **Vi** and the BLSS **redo** command editor (described in *history(BRM)*) are identical in this respect; and most **redo** commands are identical to the corresponding **vi** commands.
2. Also called **ex** commands—because they come from the **ex** editor, a line-oriented editor on which **vi** is based.
3. Usually ^C. The interrupt, erase, and kill characters are discussed in *BUG* Section 1.4.
4. **Vi** itself uses this notation to display any control characters your file may contain.

Entering and exiting vi. The following commands are used to enter **vi** from UNIX and to exit **vi** when you are done. You can exit with or without saving your changes.

vi [*file*] Enter **vi** from UNIX: edit or create the specified file. (If no file name is given, use the command :**w** *file* shown below in order to exit.)

:**w** Save ('write') the changes made to the file during the **vi** session. *Note*: **Vi** actually makes its changes to a temporary copy of your file. Your file itself is not changed until you give a :**w** command.

:**w** *file* Save the edited text file into the specified *file* (which can be different from the one you started to edit).

:**q** Quit **vi** (only after :**w** or making no changes).

:**q!** Quit **vi** without saving any changes to your file.

:**wq** Same as :**w** followed by :**q**.

These and all commands beginning with : are bottom-line commands. Don't forget to type a RETURN after them.

Moving the cursor. The *cursor* is a highlighted, possibly blinking, rectangle or underline character on your terminal, analogous to the printhead on a typewriter. The position of the cursor is your current position in the file you're editing. Before you can change any text, you must first move your cursor to where the changes will be made.[5]

On most terminals with arrow keys, the arrow keys can be used to position the cursor.

h Move backward (left) one character: same as the left arrow ←.
j Move down one character: same as the down arrow ↓.
k Move up one character: same as the up arrow ↑.
l Move forward (right) one character: same as the right arrow →.
w Move to the beginning of the next (punctuation-separated) word.
W Move to the beginning of the next space-or-tab-separated word.
b Move (back) to the beginning of the current (or previous) word.
B Move (back) to the beginning of the current (or previous) space-or-tab-separated word.
e Move (forward) to the end of the current (or next) word.
E Move to the end of the current (or next) space-or-tab-separated word.
0 Move to the beginning of the current line.
^ Move to the first non-space-or-tab character on the current line.
$ Move to the end of the current line.

5. The first time you use **vi**, try the motion commands on a file which already contains some text. If you don't have a text file to edit, you can make a copy of a UNIX system file and practice on that. For example:

```
% cp /etc/group myfile
% vi myfile
```

Alternatively, you can add text to an empty file using any of the text insertion commands described below in the section entitled *Adding text.*

G	Move to the last line in the file (**G** is mnemonic for *go to*).
*n***G**	Move to the *n*-th line in the file.
^D	Page down a half-screen.
^U	Page up a half-screen.
^F	Page forward (down) a full screen.
^B	Page backward (up) a full screen.

Searching. The following commands move the cursor to the beginning of what is found. The search starts at the current cursor position.

f*c*	Move forward to (find) the next occurrence of the character *c* on the current line.
F*c*	Move backward to the previous occurrence of *c* on the current line.
;	Repeat the last **f** or **F** command.
,	Repeat the last **f** or **F** command, but in the opposite direction.
/*string*	Search forward to the next occurrence of *string*.
?*string*	Search backward to the previous occurrence of *string*.
n	Repeat the previous **/** or **?** command (**n** is mnemonic for *next*).
N	Repeat the previous **/** or **?** command, but in the opposite direction.
%	Find the matching (,), [,], {, or } character.

The **/** and **?** commands are bottom-line commands; you must hit RETURN after specifying the string to search for.

Magic characters. In searches, the following *magic characters* have special meanings used for pattern matching (which we won't go into here): ^, $, \, ., *, [, / (in forward searches), and ? (in backward searches). To search for these characters, you must escape their special meanings by preceding them with a backslash character \. For example, to find the string 'x/2.5' search for 'x\/2\.5'.

Adding text. The following commands (except **:r**) and several others use *insert mode*: After giving the command, all text you type—including spaces, tabs, and RETURN's—is inserted into the file until you finish by pressing the ESCAPE key.

a	Add (append) text to the right of the cursor.
A	Add (append) text at the end of the current line.
i	Insert text to the left of the cursor.
I	Insert text at the beginning of the current line.
o	Add text starting with a new line below the current line ('open' a line in the file).
O	Add text, starting with a new line above the current line.
:r *file*	Read in the named *file*, starting below the current line.
<ESCAPE>	Exit insert mode.

For example, the command '**o**xxxx<RETURN>yyyy<ESCAPE>' adds two lines of text after the current line: 'xxxx' and then 'yyyy'. The command '**a**<RETURN><ESCAPE>' adds a carriage return immediately after the cursor—thereby breaking the current line in two.

The following control characters allow you to make corrections while in insert mode:

^H   Erase the last character.
^W   Erase the last word.
^D   Backtab over leading tabs inserted automatically by **vi**, if any.
<ERASE>
         Your erase character: erases the last character (same as **^H**).
<KILL> Your kill character:[6] kills the current input line.

Note: **vi** allows you to create long lines—longer than the 80 character line length of most terminals. In this case, the text *wraps around* the screen: characters after the 80th are displayed on the next line of the screen, but they are in fact on the same line of the file (as you can verify by using cursor positioning commands such as ^ and $). Except when you have a special need (such as preparing a dataset with many columns for the **read** {**autodims**} command or using **editdata** on a dataset with many columns), it's a good idea to type a RETURN before you reach the column 80 to prevent unnecessary wrap-around.

Changing and deleting text.

x   Delete (x out) the character under the cursor.
r*c*   Replace the character under the cursor with any character *c* (including tab or RETURN).
R   Replace characters by typing over them—continues until you type an ESCAPE.
s   Substitute text for the character under the cursor (enters insert mode).
dw   Delete to the end of the word.
dd   Delete the current line.
d$ Delete to the end of the current line.
D   Same as **d$**.
cw   Change text to the end of the current word (deletes that text and enters insert mode).
cc   Change the current line. (The old line disappears and you enter insert mode.)
c$ Change text to the end of the current line.
C   Same as **c$**.
J   Join together the current line and the next line.
~   Change the character under the cursor from lowercase to uppercase—or vice versa—and advance to the next character. Thus, repeated **~**'s change the case of an entire word.
.   Repeat the previous change (bottom-line commands not included).

When you delete lines, on some terminals **vi** may put @ characters in the left-hand margin to denote this. See the **^R** command, below, for a remedy.

The **d** and **c** commands (and the **y** command, described below) can be used with most of the cursor-movement commands described previously. For

---

6. Usually **^U**. See *BUG* Section 1.4.

example: **d0** deletes back to the beginning of the line; **c0** changes text back to the beginning of the line; **dW** deletes a space-or-tab-separated word; **cf**x changes through the next x character; **y$** yanks text through the end of the line; etc.

Undo commands. The following commands undo changes:

u   Undo the most recent change (including a **u** or **U** command).
U   Undo the last set of changes on the current line.

Repeat counts. Most **vi** commands can be preceded by counts, which specify a number of times to repeat the command. For example: **10k** means move up 10 lines; **7s** means substitute text for the next 7 characters; **8cw** means change 8 words; **5dd** means delete 5 lines; **6.** means repeat the previous change 6 times; etc.

Moving and copying text. **Vi** maintains an *unlabeled buffer* in which you can store pieces of text.

yy   Copy (yank) the current line into the buffer.
*n***yy**   Copy n lines (starting at the current line) into the buffer.
p   Append (put) the buffer contents after the cursor.
P   Insert (put) the buffer contents before the cursor.

Text which is deleted (via any **c**, **d**, **s**, **x**, or similar command) is automatically placed in the buffer. Thus you can move lines of text by deleting them and inserting them elsewhere with **p** or **P**.

Labeled buffers. In addition to the unlabeled buffer, there are 26 labeled buffers with names *a* through *z*. Any of the **c**, **d**, **x**, **y**, **p**, or **P** commands can be preceded by **"**x (where **"** is the double-quote character), which means to use the buffer with label x instead of the unlabeled buffer. For example, **"a10dd** puts the 10 deleted lines of text into buffer **a**. Then **"ap** puts the contents of that buffer into the file.

Other commands.

^G   Show the current line number and other status information.
^L   Clear and redraw the screen.
^R   Redraw the screen only as necessary to remove @ symbols which denote deleted lines.
ma Mark the current line with label a, where a is any lowercase letter.
'a Move to the line marked with the label a (where **'** is the single-quote character).

Commands that apply to a range of lines.

:'a**,'**b **s/**xxx**/**yyy**/g**
          Change the string xxx to the string yyy (**s** is mnemonic for *substitute*). The change is made to all occurrences of the string xxx which appear in the file between the line marked a and the line marked b, inclusive. Be careful to escape any magic characters in the xxx string with backslashes '\'.

:*'a,'b* **d**

> Delete lines from the line marked *a* through the line marked *b*, inclusive.

:*'a,'b* **co .**

> Copy lines (from the line marked *a* through the line marked *b*, inclusive) to the position immediately below the current line.

:*'a,'b* **mo .**

> Move lines (from the line marked *a* through the line marked *b*, inclusive) to the position immediately below the current line.

In the commands above, the notations *'a*, *'b*, and '.' can be replaced by any of the following: an absolute line number (such as **1** or **10** or **47**); the symbol **$** (which denotes the last line in the file); the symbol '.' (which denotes the current line); the notation **.**–*n* (which denotes the line *n* lines above the current line); or the notation **.+**n (which denotes the line *n* lines below the current line).

For example, the command **:1,$s/xxx/yyy/g** changes all occurrences of the string *xxx* to *yyy* throughout the file. The command **:.–2,.+3mo$** moves 6 lines (from the current line minus 2 through the current line plus 3) to the end of the file.

Other features. This appendix is far from a complete description of **vi**. We have omitted commands in every category discussed here, and we have not discussed any of the following features: *regular expressions* and *magic characters*, which are used for matching patterns in searches and substitutions; *options*, which can be used to customize **vi**'s behavior to your own taste; and *macros* and *abbreviations*. Most of the UNIX references mentioned at the end of Appendix A also discuss **vi**; but for complete information, refer to the long **vi** and **ex** writeups in the UNIX reference manuals. For example, in the BSD UNIX manual they are entitled *An Introduction to Display Editing with Vi* and *Ex Reference Manual*, and they appear in the volume labeled *Supplementary Documents*.

For automatic file editing—such as simple, repeated changes to many files— the UNIX commands **ed**, **sed**, and **tr** are preferable to **vi**.

BIBLIOGRAPHY

This is a list of books referred to in this text, plus some other local favorites. With one exception, journal articles, technical reports, and theses are excluded. We have attempted to include one or more titles at all levels in each area of application pertinent to BLSS.

D. F. Andrews and A. M. Herzberg (1985). *Data: A Collection of Problems from Many Fields for the Student and Research Worker.* Springer-Verlag, New York. A collection of 71 datasets, with descriptions of each, from many sources and contributors.

David A. Belsley, Edwin Kuh, and Roy E. Welsch (1980). *Regression Diagnostics: Identifying Influential Data and Sources of Collinearity.* Wiley, New York. Graduate level monograph.

Peter J. Bickel and Kjell A. Doksum (1977). *Mathematical Statistics: Basic Ideas and Selected Topics.* Holden-Day, San Francisco. Introduction to mathematical statistics at the advanced undergraduate or beginning graduate level.

Peter Bloomfield (1976). *Fourier Analysis of Time Series: An Introduction.* Wiley, New York. Undergraduate text in frequency domain time series analysis.

David R. Brillinger (1975). *Time Series: Data Analysis and Theory.* Holt, New York. (Expanded ed., 1981, Holden-Day, San Francisco.) Graduate text in frequency domain time series analysis.

Robert Goodell Brown (1963). *Smoothing, Forecasting and Prediction of Discrete Time Series.* Prentice-Hall, Englewood Cliffs, N.J.

K. A. Brownlee (1960, 1965). *Statistical Theory and Methodology in Science and Engineering.* Wiley, New York.

John M. Chambers (1977). *Computational Methods for Data Analysis.* Wiley, New York. Intermediate level survey of computational methods and algorithms pertinent to many areas of statistics.

Cuthbert Daniel and Fred S. Wood (1971, 1980). *Fitting Equations to Data: Computer Analysis of Multifactor Data for Scientists and Engineers.* Wiley, New York.

J. J. Dongarra et al. (1979). *LINPACK Users' Guide.* SIAM, Philadelphia. Manual for LINPACK, a public domain linear algebra subroutine library written in Fortran.

N. R. Draper and H. Smith (1966, 1981). *Applied Regression Analysis.* Wiley, New York. Undergraduate text; assumes knowledge of elementary statistics.

David Freedman, Robert Pisani, and Roger Purves (1978). *Statistics.* Norton, New York. Elementary (nonmathematical) statistics text; contains verbal descriptions of many important concepts from probability and statistics.

Keith A. Haycock and David R. Brillinger (1985). *LIBDRB: A subroutine library for elementary time series analysis.* Technical Report No. 48, Department of Statistics, University of California, Berkeley.

Paul G. Hoel (1947, 1954, 1962, 1971, 1984). *Introduction to Mathematical Statistics.* Wiley, New York. Undergraduate text; assumes knowledge of calculus.

Lambert H. Koopmans (1981). *An Introduction to Contemporary Statistics.* Duxbury, Boston. Introductory (noncalculus) text with unusually broad coverage: not only classical but also EDA, nonparametric, and other methods.

Charles L. Lawson and Richard J. Hanson (1974). *Solving Least Squares Problems.* Prentice-Hall, Englewood Cliffs, N.J. Numerical methods pertinent to least squares and linear algebra.

Edwin Mansfield (1980, 1983, 1987). *Statistics for Business and Economics.* Norton, New York. Introductory (noncalculus) text for business and economics majors. Also, *Basic Statistics with Applications* (1986), by the same author.

Donald R. McNeil (1977). *Interactive Data Analysis.* Wiley, New York. EDA statistics presented at the undergraduate level; describes algorithms for some routines.

William Mendenhall (1964, 1967, 1971, 1975, 1979, 1983, 1987). *Introduction to Probability and Statistics.* Duxbury, Boston. Introductory (noncalculus) text.

Donald F. Morrison (1983). *Applied Linear Statistical Methods.* Prentice-Hall, Englewood Cliffs, N.J. Undergraduate text; includes regression models, time series models, analysis of variance and covariance.

Lincoln E. Moses (1986). *Think and Explain with Statistics.* Addison-Wesley, Reading, Mass. Introductory (noncalculus) text; emphasis on verbal explanations, intuition, and common sense.

B. T. Smith et al. (1972, 1976). *Matrix Eigensystem Routines — EISPACK Guide.* Springer-Verlag, New York. Manual for EISPACK, a public domain eigenvalue-eigenvector subroutine library written in Fortran.

John W. Tukey (1977). *Exploratory Data Analysis.* Addison-Wesley, Reading, Mass. The bible of EDA.

Paul F. Velleman and David C. Hoaglin (1981). *Applications, Basics, and Computing of Exploratory Data Analysis.* Duxbury, Boston. 'The ABC's of EDA.'

INDEX

Symbols are listed together at the front of the index. Not all options are listed—only those which are discussed in the text separately from the command(s) to which they apply.